ALLE ZEIT WACH 1842

Progress in Sensory Physiology

Editors: H. Autrum, D. Ottoson,
E.R. Perl, R.F. Schmidt, H. Shimazu, W.D. Willis
Editor-in-Chief: D. Ottoson

Volume 9
1989. 71 figures. V, 227 pages
ISBN 3-540-50282-3

Contents: *R.D. Foreman:* Organization of the Spinothalamic Tract as a Relay for Cardiopulmonary Sympathetic Afferent Fiber Activity. – *G.N. Akoev, G.N. Andrianov:* Synaptic Transmission in the Mechano- and Electroreceptors of the Acousticolateral System. – *J. Syka:* Experimental Models of Sensorineural Hearing Loss – Effects of Noise and Ototoxic Drugs on Hearing. – *Y. Oomura*: Sensing of Endogenous Chemicals in Control of Feeding. – *J.E. Sánchez-Criado, O.A. Mora, A. Gallego*: Structure and Function of the Vomeronasal System – The Vomeronasal Organ as a Priming Pheromone Receptor in Mammals.

Volume 10
M. Steriade, D. Paré, B. Hu, M. Deschênes
The Visual Thalamocortical System and Its Modulation by the Brain Stem Core
1990. 27 figures. VII, 124 pages.
ISBN 3-540-51380-9

Contents: Prologue. – Basic Circuits in the Visual Thalamocortical Systems and Their Physiological Aspects. – Regulatory Systems of the Brain Stem Core, Basal Forebrain, and Hypothalamus. – Neurotransmitters. – State Dependency of Visual Thalamic and Cortical Activities – References.

Volume 11
P.J. Snow, P. Wilson
Plasticity in the Somatosensory System of Developing and Mature Mammals – The Effects of Injury to the Central and Peripheral Nervous System
1991. 151 figures. XVI, 482 pages.
ISBN 3-540-52573-4

Contents: Introductory Remarks. – Plasticity in the Peripheral Somatosensory Nervous System. – Plasticity and the Mystacial Vibrissae of Rodents. – Plasticity and the Spinal Dorsal Horn. – Plasticity and the Dorsal Column Nuclei. – Plasticity and the Somatosensory Thalamus. – Plascticity and the Somatosensory Cerebral Cortex. – Concluding Remarks. – References.

Progress in

Sensory Physiology 12

Editors: H. Autrum D. Ottoson
E. R. Perl R. F. Schmidt H. Shimazu W. D. Willis

Editor-in-Chief: D. Ottoson

With Contributions by O. R. Anderson
A. B. Grossman G. M. Innocenti G. A. Ojemann
T. B. van Wimersma Greidanus

With 64 Figures

Springer-Verlag
Berlin Heidelberg New York
London Paris Tokyo
Hong Kong Barcelona
Budapest

Editor-in-Chief

Professor Dr. David Ottoson
Karolinska Institutet, Fysiologiska Institutionen II
Solnavägen 1, 10401 Stockholm 60, Sweden

Editors

Professor Dr. Hansjochem Autrum
Zoologisches Institut der Universität München
Luisenstraße 14, 8000 München 2, Germany

Professor Dr. Edward R. Perl
Department of Physiology
University of North Carolina at Chapel Hill
Chapel Hill, NC 27514, USA

Professor Dr. Robert F. Schmidt
Physiologisches Institut der Universität
Röntgenring 9, 8700 Würzburg, Germany

Professor Dr. Hiroshi Shimazu
Department of Neurophysiology
University of Tokyo, Institute of Brain Reserach
7.3.1. Hongo, Bunkyo Ku, Tokyo, Japan

Professor Dr. William D. Willis
The Marine Biomedical Institute
University of Texas Medical Branch
Galveston, TX 77550, USA

ISBN-13:978-3-642-75966-6 e-ISBN-13:978-3-642-75964-2
DOI: 10.1007/978-3-642-75964-2

LC card number: 85-647743 (ISSN 0721-9156)

Softcover reprint of the hardcover 1st edition 1991

Phototypesetting: Thomson Press (India) Ltd. New Delhi, India
2121/3130-543210 – Printed on acid-free paper

Contents

Opioid Regulation of Pituitary Function

T.B. van Wimersma Greidanus[1] and A.B. Grossman[2]

[1] Department of Pharmacology, Rudolf Magnus Institute, Vondellaan 6, 3521 GD Utreçht, The Netherlands
[2] Department of Endocrinology, The Medical College of St. Bartholomew's Hospital, West Smithfield, London EC1A 7BE, United Kingdom

Abbreviations. DADLE, D-Ala-D-Leu-enkephalin; DAGO, Tyr-D-Ala-Gly-MePhe-NH(CH_2)OH; DAMME, D-Ala^2-MePhe^4-Met(o)-ol-enkephalin.

1 Introduction

It has long been known that administration of opiates such as morphine induces pronounced changes in the secretion of hormones from the pituitary gland in both animals and man. The discovery of endogenous peptides with opiate-like activity, the demonstration of various (sub)types of opiate receptors and the development of specific agonists and antagonists for these opiate receptor subtypes stimulated research into the effects of opiates and opioid peptides on pituitary function. In this respect not only have the pharmacological effects of opiate-like substances on hormone release from the pituitary been studied, but also the regulation of the secretion of hormones from this endocrine gland by endogenous opioids has been investigated (reviews: Herz and Millan 1984; Millan and Herz 1985; Maysinger 1985). In the various sections of this review, these aspects will be discussed for each hormone of the pituitary gland.

2 Posterior Pituitary Hormones

2.1 Vasopressin

Concerning the influence of opiates and opioid peptides on the secretion of vasopressin, both stimulatory and inhibitory effects have been described, and some disagreement exists – or at least has existed – on the nature of these effects. We have already mentioned in previous reviews (van Wimersma Greidanus and ten Haaf 1984; van Wimersma Greidanus et al. 1985) that the substances and doses used, time course, route of administration (e.g. central vs systemic), presence and/or type of anesthesia, in vivo or in vitro situation, stimulated or non-stimulated condition and, last but not least, the animal species studied, are important factors in modulating the effects of opiates and opioid peptides on vasopressin secretion.

2.1.1 *Morphine and Related Substances*

More than 40 years ago one of the earliest reports (De Bodo 1944) on the effect of morphine on vasopressin release appeared. It was suggested that morphine stimulated the secretion of vasopressin. This suggestion was largely based upon the observation that, in *dogs*, morphine administration resulted in inhibition of water diuresis. Because morphine inhibited water diuresis neither in diabetes insipidus dogs (De Bodo 1944) nor in dogs with a transsection in the supraoptic tract (Duke et al. 1951), it was assumed that the antidiuretic effect of morphine was due to a direct effect on the supraoptic cells causing an increased release of antidiuretic hormone (vasopressin) from the posterior lobe of the pituitary. In 1980 Montastruc et al. described a morphine-induced, naloxone-sensitive antidiuresis in normal dogs, but not in diabetes insipidus dogs during regular diuresis. They also interpreted these results as being suggestive of a major role for vasopressin. However, alternative theories have been put forward to explain the antidiuretic effects of morphine and related drugs with opiate activity in dogs. For example, it has been proposed that morphine reduces the number of active nephrons (Handley and Keller 1950) or that hemodynamic changes are responsible for the inhibitory effect of opiates on water diuresis in dogs. Moreover, more recently it was reported by Rockhold et al. (1983) that intracarotid injections of morphine in pentobarbital-anesthetized dogs resulted in a rise in plasma vasopressin levels as measured by radioimmunoassay. These authors postulated a causal relationship between the documented morphine-induced fall in arterial blood pressure and the concomitant rise in plasma vasopressin (arginine vasopressin) levels. However, it has recently been demonstrated by Hellebrekers et al. (1987) that, following intravenous administration of the μ-(opiate) receptor type agonist methadone to conscious dogs, increased vasopressin levels in the systemic circulation are observed. From the magnitude of this naloxone-reversible response and from the lack of changes in biochemical and hemodynamic parameters sufficient to explain the rise in plasma vasopressin levels, these latter authors concluded that methadone exerts a direct stimulatory influence on the systemic release of vasopressin in dogs.

In *rabbits*, administration of morphine into the third ventricle resulted in a sustained elevation of vasopressin levels in plasma (Firemark and Weitzman 1979). Pretreatment with the opiate antagonist naloxone failed to prevent the rise in plasma levels of vasopressin following morphine administration, suggesting that naloxone-sensitive opiate receptors are not directly involved in vasopressin release in the rabbit.

In *rats*, both diuresis and antidiuresis have been reported to occur following the administration of morphine (Giarman and Condouris 1954; Fujimoto and Hisada 1978; Fujimoto 1971; Marchand 1970). The antidiuretic effect of morphine appears to be independent of vasopressin release, and several publications support the notion that factors other than opiate-induced vasopressin secretion are involved in this effect (Grell et al. 1985; Huidobro and Huidobro-Toro 1979; Wilson and Ngsee

1982). Although a few publications report an increase of plasma vasopressin levels following morphine administration (Aziz et al. 1981; Haldar 1982a), a number of reports have suggested that the most consistent response to administration of morphine in rats is inhibition of vasopressin release (Aziz et al. 1981a; Lutz-Bucher and Koch 1981; Miller 1975; Nutt and Jasinski 1974; van Wimersma Greidanus et al. 1979; Evans et al. 1988a). Subcutaneous administration of the narcotic antagonists oxilorphan and butorphanol resulted in a diuresis (Miller 1980; Nutt and Jasinski 1974) accompanied by decreased urinary excretion of radioimmunoassayable vasopressin (Miller 1980). These data suggest, according to Miller (1980) and to Nutt and Jasinski (1974), that the narcotic antagonists caused diuresis by inhibiting vasopressin release from the posterior pituitary. However, morphine-related substances such as butorphanol have both opioid agonistic and antagonistic activities and hence their effects on posterior pituitary function cannot be interpreted as pure antagonistic actions. Kamoi et al. (1979) also studied the effects of opiates and opiate antagonists on the osmoregulation of vasopressin release. They administered butorphanol, naloxone and/or morphine subcutaneously to rats and concluded from their results that opiate agonists inhibit vasopressin secretion, primarily by raising the setpoint of the osmostat. Huidobro and Huidobro-Toro (1979) investigated the antidiuretic effect of morphine and vasopressin in normal and Brattleboro rats. They found that although both compounds have marked antidiuretic properties, they differ markedly in their effect on the composition of urine. In addition, they found that morphine produces antidiuresis in Brattleboro rats despite the severe diabetes insipidus due to the absence of vasopressin (Huidobro-Toro 1980). Thus, the antidiuretic effect of morphine may not involve vasopressin. Kamoi et al. (1979) and van Wimersma Greidanus et al. (1979) both reported that morphine induced a decrease of radioimmunoassayable vasopressin levels after peripheral or intracerebroventricular administration. This decrease of vasopressin levels after administration of morphine was observed in nonstimulated animals as well as in rats in which the hormone level was elevated by water deprivation or by administration of hypertonic saline. The suppressive effects of morphine on plasma vasopressin levels could be blocked by administration of naltrexone (van Wimersma Greidanus et al. 1979) or naloxone (Kamoi et al. 1979). Similar results were obtained by Lutz-Bucher and Koch (1981). Interestingly, administration of these opiate antagonists on their own generally does not effect plasma vasopressin levels (Kamoi et al. 1979; van Wimersma Greidanus et al. 1979, 1981; ten Haaf et al. 1986, 1987). Aziz et al. (1981a) also studied the effect of intracerebroventricular injections of morphine on vasopressin release in rats. They found that the vasopressin response to morphine was both dose- and time-dependent. Higher doses produced short latency stimulation of vasopressin secretion followed by a long-term fall, whereas the lower dose resulted only in a long-latency inhibition. They also reported that the most consistent response to morphine administration was inhibition of vasopressin release. This inhibition was counteracted by naloxone and was independent of the effects of morphine on the cardiovascular system. In addition to its suppressive effect on the release of vasopressin into the systemic circulation, morphine decreases vasopressin concentrations in portal plasma (Koenig et al. 1986).

Interestingly, Clarke et al. (1980) reported that morphine selectively inhibits the firing of phasically discharging vasopressin neurons, pointing to a reduction in vasopressin release in response to the intracerebroventricular injection of morphine. Pittman et al. (1980) also described a depressant action of morphine on paraventricular neuronal activity recorded from slices of rat hypothalamus in vitro, but this effect was more pronounced on unidentified neurons than on vasopressin neurons, and may reflect inhibition of CRF-41 (see below). Miller (1981) studied the effect of morphine on the release of vasopressin in perfused, isolated hypothalamic tissue. In this system morphine induced a decline in vasopressin release, whereas naloxone induced an enhancement. Results of other in vitro experiments also suggest that morphine inhibits electrically evoked vasopressin secretion from rat medial basal hypothalamus (Knepel and Reimann 1982), vasopressin secretion from electrically-stimulated rat neurointermediate lobes (Iversen et al. 1980; Clarke and Patrick 1983) and the hormonal output from neurointermediate or posterior lobes of rat pituitaries (Lutz-Bucher and Koch 1980). In addition, Wright and Clarke (1983) described a suppressive effect of morphine on electrically evoked, but not on the basal release, of neurohypophyseal hormone release from posterior pituitaries in vitro. However, Dencker Christensen and Fjalland (1982), using an in vitro system comparable to that of Iversen et al. (1980), were not able to demonstrate an effect of morphine on the evoked release of vasopressin. In contrast, Haldar (1982) reported that while morphine did not release vasopressin into the medium when incubated with isolated neural lobes, a gradual increase in release was observed when incubated with intact, hypothalamo-neurohypophyseal tissue. Recently, Lacoumenta et al. (1988) reported that sufentanil, a potent μ-receptor opiate in clinical anesthetic use, which attenuates stress-induced vasopressin release in cardiac surgery, inhibited osmotically stimulated vasopressin release from rat hypothalami in vitro.

2.1.2 *Endorphins*

Intravenous as well as intracerebroventricular administration of β-endorphin to *rabbits* induced a sustained and pronounced rise in plasma vasopressin levels (Firemark and Weitzman 1979; Weitzman et al. 1977).

Similar to the effects of morphine, the endogenous opioid β-endorphin induces antidiuresis in *rats*. On a molar base, β-endorphin is much more potent than morphine in producing oliguria (Huidobro-Toro et al. 1979; Tseng et al. 1978). As for morphine, it has been questioned whether this antidiuretic effect is mediated by a β-endorphin-induced stimulation of vasopressin from the posterior pituitary. Since the antidiuretic effect of β-endorphin has also been found in Brattleboro rats lacking vasopressin (Huidobro-Toro 1980), this appears unlikely. Studies performed by van Wimersma Greidanus et al. (1979, 1981) clearly indicated that in rats intracerebroventricular, but not systemic, administration of β-endorphin resulted in suppression of basal as well as of (by water deprivation) stimulated vasopressin levels in plasma (ten Haaf et al. 1986). Furthermore, vasopressin release during footshock-induced stress (Knepel et al. 1982) and vasopressin release induced by histamine were inhibited by β-endorphin (ten Haaf et al. 1986). Systemic but not intracerebroventricular administration of naloxone counteracted the suppressive action of β-endorphin on vasopressin secretion, whereas central administration of naltrexone partially antagonized this effect of β-endorphin (van Wimersma Greidanus et al. 1981; ten Haaf et al. 1986). Removal of N-terminal tyrosine residue – tyrosine is an essential amino acid for interaction with opiate receptors – resulted in disappearance of the suppressive effect of β-endorphin on vasopressin release (van Wimersma Greidanus et al. 1981).

A weak but significant decrease of basal but not of stimulated release of vasopressin from the rat neurointermediate lobe by β-endorphin in vitro was reported by Maysinger et al. (1984), whereas Iversen et al. (1980) reported an inhibitory effect on electrically evoked vasopressin release from neurointermediate lobes, and Dencker Christensen and Fjalland (1982) and Weitzman et al. (1977) were unable to observe any effect of β-endorphin in this latter situation. Using rat medial basal hypothalamus in vitro, Knepel and Reimann (1982) found that β-endorphin induced a marked decrease in vasopressin release evoked by electrical stimulation.

2.1.3 Enkephalins

Fujimoto and Hisada (1978) reported a significant diuresis following intracerebroventricular administration of Met-enkephalin or Leu-enkephalin. In contrast, Bisset et al. (1978) reported that administration of Leu-enkephalin produced an antidiuretic response. They also showed that injection into the cerebral ventricles was more effective than intravenous injection, indicating a central site of action. They suggested that the antidiuretic response is mediated by release of vasopressin. This suggestion could not be confirmed by Aziz et al. (1981b), who determined vasopressin concentrations in plasma of rats by radioimmunoassay following intracerebroventricular administration of graded doses of Met-enkephalin. Lower doses (1–10 μg) did not induce changes in vasopressin levels over a short time interval (5–10 min). However, very high doses (10–100 μg) induced a fall in vasopressin levels, although in the case of 100 μg this was preceded by a transient increase. Also, Iovino et al. (1983) described a dual effect of a Met-enkephalin analog on plasma vasopressin levels in rats. In contrast to the findings of Aziz et al. (1981b) they observed lower vasopressin levels 60 s after intracerebroventricular administration of 10 μg Met-enkephalin, followed by an increase at 20 min after administration. Since they did not observe an effect upon systemic administration, they suggest that the mechanism by which enkephalin affects vasopressin release is due to central opiate receptors, perhaps mediated by activation of the renin-angiotensin system. Summy-Long et al. (1981a, b) did not observe an effect of intraventricularly administered Leu-enkephalin (100 μg) on basal vasopressin levels, but found an inhibition of both the hypertonic sodium chloride-induced and the angiotensin-induced (Keil et al. 1984) increase of plasma vasopressin. A suppressive effect of intracerebroventricularly injected Met-enkephalin on plasma concentrations in spontaneously hypertensive rats was reported by Rockhold et al. (1981). van Wimersma Greidanus et al. (1983) reported that intracerebroventricular administration of Leu-enkephalin or Met-enkephalin failed to affect basal vasopressin levels 30 min after injection.

Recently, Wang et al. (1989) examined the effect of administration of enkephalins on vasopressin secretion in conscious *sheep*. Intracerebroventricular but not intravenous infusion of the Met-enkephalin analog DAMME (FK 33-824) significantly increased plasma concentrations of vasopressin, whereas Met-enkephalin and Leu-enkephalin were ineffective.

In vitro it appeared that neither Leu-enkephalin nor Met-enkephalin, nor their analogs, affected basal or evoked vasopressin release from rat neurointermediate lobes and neurohypophyses (Bicknell and Leng 1982; Dencker Christensen and Fjalland 1982; Nordmann et al. 1986; Pitzel and König 1984). In contrast, a few studies reported a slight decrease of spontaneous (Maysinger et al. 1984) or stimulated (Iversen et al. 1980) vasopressin release following enkephalin administration. In addition, Leu-enkephalin was found to antagonize vasopressin secretion induced by K^+ and veratridine depolarization, whereas it failed to affect isoproterenol-stimulated release of vasopressin from perifused neurointermediate lobes of rat pituitaries (Al Zein et al. 1984). According to Lightman et al. (1982), the Leu-enkephalin analog DADLE inhibited electrically stimulated vasopressin release from the rat neurointermediate lobe. This effect was markedly influenced by the calcium concentration of the medium. Finally, it was reported that Leu-enkephalin exerted a weak depressant action on the excitability of a small percentage of phasically active supraoptic neurons (Arnauld et al. 1983).

2.1.4 Dynorphins, Other Prodynorphin Derivatives and κ-Receptor Agonists

Recently it has become evident that so-called κ-agonists, i.e. opiates and opioid peptides characterized as having agonistic activity at the opioid receptor of the κ-type, have marked diuretic effects in *rats* (Leander 1983a), *mice* (Rathbun et al. 1983) and *dogs* (Slizgi et al. 1984). Systemic administration of ethylketocyclazocine (a prototype κ-receptor agonist) was reported by Slizgi and Ludens (1982) to result in a dose-dependent increase in urine formation and reduction of plasma vasopressin levels in conscious rats. The authors proposed that this drug induces its diuretic effect by inhibition of vasopressin secretion, as well as by attenuation of the vasopressin response in the kidney. This proposition was strengthened by their subsequent studies in Brattleboro rats (Slizgi and Ludens 1986). Administration of U-50488, a chemically novel and highly selective κ-receptor agonist, has also been shown to induce a diuretic effect in rats (Leander 1983b; von Voigtlander et al. 1983) and dogs (Slizgi et al. 1984), but not in Brattleboro rats homozygous for diabetes insipidus (Leander 1983b), and to cause a fall in plasma vasopressin levels following intracerebroventricular (Carter and Lightman 1984a) or systemic (Leander et al. 1985; Oiso et al. 1988) application in rats. Carter and Lightman (1984b) found that local injections of U-50488 into the nucleus of the solitary tract induced a marked pressor response which was associated with a significant rise in circulating vasopressin levels. In addition, Leander et al. (1987) reported that other full and partial κ-agonists induced a diuresis, these diuretic effects correlating with plasma

vasopressin levels in water-deprived rats. According to these authors, κ-agonists induce their diuretic effects by suppression of plasma vasopressin levels. However, according to Ashton et al. (1989) this diuresis did not appear to be mediated by inhibition of vasopressin secretion, but could be attenuated by prior adrenal demedullation. Dynorphin, the assumed endogenous ligand for κ-receptors (Oka et al. 1982; Quirion and Pilapil 1984), appears to be a potent suppressor of basal and water deprivation-elevated plasma levels following intracerebroventricular but not systemic administration (ten Haaf et al. 1987). Also Leu-morphine, like dynorphin a κ-receptor agonist derived from the prodynorphin precursor molecule, is a potent inhibitor of vasopressin secretion following intracerebroventricular administration under basal and various stimulated conditions (Yamada et al. 1988). Recently Yamada et al. (1990) showed that the highly selective κ-opioid receptor agonist U-62066E markedly inhibited the elevated vasopressin response to osmotic stimulation and acute haemorrhage.
In vitro studies reveal that the influence of κ-agonists on vasopressin release is not clear. U-50488 modified neither basal nor stimulated vasopressin secretion from isolated neurointermediate lobes of the pituitary in an in vitro perfusion system, indicating that a κ-receptor mechanism in the hypothalamus rather than in the pituitary operates in the control of vasopressin release (Carter and Lightman 1984a). However, Zhao et al. (1988) reported that U-50488 inhibited electrically evoked vasopressin release from rat neural lobes, and suggested that κ-receptors sensitive to U-50488 mediate opioid inhibition of secretion from vasopressin nerve terminals. It has been reported that dynorphin inhibits spontaneous release of vasopressin, but stimulates the electrically-evoked secretion of this hormone from rat neurointermediate lobes (Maysinger et al. 1984). Knepel et al. (1983) also showed that dynorphin increases vasopressin release from stimulated neurointermediate lobes, whereas Bicknell et al. (1985a) did not find any significant effect. van Wimersma Greidanus et al. (1988) investigated the effect of dynorphin on vasopressin release in vitro; they found a tendency towards an inhibition of electrically stimulated release which did not reach statistical significance, with no effect on the basal release of vasopressin. Thus, as found for other opiates and opioid peptides, disagreement exists on the effect of dynorphin on vasopressin release in vitro. Differences in experimental conditions and doses used are, among other factors, thought to cause these discrepancies. Knepel et al. (1983) and Maysinger et al. (1984) showed that the in vitro effects of dynorphin strongly depend on the doses used and on the experimental conditions.

2.1.5 Human Studies

Studies in man reveal that opiate agonists and antagonists induce changes in vasopressin secretion, with the most consistant response being an opiate-induced inhibition of vasopressin, as was noted in the rat.
As in rats, morphine has a marked antidiuretic effect on healthy people given water, whereas nalorphine produces antidiuresis in some subjects only (Schnieden and Blackmore 1955). Conversely, the narcotic antagonists oxilorphan, butorphanol and cyclazocine induce diuresis in normally hydrated subjects (Miller 1980; Nutt

and Jasinski 1974). Oxilorphan even augmented urine flow and decreased urine osmolality in water-deprived subjects (Nutt and Jasinski 1974). These authors suggest that the diuretic action of oxilorphan is due to inhibition of vasopressin release, although an antagonism between vasopressin and morphine at a renal level could not be ruled out. However, Miller (1980) showed that oxilorphan suppressed urinary vasopressin excretion when the osmotic threshold for vasopressin release was exceeded. Although these studies were interpreted as showing that opiate antagonists inhibit vasopressin release, it is clear that these drugs may well be acting as opiate agonist drugs at non-μ-receptors in naive subjects. In addition, it has been shown that in patients given a neuroleptanesthetic followed by postoperative intramuscular narcotics, plasma vasopressin levels were more than twice as great as in patients given an epidural anesthetic followed by epidural fentanyl (von Bormann et al. 1983). Small doses of fentanyl may stimulate vasopressin secretion during anesthesia (Lehtinen et al. 1984), whereas large doses of opiates are able to prevent the vasopressin response to surgical stress. This may indicate that the action of fentanyl, and maybe other opiates, on vasopressin release are bidirectional, although changes in blood volume and pressure must also be taken into account.

β-Endorphin administered intravenously to healthy men had no stimulatory effect on the release of vasopressin (Reid et al. 1981a), but the methods used and the low basal levels of vasopressin did not allow the determination of a possible inhibitory effect.

Infusion of the long-acting analog of Met-enkephalin DAMME (FK 33-824) resulted in water diuresis in dehydrated or hypertonic saline-loaded subjects by inhibiting vasopressin release (Grossman et al. 1980). Such an infusion also inhibited the rise of vasopressin in response to tilt in normal volunteers (Lightman and Forsling 1980a), but DAMME was not found to alter the vasopressin response to a hypertonic stimulus (Lightman et al. 1980). Brownell et al. (1980) described a reduction of urinary concentrations of vasopressin following administration of DAMME, suggesting an inhibition of vasopressin secretion by Met-enkephalin in humans, while the enkephalin analog Met-kephamid suppressed plasma vasopressin (Zerbe et al. 1982). More recently, the selective κ-opioid agonist U-62,066E has been shown to produce a water diuresis in humans, which may have resulted from suppression of plasma vasopression levels (Peters et al. 1987). Thus, the majority of the data in man suggest that opioids and opiates inhibit the release of vasopressin.

2.1.6 Opiate Antagonists

In various species naloxone, an opiate antagonist, has been investigated for its effects on vasopressin release in order to determine the role of endogenous opioids on posterior pituitary function. Most of the described effects of opiates and opioid peptides on vasopressin release could be counteracted by administration of naloxone. This opiate antagonist by itself reduces diuretic responses induced by water and alcohol, but is inactive in normally hydrated *rats* (Guiol et al. 1984). In higher doses naloxone decreased urine flow in Brattleboro rats, whereas lower doses failed to do so. Auriac et al. (1985) observed a reduced diuresis during the 2 h of naloxone injection in normal as well as in homozygous diabetes insipidus rats from the Brattleboro strain, suggesting that the antidiuretic properties of naloxone are independent of vasopressin secretion.

On the other hand, naloxone does not prevent GABA-induced inhibition of the angiotensin-stimulated release of vasopressin (Summy-Long et al. 1986), and the vasopressin concentrations were unaffected by naloxone in rats treated with either saline (placebo) or histamine, nicotine, isoproterenol or hypertonic saline (Summy-Long et al. 1986). Similar results have been obtained by van Wimersma Greidanus et al. (1979, 1981) and ten Haaf et al. (1986, 1987). Furthermore, the opiate antagonist naltrexone did not affect basal plasma vasopressin or the elevated vasopressin concentrations caused by osmotic stimulation, hypovolemia or nicotine (Rosella-Dampman et al. 1983). However, in spontaneously hypertensive rats from the Wistar–Kyoto strain, naloxone increased plasma concentrations of vasopressin (Rosella-Dampman et al. 1985; Rockhold et al. 1986), indicating that naloxone-sensitive endogenous opioid mechanisms may play a regulatory role in the control of vasopressin secretion in these animals. Surprisingly, naloxone reduced the vasopressin response to hypoxia in male rats (Forsling and Aziz 1983) and increased plasma vasopressin in female rats in diestrus but not in pregnant rats at 20 or 21 days of gestation or before or during parturition (Hartman et al. 1986a). Naloxone also induced a small but significant increase of vasopressin levels in the plasma of lactating rats (Bicknell et al. 1985c). Naloxone further elevated plasma vasopressin release in response to osmotic stimulation in immature rats but not in adult animals (Hartman et al. 1986b). Naloxone significantly potentiated the vasopressin response to immobilization, but in female rats only (Carter et al. 1986). In an elegant study recently performed by Forsling et al. (1988) comparing the vasopressin response to intraperitoneal and to intravenous administration of hypertonic saline, naloxone was shown to both increase resting vasopressin levels and effectively potentiate vasopressin release in response to hypertonic saline by reducing the osmotic threshold for hormone release. Yamada et al. (1989) demonstrated that iv injection of naloxone (0.5 mg/kg) raised the basal plasma vasopressin level, enhanced vasopressin secretion after 72 h water deprivation, prolonged vasopressin secretion induced by angiotensin and increased carbachol-induced vasopressin secretion. In contrast, in dehydrated *sheep* vasopressin secretion was not affected by administration of naloxone (Thornton and Parrott 1989). Also, in conscious normal sheep intracerebroventricular infusion of naloxone failed to change plasma concentrations of vasopressin (Wang et al. 1989), but this opiate antagonist attenuated the increase in plasma vasopressin induced by DAMME.

Incubation with naloxone did not change vasopressin release from electrically stimulated isolated rat neurohypophysis in vitro (Dencker Christensen and Fjalland 1982; Bicknell and Leng 1982; van Wimersma Greidanus et al. 1988; van de Heijning et al. 1990), whereas according to Knepel et al. (1983) naloxone increases the evoked vasopressin release from rat pituitary neurointermediate lobes, but not from the isolated neural lobes. Maysinger et al. (1984) also found that naloxone induced an increase in the evoked (but not spontaneous) release of vasopressin from neurointermediate lobes of female rat pituitaries in an in vitro superfusion system.

In *man*, naloxone is without effect on basal vasopressin levels (Lightman and Seckl

1987), and did not alter the vasopressin response to a hypertonic stimulus (Lightman et al. 1980). However, naloxone reduced resting levels of vasopressin and inhibited the rise in vasopressin induced by an orthostatic stimulus according to Lightman and Forsling (1980b). The vasopressin response to insulin-induced hypoglycemia has been reported to be either unchanged (Lightman and Seckl 1987; Chiodera and Coiro, 1990; Johnson et al. 1990) or blocked (Grossman and Forsling, unpublished observations) by high-dose naloxone infusion. In general, opiate-induced inhibition of vasopressin in man is relatively resistant to naloxone antagonism (Grossman et al. 1980; Brownell et al. 1980).

2.1.7 Interaction Between Endogenous Vasopressin and Opioid Systems

Since the hypothalamus contains a considerable amount of β-endorphin, and immunohistochemical studies have indicated that the paraventricular nucleus and, to a lesser extent, the supraoptic nucleus contain β-endorphin-like innervation (Dupont et al. 1980; Khachaturian et al. 1985), it is tempting to assume that endogenous β-endorphin may regulate the release of vasopressin at the hypothalamic level. Since dopamine and catecholamines may also regulate the release of vasopressin, it is possible that noradrenergic and dopaminergic pathways interact with opioid pathways in the control of vasopressin secretion (Bicknell et al. 1987; Lightman et al. 1982, 1983a; Forsling et al. 1988).

Recently, it has been proposed that opioid receptors of the κ-subtype are involved in the release mechanism of vasopressin (Gerstberger and Barden 1986). The demonstration that the opioid receptors on pituicytes in the neural lobe (Lightman et al. 1983b) are of the κ-subtype (Bunn et al. 1986; Herkenham et al. 1986), suggests an important role for the assumed endogenous κ-agonist (dynorphin) in vasopressin secretion. In addition, prodynorphin and products of this opioid precursor such as dynorphin and α-neo-endorphin have been visualized in populations of neurons in the brain and in fibers in the posterior lobe of the pituitary (Ito et al. 1981; Khachaturian et al. 1985; Palkovits et al. 1983; Watson et al. 1981; Weber et al. 1982). Both dynorphin and α-neo-endorphin have been identified within the paraventricular, supraoptic, suprachiasmatic and accessory nuclei of the hypothalamus, the sites of origin of vasopressin-containing neurons. In fact, dynorphin and vasopressin occur in the same hypothalamic cells of rats (Watson et al. 1982), and dynorphin has been shown to be present within vasopressin neurosecretory granules in the pituitary of rats (Whitnall et al. 1983). A more recently performed re-examination confirmed co-storage of immunoreactive dynorphin-(1–8)-like material with vasopressin (Gaymann and Martin 1987). This suggests that endogenous dynorphin plays an important physiological role in the regulation of vasopressin release. In this respect it is interesting to note that stress and hypovolemic hypotension produce site- and time-dependent changes in the distribution of dynorphin, Leu-enkephalin and vasopressin in the rat central nervous system (Feuerstein et al. 1985) and that osmotic stimulation results in a parallel increase in mRNAs for both prodynorphin and provasopressin in hypothalamic nuclei (Sherman et al. 1986).

The identification of an extensive enkephalinergic system within the hypothalamo-neurohypophyseal axis has led to the speculation that naturally occurring enkephalins are involved in the regulation of the posterior pituitary (Micevych and Elde 1980; Rossier 1982; Rossier et al. 1979). In this respect it is interesting to notice that the enkephalins coexist with vasopressin (and oxytocin) in nerve terminals of rat neurohypophysis (Martin and Voigt 1981). It appears that vasopressin-containing neurons are often immunoreactive for Leu-enkephalin (whereas Met-enkephalin is invariably associated with oxytocin-containing nerve terminals) (Adachi et al. 1985; Martin and Voigt 1981). The data suggest that enkephalins and dynorphins are co-peptides of the neurohypophyseal neurosecretory system, and it is postulated that the secreted enkephalins may act on autoreceptors interfering with the release of oxytocin and/or vasopressin. Recently it was reported (Blanco et al. 1989) that Met-enkephalin treatment caused an increase in number of immunostained vasopressin neurons in the rat supraoptic nucleus. This effect was more pronounced in females than in males. Although this does not really clarify the issue of the regulation of vasopressin release by opioids, it supports the view that dynorphin and enkephalin pathways to the posterior lobe are functionally important (Rossier 1982).

2.1.8 Conclusions

The main products of the three opioid precursors pro-opiomelanocortin, prodynorphin and pro-enkephalin – β-endorphin, dynorphin and enkephalin respectively – are presumed to control the secretion of neurohypophyseal hormones. In particular, concomitantly stored and presumably co-released enkephalins and dynorphins may act as regulators of the secretion of these principles (Martin et al. 1983). Activation of different types of opiate receptors by opioid peptides may differentially and selectively modulate the release of vasopressin and oxytocin from the posterior pituitary, possibly by an interaction at the level of the pituicytes (van Leeuwen et al. 1983).

Although similar effects have not been found under all circumstances and in all species, the general opinion is that opioid peptides have a predominantly inhibitory influence on the release of vasopressin in rats and man, whereas in dogs and rabbits the picture may be different, tending to a stimulatory influence. It has been shown (Mansour et al. 1986) that a marked difference exists between species in the amount of opiate receptors of the various subtypes in the neural lobe of the pituitary as well as in the processing of prodynorphin (Dores and Akil 1987).

From our own findings and from the data available in the literature on the effects of opioid peptides on vasopressin secretion and on the distribution of the various opioid systems in brain and pituitary, we tend to assume that in rats the inhibitory influence of β-endorphin on vasopressin release takes place at the level of the hypothalamus. Taking into account the conflicting data on the effects of prodynorphin derivatives on vasopressin release in vitro, we also tend to assume that the primary site of action of these opioids is at the hypothalamic rather than at the pituitary level. This is despite the fact that κ-receptors are localized on nerve terminals and/or processes of pituicytes in the neural lobes of the rat pituitary (Bunn

et al. 1986; Herkenham et al. 1986) and the observation that dehydration reduces κ-opiate receptor binding in the neurohypophysis of the rat (Brady and Herkenham 1987; Brady et al. 1988). In our opinion, the κ-receptor in the posterior pituitary is more directly involved in the secretion of oxytocin (see below) than in that of vasopressin. Also, the observations that the electrical activity of putative vasopressinergic neurons is influenced by opioids and opiate antagonists (Arnauld et al. 1983; Yamashita et al. 1988), and that water deprivation specifically activated dynorphin mRNA in magnocellular neurons of the paraventricular nucleus (Sherman et al. 1986), support the view of a central action of these principles. In fact, it has been suggested by Summy-Long et al. (1990) that the co-release of a neuromodulator (dynorphin peptides) with vasopressin attenuates the inappropriate secretion of oxytocin during dehydration. Experimental evidence obtained by Summy-Long (1990) and by us (van de Heijning et al., submitted) strongly supports this hypothesis of an opioid peptide (dynorphin) co-released with vasopressin that inhibits oxytocin release by acting on κ-receptors at the oxytocin nerve terminals in the posterior pituitary. Consequently this cross-inhibition results in a preferential release of vasopressin as response to appropriate stimulation.
Pro-enkephalin derivatives (almost exclusively Met-enkephalin) should affect, according to our assumption, the release of oxytocin rather than that of vasopressin in rats. It is clear that the use of more specific agonists and antagonists for the various opiate receptor subtypes is needed to answer more precisely the question as to which of the opioid peptides and which receptor subtypes play(s) a role in the regulation of vasopressin secretion. In addition, marked differences in species, in sex and in the state of activation of the neurohypophyseal (vasopressin) system are complicating factors.

2.2 *Oxytocin*

2.2.1 *Morphine and Related Substances*

One of the first reports dealing with the effect of opiates on oxytocin secretion was published by Haldar and Sawyer (1978). It was found that morphine as well as its analogs butorphanol and oxilorphan inhibited initial milk yield in lactating *mice* due to an inhibition of oxytocin release. This observation was later confirmed and extended by the finding of an inhibitory influence of morphine on the release of oxytocin induced by suckling or by administration of acetylcholine (Haldar et al. 1982). The inhibitory effects were antagonized by naloxone. Clarke et al. (1979) also provided evidence that opiates inhibit carbachol-induced as well as electrically or osmotically-induced release of oxytocin in *rats*, and that the opiates act on or close to axon terminals within the neurohypophysis. Naloxone reversed this opiate-induced inhibition of release of oxytocin (Wright et al. 1983). Lutz-Bucher and Koch (1981) also reported an inhibitory effect of morphine on experimentally-induced release of posterior pituitary hormones. Morphine completely abolished the increase in levels of oxytocin in plasma of rats treated with hypertonic salt or 2-deoxy-D-glucose. Clarke and Wright (1984) demonstrated that in lactating rats pentazocine, pethidine and morphine markedly suppressed oxytocin release at doses sufficient to produce analgesia, suggesting that secretion of the hormone is substantially reduced during opiate-induced analgesia. Furthermore, Russell and Spears (1983) observed an inhibition of suckling-induced oxytocin secretion in conscious lactating rats following morphine administration, as well as a disruption in maternal behavior (Russell et al. 1989). Pethidine and morphine appeared to retard parturition in the rat and, since parturition is propelled by secretion of oxytocin, this further points to an opiate-induced inhibition of oxytocin secretion (Gosden et al. 1985; Leng and Russel 1987; Evans et al. 1989b; Russell et al. 1989). Also, in vitro morphine inhibited the basal (Lutz-Bucher and Koch 1980) and the electrically evoked release of oxytocin in a naloxone-reversible fashion (Clarke and Patrick 1983).

2.2.2 *Endorphins*

As was found for morphine, intracerebroventricularly administered β-endorphin also reduced oxytocin secretion in response to an osmotic challenge (Wright and Clarke 1983; Wright et al. 1983), to acetylcholine administration or suckling (Haldar et al. 1982). Intracerebroventricular administration of β-endorphin also suppressed basal levels of oxytocin in plasma of male *rats* (van Wimersma Greidanus and ten Haaf 1985). Likewise, destyrosine-β-endorphin and γ-endorphin decrease basal oxytocin levels (van Wimersma Greidanus and ten Haaf, 1985), although these compounds do not bind to classical opiate receptors. Thus, opiate agonists such as morphine and related substances as well as endorphins suppress the release of oxytocin in vivo.
Using isolated neurointermediate lobes in vitro, no effect of β-endorphin was observed on spontaneous or electrically evoked oxytocin release by Maysinger et al. (1984) and by Bicknell et al. (1985a). However, Vizi and Volbekas (1978) found β-endorphin to increase oxytocin release from isolated neural lobe preparations.

2.2.3 *Enkephalins*

Enkephalins such as the analogs DADLE, DAGO, D-Ala-Met-enkephalin and Met-kephamid have been shown to suppress osmotically evoked release of oxytocin in lactating *rats* (Wright et al. 1982; Wright and Clarke 1983, 1984); D-Ala-Met-enkephalin also produced a naloxone-reversible inhibition of oxytocin release induced by acetylcholine or suckling (Haldar et al. 1982). Leuenkephalin has been documented to inhibit angiotensin-stimulated, but not basal, oxytocin release following intracerebroventricular administration of very high doses (100 μg) (Keil et al. 1984). van Wimersma Greidanus and ten Haaf (1985) reported a suppressive effect of Met-enkephalin and Leu-enkephalin on basal levels of oxytocin following intracerebroventricular injection of relatively high doses (5 μg).
In vitro spontaneous oxytocin release from neurointermediate lobes remained unchanged after Leu-enkephalin administration (10^{-6} *M*), while the electrically

evoked release was stimulated by this treatment; 10^{-7} *M* Leu-enkephalin had no effect (Maysinger et al. 1984). DADLE inhibited oxytocin release from electrically stimulated rat neurointermediate lobes in vitro (Bicknell and Leng 1982; Bicknell et al. 1985a). D-Ala-Met-enkephalin produced a significant decrease in the amount of oxytocin released from an incubated hypothalamo-neurohypophyseal system, whereas the release of oxytocin into the incubation medium was not altered by D-Ala-Leu-enkephalin (Schriefer 1985).

Since DAGO has a greater affinity for μ-receptor sites, whereas DADLE and Met-kephamid act preferentially on δ-receptors, it has been suggested (Wright and Clarke 1984) that the inhibitory action of enkephalins on oxytocin release may be mediated through both μ- and δ-receptors. Concerning the site of this inhibitory action, it seems most likely that it takes place at the level of the oxytocin nerve terminals, but other sites, such as the oxytocinergic cell bodies in the magnocellular nuclei of the hypothalamus, cannot be excluded (Wright and Clarke 1984). The observation that morphine and DADLE inhibit neuronal firing in the supraoptic nucleus in vitro (Wakerley et al. 1983) supports this view.

2.2.4 Dynorphins, Other Prodynorphin Derivatives and κ-Receptor Agonists

In 1982 and 1983 it was reported that intracerebroventricular administration of dynorphin, the assumed physiological agonist for opiate receptors of the κ-subtype, suppressed osmotically induced oxytocin release in lactating *rats* (Wright et al. 1982, 1983; Wright and Clarke 1983). Recently, van de Heijning et al. (1989) found that intracerebroventricular administration of dynorphin-(1–13) suppressed basal levels as well as water deprivation-induced elevated levels of oxytocin in plasma of male rats. Also, synthetic κ-agonists such as U-50488 inhibited electrically stimulated oxytocin release in lactating rats anesthetized with urethane (Coombes and Russell 1987). Grell et al. (1989) reported a decrease in plasma oxytocin levels in male rats following administration of the κ-opioid receptor agonist bremazocine.

In vitro, dynorphin-(1–8) was reported by Falke (1988) to inhibit the potassium depolarization-stimulated release of oxytocin from isolated rat neurosecretosomes, while Bondy et al. (1988) demonstrated that dynorphin-(1–8) and dynorphin-(1–17) produced a substantial reduction in oxytocin release from male rat neurointermediate lobes evoked by low-frequency, but not high-frequency, electrical stimulation. Also the selective κ-agonist U-50488 inhibited the electrically evoked release of oxytocin from male rat neural lobes. Surprisingly, it was reported by Maysinger et al. (1984) that dynorphin-(1–17) did not affect spontaneous oxytocin release from rat neurointermediate lobes but increased the electrically evoked release. This stimulatory effect on oxytocin release was observed following administration of a 10^{-7} *M* concentration of dynorphin-(1–17), but not using 10^{-6} *M*. In addition, the destyrosine derivative of dynorphin-(1–17) appeared to be without effect. Bicknell et al. (1985a) investigated the effects of various agonists on electrically stimulated oxytocin secretion and were unable to observe an effect of dynorphin-(1–13) or the synthetic κ-agonist ethylketocyclazocine. According to Grell et al. (1989), the κ-opioid receptor agonist bremazocine inhibited the secretion of oxytocin from isolated neurohypophyses by 60%.

From one study in *man*, it appears that intravenous bolus administration of β-endorphin does not affect basal oxytocin levels. However, an inhibitory role for β-endorphin could not be excluded because of the low levels of oxytocin and the restrictions of the assays applied (Reid et al. 1981a).

2.2.5 Opiate Antagonists

Naloxone and other opiate receptor antagonists have been used in vivo and in vitro in order to investigate the possible role of endogenous opioids in the regulation of oxytocin secretion. It has been suggested that oxytocin release is inhibited by endogenous opioids, based on the results of experiments in which the effect of naloxone was examined without administration of opiates or opioid peptides. In fact, it has been shown that the intramammary pressure response of *rats* is increased after administration of naloxone, and that carbachol and electrical stimulation of isolated rat neurohypophyses became more effective in releasing oxytocin after naloxone (Bicknell and Leng 1982; Clarke and Patrick 1983; Clarke et al. 1979; Maysinger et al. 1984; Wammack and Racké 1988; van Wimersma Greidanus et al. 1988; van de Heijning et al. 1990). Thus, it is assumed that endogenous opioids tonically suppress the release of oxytocin. This view is also supported by the observation by Haldar and Bade (1981) and Summy-Long et al. (1984, 1986) that opiate antagonists significantly reduced the inhibition of oxytocin release by heat stress in lactating mice and that these drugs further increased oxytocin release during water deprivation and hemorrhage and after histamine, nicotine or an osmotic challenge. Also, Carter and Lightman (1987a) reported that naloxone enhanced oxytocin secretion during days 1–3 of a 9-day period of drinking hypertonic saline. Flanagan et al. (1988) described a potentiation by naloxone of the stimulation of oxytocin secretion by cholecystokinin and lithium chloride, and Russell et al. (1988) reported that naloxone selectively increased oxytocin secretion after stress. This finding is in agreement with the observation that naloxone stimulated secretion of oxytocin, but not of arginine vasopressin or prolactin in response to immobilization stress (Carter and Lightman 1987a; Carter et al. 1986; Samson et al. 1985). However, basal oxytocin levels in rat plasma or spontaneous release of oxytocin from neurohypophyses, neural lobes or neurointermediate lobes in vitro are hardly affected by opiate antagonists (Carter and Lightman 1987; Haldar and Bade 1981; Maysinger et al. 1984; Rosella-Dampman et al. 1985; van Wimersma Greidanus et al. 1985, 1988; van de Heijning et al. 1990). Thus, endogenous opioids play a role in the phenomenon of stress-induced inhibition of oxytocin release. In addition, it has been suggested that the opioid modulation of oxytocin release may play a role in the interval between successive births during normal parturition in rats (Bicknell et al. 1983a; Leng et al. 1985). It has also been suggested that the inhibitory action of relaxin on oxytocin secretion is mediated through an opioid system, and that this relaxin-opioid-oxytocin system may be important in the control of the timing of birth (Jones and Summerlee 1986). Moreover, changes in the levels of opioids and opiate receptors in the preoptic area during pregnancy and lactation (Hammer and Bridges 1987) further implicate endogenous opioids in these oxytocin-dependent processes in reproduction. Recently, the results of a series of experiments performed by Bicknell et al. (1988a, c) clearly demonstrated that opioid receptors mediate powerful inhibitory inputs to the electrical and secretory activities of magnocellular oxytocin neurons.

In *rabbits*, naloxone did not alter oxytocin secretion during coitus (Todd and Lightman 1986); in the *goat*, however, although naloxone had no effect on basal plasma concentrations of oxytocin it markedly potentiated oxytocin secretion following vaginocervical stimulation (Seckl and Lightman 1987). Moreover, in dehydrated *sheep* oxytocin secretion was not affected by administration of naloxone (Thornton and Parrott 1989).

In normal *man*, intravenously administered naloxone appeared to have no effect on basal oxytocin levels (Honor et al. 1986), but induced a significant rise in oxytocin levels during insulin-induced hypoglyemia (Dunne et al. 1987; Lightman and Seckl 1987; Seckl et al. 1988) and during smoking. Since naloxone did not influence oxytocin levels during infusion of hypertonic saline it was concluded by Seckl et al. (1988) that in man a stimulus-dependent opioid-mediated inhibition of oxytocin secretion exists.

2.2.6 Interaction Between Endogenous Oxytocin and Opioid Systems

Enkephalins coexist with oxytocin and vasopressin in nerve terminals of rat neurohypophysis (Martin and Voigt 1981). It appears that Met-enkephalin is invariably associated with oxytocin-containing nerve terminals (whereas vasopressin-containing neurons often are immunoreactive for Leu-enkephalin) (Gaymann and Martin 1987).

The hypothalamus also contains a considerable amount of enkephalin and β-endorphin (Nilaver et al. 1979). Immunohistochemical studies indicate that the paraventricular nucleus and, to a lesser extent, the supraoptic nucleus contain β-endorphin-like innervation. This suggests that opioid peptides like β-endorphin may regulate the release of neurohypophyseal hormones at the hypothalamic level. Since dopamine may regulate the (reflex) release of oxytocin (Clarke and Patrick 1983; Moos and Richard 1979, 1982; Seybold et al. 1978; Vizi and Volbekas 1980) it is possible that opiates and opioid peptides exert their effects on the release of oxytocin via dopaminergic pathways. However, opioid-noradrenergic interactions in the neurohypophysis may also occur. According to Zhao et al. (1988), opioid peptides within the neurohypophysis suppress secretion of oxytocin and, to a lesser extent, vasopressin partly through inhibition of the release of facilitatory norepinephrine. Thus, direct and indirect effects of opioids on the posterior pituitary are involved in addition to a regulatory influence at the hypothalamic level.

2.2.7 Conclusions

Various reports provide electrophysiological data supporting the view that endogenous opioids act at the final common pathway or at the neurosecretory terminals to inhibit oxytocin release (Shibucki et al. 1988; Summy-Long et al. 1984, 1986; Way et al. 1987), but an (additional) central site of action cannot be excluded (Bicknell et al. 1988a). In an earlier review we suggested that pituitary as well as central (hypothalamic) sites of opioid action are involved in the (inhibitory) regulation of oxytocin secretion (van Wimersma Greidanus and ten Haaf 1985) and our view has not basically altered. Studies using chronic morphine treatment and studying morphine dependence, cross-tolerance and withdrawal effects in relation to oxytocin secretion confirm this idea (Bicknell et al. 1985c, d, 1985b, c; Leng et al. 1987; Rayner et al. 1988). During recent years data have become available that at the pituitary level κ-receptor agonists rather than δ-receptor agonists regulate oxytocin secretion, at least in rats. According to Carter and Lightman (1987b) μ and κ-opiate receptor types are differentially involved in the opioid regulation of stress-induced oxytocin secretion, depending on the type of stress. Also, recent data obtained by Evans et al. (1989a) strongly suggest that μ- and κ-opiate receptor agonists inhibit oxytocin (and vasopressin) secretion. Moreover, it has been put forward that opioids released from the neurosecretory terminals in the neural lobe are involved in the inhibition of oxytocin secretion and that opioids from the neighboring intermediate lobe can effectively be ruled out of consideration (Bicknell et al. 1983b, 1985a, b).

2.3 General Conclusions: Posterior Pituitary Hormones

In previous reviews we reported the various experimental findings as described in the literature on the effects of opiates on the release of vasopressin and oxytocin from the posterior lobe of the pituitary (van Wimersma Greidaus 1987; van Wimersma Greidanus and ten Haaf 1984; van Wimersma Greidanus et al. 1985). We concluded that over the years the literature has inclined towards the view that there is a suppressive influence of opiates and opioid peptides on vasopressin release. In particular, over the past few years an increasing amount of evidence has been obtained that these substances inhibit rather than stimulate the release of hormones from the posterior pituitary. The disagreement occasionally expressed in the literature as to whether morphine and related substances inhibit or stimulate the secretion of vasopressin may in large part be caused by the variety of animal species studied, by the differences in routes of drug administration, and by the different experimental situations.

Interestingly, no disagreement whatsoever exists on the effect of opiate-like substances on the release of oxytocin. Virtually all studies performed in this area reveal that opioids and opiate-like substances suppress the oxytocin secretion. It may be concluded that various opioid peptides and opiate receptor agonists suppress oxytocin release from the posterior pituitary in rats and mice. In other species only few data are available. As a result of the studies on the influence of opiate antagonists on oxytocin release, it can also be concluded that endogenous opioids are involved in the regulation of hormonal secretion from the posterior pituitary. The general divergence of opinion concerning the influence of opiates and opioid peptides on the release of either vasopressin or oxytocin implies that opioids may differentially affect the release of vasopressin and oxytocin, an assumption which is based also on direct experimental evidence, reported by Clarke and Patrick (1983) and by Summy-Long et al. (1984) among others. In addition, the differential regulation of posterior pituitary function by opioids may depend on age and sex as well as on the state of activation of the hypothalamo-neurohypophyseal system (Summy-Long et al. 1984; Carter et al. 1986; Hartman et al. 1986a, b).

In general, it is believed that, at least in rats, the predominant inhibitory influence of opioids on vasopressin release takes place at the hypothalamic level. As far as the suppressive effect of opioids on oxytocin secretion is concerned, sites of action at the level of the neural lobe as well as the hypothalamus are assumed to be involved. It is hypothesized that the source of opioid peptides that attenuate oxytocin secretion may also be in the neural lobe, and enkephalin and dynorphin, which are depleted from the lobe along with vasopressin after salt loading (Höllt et al. 1981; Rossier et al. 1979; Zamir et al. 1985) and water deprivation (Mata et al. 1977; Zamir et al. 1985), have been put forward as likely candidates for inhibitors of oxytocin release in this respect (Hartman et al. 1986b). Since it has also proposed that κ-receptors rather than δ-receptors are involved, the endogenous ligand for κ-receptors, i.e. dynorphin, remains the most likely candidate. In this respect it is interesting to notice that dehydration results in a change in κ-opiate receptor binding in the neurohypophysis (Brady and Herkenham 1987; Brady et al. 1988). The finding that κ-receptors are localized on nerve terminals and/or processes of pituicytes in the neural lobe of the rat pituitary (Bunn et al. 1986; Falke and Martin 1988; Gerstberger and Barden 1986; Herkenham et al. 1986) further favors this idea. Thus, dynorphin released from (activated) vasopressin neurons may be responsible for the attenuation of oxytocin secretion (Hartman et al. 1987; Bondy et al. 1988; Van de Heijning et al. 1990; see also Sect. 2.1.8). It may be that the biological significance of this interaction and differential endogenous opioid inhibition at the level of the neural lobe represents a mechanism to selectively modulate the output of the two neurohypophyseal principles to specific stimuli (Summy-Long et al. 1984; Bondy et al. 1988). In addition, the differential response may be regulated by opioid input at the level of the magnocellular nuclei in the hypothalamus.

3 Anterior Pituitary Hormones

Just as for the release of hormones from the posterior pituitary, it has been known for some decades that narcotic analgesics influence the function of the anterior pituitary. Earlier reviews dealing with this subject (Smith 1972; George 1971; George and Lomax 1972; de Wied et al. 1974; Meites et al. 1979) appeared at times when the activity of anterior pituitary function was sometimes still measured by indirect methods.

With the availability of well developed radioimmunological techniques it has become possible to assess the influence of morphinomimetics and of opioid peptides on anterior pituitary function. Other reviews have also recently been published on this issue (Szekely 1983; Morley 1983; Grossman 1983; Herz and Millan 1984; Millan and Herz 1985; Maysinger 1985).

3.1 Growth Hormone

3.1.1 Morphine and Related Substances

Narcotic analgesics, mainly morphine, have been shown to cause an increase in radioimmunoassayable growth hormone (GH) upon acute administration in *rats* (Kokka et al. 1973; Bruni et al. 1977; Bluet-Pajot and Schaub 1977; Spiegel et al. 1982; Pechnick et al. 1985; Chihara et al. 1978; Cocchi et al. 1977). The very short delay of the GH response to morphine suggests that the opiate acts directly on receptor(s) located within the hypothalamo-hypophyseal system (Bluet-Pajot and Schaub 1977). van Vugt and Meites (1980) showed that systemic administration of morphine resulted in a dose-dependent increase in circulating GH levels. This effect appeared to be counteracted by concurrent administration of the opiate antagonist naloxone. In addition, it appeared that naloxone given alone significantly lowered serum GH concentrations in the rat (van Vugt and Meites 1980; Bruni et al. 1977; Shaar et al. 1977). However, this effect of naloxone may depend on the dose used (Shaar et al. 1977) and it has also been reported that naloxone did not affect the pulsatile release of GH (Martin et al. 1979). However, the suckling-induced rise of GH can be inhibited by naloxone (Riskind et al. 1984). According to Lien et al. (1979), naloxone counteracted the effect of morphine on GH, but did not affect nonstimulated GH levels.

Similar to morphine, the opiate receptor agonist methadone stimulated GH release in adult male rats following subcutaneous injection. Not only the opiate antagonist naloxone but also naltrexone blocked this stimulatory effect of both morphine and methadone (Shaar and Clemens 1980). When administered alone, both naloxone and naltrexone were reported to reduce serum GH concentrations (Shaar and Clemens 1980). Neither the dophaminergic agonist apomorphine nor the α-adrenergic receptor agonist clonidine significantly affects the stimulatory effect of morphine on GH release. Moreover, morphine caused GH release in rats pretreated with α-methyl-*p*-tyrosine, a catecholamine synthesis inhibitor. These data suggest that opiates stimulate GH release via a mechanism independent of α-adrenergic or general catecholaminergic influences (Shaar and Clemens 1980). However, pilocarpine completely blocked the effect of morphine on GH release, whereas physostigmine, a cholinesterase inhibitor, reduced the stimulatory effect of morphine on GH secretion (Shaar and Clemens 1980), suggesting that cholinergic systems may be involved in the opiate-induced stimulation of GH release, although the mechanism by which this occurs is still not clarified.

In addition to opiate agonists (morphine, methadone), partial agonists such as levallorphan, nalorphine and pentazocine also raised GH levels. Some of these partial agonists appeared to suppress the morphine-induced elevation of GH, presumably because of their partial antagonist capacity.

In the immature male *pig* levels of GH were increased after morphine treatment and this increase was delayed by naloxone. Naloxone alone did not affect GH secretion (Trudeau et al. 1988).

In *goats*, intravenous administration of morphine (10 mg) raised plasma GH levels, which returned to basal concentrations within 40 min. A similar, although somewhat delayed, response was observed following subcutaneous administration of morphine. Naloxone completely inhibited the GH response to morphine (Hart and Cowie 1978).

In male *human* volunteers, intravenous methadone and nalorphine, but not morphine or pentazocine, have been reported to elevate GH. These responses were reversed by low doses of naloxone. Infusions of naloxone did not alter basal GH levels (Delitala et al. 1981a, 1983b). GH levels did not change in female patients after intravenous administration of morphine as pre-anesthetic medication (Tolis et al. 1975), nor in three patients with pain caused by cancer (Foley et al. 1979). Naloxone did not significantly change GH plasma levels in normal male volunteers (Volavka et al. 1980). Recently, it has been reported that the μ-opiate receptor agonist fentanyl caused a moderate, dose-dependent increase in plasma GH concentrations, peaking 45–75 min after an intravenous bolus injection into male subjects (Hoehe et al. 1988). However, further analysis of the data revealed marked individual differences in GH responses to fentanyl administration.

3.1.2 Endorphins

In 1977 it was shown by Rivier et al. that β-endorphin, in doses from 30 μg to 100 μg, as morphinomimetics, stimulated GH secretion in *rats*. Dupont et al. (1977) reported that β-endorphin injected intraventricularly in rats that had been treated with antiserum to somatostatin led to a 6–10 fold increase of plasma GH levels. Other workers have also described a stimulatory effect of β-endorphin on GH release in rats (Chihara et al. 1978; Collado-Escobar et al. 1986). α-Endorphin and γ-endorphin failed to produce any significant change in serum GH concentrations (Chihara et al. 1978, confirming earlier data obtained by Rivier et al. 1977) showing that α-endorphin was devoid of GH-releasing activity. Neither morphine nor β-endorphin influenced release of GH by cultured rat pituitary cells in vitro, suggesting a central nervous system site of action of these substances in their GH-releasing effect in vivo (Rivier et al. 1977). However, Khorram and McCann (1986) could not confirm the finding of a stimulatory effect of β-endorphin on GH release. Lower doses (20 ng) of intracerebroventricularly administered β-endorphin did not affect GH levels in female rats, whereas higher doses (150 ng) of this opioid peptide tended to increase GH levels within 15 min; however, this increase did not reach statistical significance.

It has been suggested that pituitary β-endorphin is involved directly in the GH rise in response to suckling in the rat. This suggestion is in part based upon the observation that naloxone blocked this rise and that inhibition of the rise in circulating β-endorphin release in response to suckling also prevented the GH response (Riskind et al. 1984).

In *humans*, Reid et al. (1981b) found that synthetic human β-endorphin, administered as a 2.5-mg intravenous bolus, did not appear to alter basal levels of GH in male and female volunteers. Foley et al. (1979) also reported that intravenous administration of β-endorphin did not affect plasma GH levels, while intracerebroventricular administration suppressed plasma GH.

3.1.3 Enkephalins

Bruni et al. (1977) reported that in male *rats* intraperitoneal administration of 5 mg/kg Met-enkephalin resulted in a significant rise in GH serum concentrations in one experiment, but not in another. Concurrent administration of naloxone (0.2 mg/kg) with Met-enkephalin reversed the stimulatory effect of the opioid peptide on GH release (Bruni et al. 1977; van Vugt and Meites 1980). Met-enkephalin increased GH-levels in rats pre-treated with antiserum to somatostatin, but to a lesser extent than β-endorphin (Dupont et al. 1977). Also, Leu-enkephalin, administered at intraventricular doses of 50–100 μg, produced a dose-related increase in plasma GH in infant rats (Cocchi et al. 1977). Pretreatment of the pups with naloxone (2 mg/kg) almost totally suppressed the GH-releasing effect of Leu-enkephalin. In contrast to these data, Met-enkephalin was found to be ineffective in stimulating GH release following intravenous injection of 300 μg into anesthetized adult rats by Rivier et al. (1977). These latter authors also reported that Met-enkephalin is devoid of pituitary in vitro activity. Shaar et al. (1977) reported in immature female rats a subcutaneous injection of 1–10 mg/kg D-Ala^2-Met-enkephalin amide or D-Ala^2-Leu-enkephalin amide, but not of Met-enkephalin amide, resulted in significant dose-dependent elevations of serum GH levels. These effects could be antagonized by naloxone, which displayed maximal antagonistic activity at 0.2 mg/kg. Met-enkephalin, in a dose of 10^{-9} *M*, led, according to Chihara et al. (1978) only to a meager, nonsignificant change in serum GH. However, D-Ala^2-Met-enkephalin amide stimulated GH release significantly. This stimulatory effect was also observed in rats treated with somatostatin-antiserum.

The long-acting analog of Met-enkephalin, DAMME (FK 33-824), given intravenously to healthy male volunteers in a dose of 0.25 mg, raised serum GH levels (Stubbs et al. 1978). This effect occurred within 15 min with a peak at approximately 45 min. Pretreatment with 0.4 mg naloxone attenuated the response to DAMME. Delitala et al. (1981b) also reported that DAMME produced a clear elevation in GH. By using dopamine and the dopamine antagonist domperidone, they showed that it is unlikely that dopamine is involved in the influence of enkephalins of GH secretion in *man*. There is also evidence that neither α_1- nor α_2-adrenoreceptors are involved (Bouloux et al. 1986). In patients with cortisol deficiency due to either Addison's disease or bilateral adrenalectomy for Cushing's disease, DAMME (0.5 mg intramuscularly) induced an increase in GH similar to that described in normal subjects (Allolio et al. 1982). However, in cases of long-standing hypercortisolism, the GH response to DAMME was completely abolished. It is likely that this reflects the effects of glucocorticosteroids on the pituitary, as GH responses to most stimuli are attenuated in Cushing's syndrome (Allolio et al. 1982).

3.1.4 κ-Receptor Agonists

Subcutaneous administration of the κ-receptor agonist ethyl-ketocyclazocine elicited an increase in serum levels of GH in male *rats* (Pechnick et al. 1985), which was completely blocked by pretreatment with naloxone (10 mg/kg) or WIN 44,441-3.

This latter drug has been reported to possess greater antagonist potency than naloxone towards κ-opiate receptors (Michne et al. 1978). Krulich et al. (1986a) tested the effects of the κ-receptor agonists bremazocine and U-50488 on GH secretion in male rats. They found that the lower doses (0.013 and 0.041 mg/kg intravenously) of bremazocine were ineffective, but that the higher doses (0.125 and 0.5 mg/kg) caused a decline of GH levels. A similar effect was observed with increasing doses of U-50488. Since the authors confirmed, in the same series of experiments, the stimulatory effect of morphine on GH release, it is evident that there is a striking difference between the effects of κ-agonists and morphine (as well as endorphins and enkephalins) in their influence on GH secretion (Krulich et al. 1986a). In this respect it is surprising that the assumed endogenous κ-receptor agonist dynorphin (Chavkin et al. 1982; Kato et al. 1981) has been shown to stimulate GH secretion following intracerebroventricular administration. Krulich et al. (1986b) also showed that pretreatment of rats with bremazocine inhibited the GH-releasing response to morphine in a dose-dependent fashion. U-50488 also inhibited the GH-releasing effect of morphine, but this κ-agonist appeared to be less potent in this respect than bremazocine. Comparable effects of the κ-agonists were obtained by Krulich et al. (1986b) when the α_2-adrenergic receptor agonist clonidine was applied to stimulate GH secretion. Interestingly, pretreatment with morphine seemed to augment the effect of clonidine. Moreover, it was reported by the same authors that neither κ-agonist had any effect on the stimulation of GH secretion by GH-releasing hormone (tHRH).

In *humans* it was shown by Pfeiffer et al. (1986) that the κ-agonists MR 2033 (racemic benzomorphan) and its levorotatory isomer MR 2034 induced a marked, relatively rapid and dose-dependent increase in plasma GH levels. Pretreatment with naloxone significantly antagonized this increase in GH.

3.1.5 Conclusions

Since the Met-enkephalin analog DAMME, which rather consistently elevated GH levels in human subjects, is a classical and selective μ-agonist, with low affinity for δ-sites and very low affinity for κ-sites, it is tempting to assume that in humans receptors of the μ-type mediate increases in GH secretion, although this does not explain the lack of efficacy of morphine itself. Alternatively, another type of opiate receptors activated by μ-opiate receptors agonists at high concentrations (e.g. δ-receptors) may be involved. The very potent effects of MR 2033 and MR 2034, on the other hand, suggest that, in addition to μ-receptors, activation of the κ-receptors results in GH release. Thus, the release of GH in humans may be controlled in a stimulatory way by (at least) two types of opiate receptors (Pfeiffer et al. 1986), but most reports emphasize the predominance of the μ-receptors.

Although the GH-releasing effect by morphine can be counteracted by naloxone in rats, it can be selectively blocked by the preferential δ-receptors antagonist ICI-154,129 (a Leu-enkephalin derivative), but not by the selective μ-antagonists naloxazone or β-funaltrexamine (Koenig et al. 1984). Thus, it is likely that activation of δ-receptors results in stimulation of GH secretion. Various data support the view that this effect occurs by an action at the hypothalamic level. Stimulation of κ-receptors, on the other hand, results in an inhibition of GH secretion in rats. The nature of this inhibitory effect is not clear. Because of the lack of effects in vitro, it does not seem to involve a direct action on the pituitary somatotrophs. Morphine-induced stimulation of GH release in the rat is not altered in the presence of somatostatin antiserum, but is abolished by coadministration with antiserum to GHRH. This suggests that stimulation is purely via GHRH. However, morphine is able to directly inhibit somatostatin release from rat hypothalamus in vitro (Lengyel et al. 1985), and studies in man demonstrate an additive effect of GHRH with DAMME on GH release, also implicating somatostatin (Delitala et al. 1989). Studies with intervening neurotransmitters in the rat and dog were also controversial, with evidence in favor of catecholamines, acetylcholine, GABA, dopamine and histamine (reviewed by Grossman and Rees 1983). However, in man only a possible interaction with acetylcholine has been suggested (Delitala et al. 1983b).

In summary, it is probable that the major opiate receptor involved in GH regulation is the μ-receptor, which is responsible for GH stimulation in many species. While in the rat this may modulate both GHRH and somatostatin, in the human the principal regulation is via somatostatin, either directly or via a cholinergic neuron. The role of endogenous opioids in the rat is uncertain, but in man they appear to be of little importance, although the response of GH to acute severe exercise in althletes (but not in normals) was blocked by naloxone (Moretti et al. 1983).

3.2 Thyroid-Stimulating Hormone

3.2.1 Morphine and Related Substances, Opioid Peptides, Opiate Antagonists

In one of the earliest publication on an effect of morphine on thyroid function, it was reported that morphine decreased ^{131}I uptake by the thyroid gland following peripheral administration for 5 consecutive days (Samel 1958).

Acute administration of morphine was later shown to decrease basal thyroid-stimulating hormone (TSH) release and to reduce thyroid function (Lomax et al. 1970; Bruni et al. 1977; Judd and Hedge 1982; Muraki et al. 1980; Sharp et al. 1981). This suppressive influence of morphinomimetics on basal putuitary-thyroid function has been shown in different species using various parameters of thyroid function. Concurrent injection of naloxone in *rats* reduced the response to morphine, whereas naloxone given alone failed to alter serum TSH concentrations (Bruni et al. 1977). However, in *mice* a single injection of various morphine derivatives stimulated thyroidal ^{131}I release (Redding et al. 1966), whereas chronic injection of these substances inhibited TSH release. Like morphine, methadone (Kuhn and Bartolome 1985), Met-enkephalin (Bruni et al. 1977), Leu-enkephalin (May et al. 1979) and β-neo-endorphin (Mitsuma and Nogimori 1983b) induced a decrease in serum TSH concentrations in *rats* that were reversible by naloxone. These morphinomimetics and opioid peptides have also been found to inhibit TSH secretion in response to cold in a naloxone-reversible way. Since cold-stimulated TSH release has been used to monitor drug effects at the hypothalamic level

(Männistö 1983), it is tempting to assume that the inhibitory effect of opiates and opioid peptides on TSH secretion is mediated by an action in the hypothalamus. This view is further supported by the observation that perfusion of the median eminence with morphine blocked cold-induced thyrotropin-releasing hormone (TRH) secretion (Arancibia et al. 1986), and that α- and β-endorphin as well as Met- and Leu-enkephalin decreased the release of TRH from incubated hypothalamic tissue (Jordan et al. 1986). On the basis of these findings it has been suggested that opiates and opioid peptides may affect TSH secretion from the pituitary by an interaction with opioid receptors located on TRH neurons or nerve endings in the hypothalamus or median eminence (Rauhala 1988).

Other studies revealed that Leu-enkephalin inhibited TRH-stimulated TSH secretion (Mitsuma and Nogimori 1983a), whereas β-neo-endorphin and morphine appeared to be ineffective (Mitsuma and Nogimori 1983b; Muraki et al. 1980; Sharp et al. 1981). Direct microinfusion of β-endorphin stimulated TSH secretion (Judd and Hedge 1982), but morphine and opioid peptides had no effect on basal TSH release from anterior pituitaries in vitro (Jordan et al. 1986; Sharp et al. 1981). However, in contrast with these data, β-endorphin was reported to stimulate basal release of TSH from anterior pituitary cells in vitro (Judd and Hedge 1983). Recently, it was reported by Rauhala (1988) that morphine, but not MR 2034, DADLE or naloxone, stimulated TSH from dispersed anterior pituitary cells in a superfusion system.

Concerning TRH-stimulated TSH release from dispersed pituitary cells, generally no effect to various opioids and opiates has been found (Judd and Hedge 1983; Sharp et al. 1981), although Leu-enkephalin was reported in a single study to inhibit TRH-stimulated TSH release (May et al. 1979). Thus, it is likely that the major opioid effects on TSH release occur at the hypothalamic rather than at the pituitary level.

The central site of action of opiates and opioid peptides on TSH secretion is further supported by data from reports studying TSH release following central administration of these drugs. β-Endorphin decreased basal TSH secretion following microinfusion of 10 μg into the third ventricle (Judd and Hedge 1982), but not after infusion into the mediobasal hypothalamus (Grandison et al. 1980) or the lateral ventricle (Velkeniers et al. 1983) in doses of 1 and 5 μg. Microinfusion of morphine into the anterior and posterior hypothalamus suppressed thyroid function (Lomax et al. 1970), but lesions in the posterior and not in the anterior hypothalamus abolished the inhibitory effect of morphine (Lomax and George 1966). According to Lomax et al. (1970), morphine may stimulate inhibitory neurons in the caudal hypothalamus which depress pituitary-thyroid function. In a recently performed series of experiments on opiates and TSH secretion, morphine was found to decrease the TSH response to cold when infused into the third ventricle or the median eminence of male rats (Männistö et al. 1984). Infusion into the anterior hypothalamus or periaqueductal gray had no effect on the response of TSH to cold, while infusion into the posterior hypothalamus significantly enhanced it. In general these effects were counteracted by pretreatment with naloxone. The authors (Männistö et al. 1984) concluded that morphine has a dual hypothalamic action on cold-stimulated TSH secretion: a periventricular inhibition and a posterior hypothalamic

stimulation. The same authors (Rauhala et al. 1987, 1988) reported recently that β-endorphin, DAMME, Met-enkephalin and the κ-receptor agonist MR 2034 all inhibited the cold-stimulated secretion of TSH when infused into the third ventricle. The effect of MR 2034 was not antagonized by systemic administration of naloxone, and the influence of β-endorphin was also not affected by this opiate antagonist in a statistically significant way. Moreover, neither Leu-enkephalin nor naloxone had any effect on the cold-stimulated secretion of TSH following intracerebroventricular administration. In addition, it appeared that β-endorphin and MR 2034 stimulated TSH secretion upon infusion of relatively low doses into the posterior hypothalamus, whereas enkephalins generally did not affect TSH following this local administration. From these results the authors (Rauhala et al. 1987, 1988) conclude that different types of opiate receptors, evidently μ- and κ-receptors, located at periventricular sites, have an inhibitory effect on TSH secretion stimulated by cold. In addition, they conclude that smaller but not larger doses of opiates stimulate TSH secretion by an action in the posterior hypothalamus.
Various regimens used to investigate development of tolerance revealed that tolerance also develops to the effect of morphine on cold-stimulated TSH secretion (Rauhala et al. 1988).
As mentioned before, opiate antagonists generally have no effect on basal TSH secretion in rats (Bruni et al. 1977; Grandison et al. 1980; Muraki et al. 1980; Judd and Hedge 1982), suggesting that endogenous opioid peptides do not play a physiological role in basal TSH secretion. Since opiate antagonists have been reported to inhibit the stress-induced decrease in TSH levels (Briski et al. 1984; Judd and Hedge 1982; Sharp et al. 1981), it may be that endogenous opioid systems are in some way involved in the decreased TSH secretion seen during stress. Since it has been reported that naloxone decreased TSH secretion during suckling in rats (Riskind et al. 1984), an opposite (TRH-mediated?) stimulatory effect of endogenous opioids on TSH secretion during suckling might be suggested.
Interestingly, it has been demonstrated (Bhargava and Das 1986) that ligands acting at the δ- and κ-receptors interact with brain TRH receptors, pointing to an interaction between TRH and opiates in the central nervous system.
In nonrestrained *rhesus monkeys* dynorphin-(1–13) was shown to have no significant effect on plasma TSH levels. Likewise, dynorphin-(1–10) amide produced no change in plasma TSH (Gilbeau et al. 1986).
Various μ-agonists, such as morphine, FK 33-824 (DAMME), methadone and dermorphin, increased circulating levels of TSH in *humans* (Stubbs et al. 1978; Delitala et al. 1983b, 1984; Morley et al. 1984; Grossman et al. 1981a), whereas in general human studies have failed to demonstrate a clear-cut effect of naloxone on circulating TSH (Morley 1983; Roti et al. 1984), although incidentally a small but significant depression in TSH has been reported (Grossman et al. 1981a). Naloxone does not markedly alter the response to TRH. Dermorphin, infused for 30 min induced a significant increase in serum TSH concentrations in euthyroid volunteers. Naloxone pre-treatment prevented this rise in serum TSH (Roti et al. 1984).
The Met-enkephalin analog DAMME, significantly elevated TSH in male volunteers upon intravenous administration (Delitala et al. 1981b). This elevation was abolished by infusion with dopamine and enhanced by domperidone. The racemic

κ-agonist MR 2033 and its levorotatory isomer MR 2034, both benzomorphan derivatives, each significantly suppressed plasma TSH in normal *men* following intravenous injection (Pfeiffer et al. 1986). This effect was not antagonized by pretreatment with naloxone. Moreover, this suppressive effect was stereospecific since the (+) isomer, MR 2035, did not affect TSH secretion.

3.2.2 Conclusions

In conclusion, it would appear that in various animal species opiates and opioid peptides suppress basal TSH release. However, morphine has a dual action on TSH secretion stimulated by cold in rats. Periventricular sites of action are involved in the suppressive effect of morphine, whereas the posterior hypothalamus is involved in its stimulatory effect. The attenuation of TSH, as induced by morphine, DAMME and Met-enkephalin, is most probably mediated by periventricular μ-receptors. MR 2034 evidently decreases TSH through periventricular κ-receptors. The stimulatory effect of morphine, β-endorphin and MR 2034 is apparently mediated by μ- and κ-receptors in the posterior hypothalamus. In humans, opiates acting as agonists for the receptors induce a weak stimulatory effect on TSH, possibly by inhibiting the release of either somatostatin (see above) or dopamine (see below). Due to the lack of reproducible and consistent effects of naloxone it seems unlikely that endogenous opioids play an important or dominant role in the physiological regulation of TSH release in animals or in humans.

3.3 Prolactin

3.3.1 Morphine and Related Substances, Opioid Peptides

It has long been accepted that morphine can stimulate prolactin release in the *rat*, and more recently further information has been provided by work on opioid peptides. The stimulation of release of prolactin by drugs such as morphine is readily reversed by even low doses of naloxone (van Vugt and Meites 1980; Pan and Teo 1989), suggesting the involvement of opiate receptors of high naloxone sensitivity, but different to those which control GH release (Lien et al. 1979) Prolactin release also rises in response to β-endorphin, Met-enkephalin (Meites et al. 1979) and dynorphin (Kato et al. 1981), although the particular response in any given study is dependent on the dosage and route of administration. Of greater interest has been the mechanism by which opiates and opioids affect prolactin release. Ferland et al. (1977), using a quantitative microfluorimetric technique, demonstrated that Met-enkephalin increased catecholamine stores in the medial palisade zone of the median eminence; they interpreted this as indicating decreased release of hypothalamic dopamine. More directly, Gudelsky and Porter (1979) found that morphine, β-endorphin and an enkephalin analog all substantially reduced dopamine concentrations in portal plasma, and these reductions were naloxone-reversible. This is compatible with data indicating that blockade of dopamine breakdown by means of pargyline blunted the effect of β-endorphin on plasma prolactin, and was further taken to indicate that opioids preferentially inhibit the release of newly synthesized hypothalamic dopamine (Van Loon et al. 1979, 1980). Iontophoresis of morphine into the arcuate nucleus markedly reduced the concentration of dopamine in portal plasma (Haskins et al. 1981). Wilkes and Yen (1980) also reported that β-endorphin inhibited dopamine efflux from both male and female rat hypothalamus in vitro by approximatey 70%. Thus, there is a substantial body of evidence that opioids may inhibit the release of hypothalamic dopamine, which is considered to be the principal factor controlling prolactin release. However, dopamine is not the only hypothalamic agent able to modify prolactin levels, and it has been unclear whether the total effect of opioids on prolactin could be explained by this mechanism. Opioids can interact with neuroamines other than dopamine, and, in particular, they may increase the turnover (and presumably release) of brain serotonin (Algeri et al. 1978); serotonin is known to be a potent prolactin-releasing factor which acts at a suprapituitary level (Garthwaite and Hagen 1979), and the prolactin-releasing activity of enkephalin is attenuated by serotonin antagonists such as metergoline and methysergide, and blocked by destruction of serotonin terminals (Spampinato et al. 1979). Direct injection of morphine into the mesencephalic dorsal raphe nucleus, the major source of brain serotonin neurons, will also stimulate prolactin release (Johnson 1982). However, neither metergoline nor methysergide is a specific serotonin antagonist, and metergoline may directly inhibit pituitary release by a nonserotonergic mechanism (Besser et al. 1980). Certainly, in the primate there is little evidence for the involvement of serotonin in opioid-mediated prolactin secretion (Wehrenberg et al. 1981; Pan and Tao 1989).

This problem was tackled in a different manner by Arita and Porter (1984), who compared the amount of dopamine required to overcome morphine- and β-endorphin-stimulated prolactin release with that following α-methyl-*p*-tyrosine, and concluded that changes in portal dopamine were insufficient to account for the total change induced by opioids.

An alternative mode of action of opioids on prolactin release is via a direct interaction with the pituitary. While early studies were uniform in demonstrating no direct effect of opiates on the in vitro release of prolactin, Enjalbert et al. (1979) reported that Met-enkephalin and β-endorphin antagonized dopamine-induced inhibition of prolactin release from fresh hemipituitaries incubated in vitro. However, subsequent studies to confirm this finding have generally been unsuccessful, and we were completely unable to show any such action using either dispersed pituitary cells or pituitary fragments in a perifusion system (Grossman and Clement-Jones 1983). Cheung (1984) found that she could reproduce the dis-inhibition of pituitary prolactin by β-endorphin in vitro only at a single concentration of β-endorphin, and at 1 h of incubation: incubation for less than or more than 1 h was ineffective. More recently, the probable solution to these apparently discrepant findings was provided in a series of studies investigating both incubation and perifusion techniques, in which a direct effect of opioids could only be demonstrated using prolonged static incubation (Bentley and Wallis 1986). Previous studies on the thyrotroph dopamine receptor had shown that this receptor only appears during incubation and is not present in fresh pituitary cells with continually replenished medium (Price et al. 1983); it appears that opiate receptors may also be subject to induction by presumptive incubation products, which does not clearly mirror the situation in vivo. Thus, there is little evidence that the acute stimulant effects of opioids in vitro are due in any major way to direct pituitary stimulation of prolactin release.

3.3.2 Opiate Antagonists

The majority of studies have clearly shown that naloxone infusion will inhibit both basal, prolactin release (in most but not all studies) and, particularly, stress-induced prolactin release in male *rats* (van Vugt and Meites 1980). As it is equally effective in hypothalami subjected to de-afferentation, the endogenous opioid involved is likely to be acting at a hypothalamic locus. Furthermore, the lack of efficacy of the quaternary derivative naloxone bromide in changing prolactin suggests a site within the blood-brain barrier (Panerai et al. 1981). In an early study, antiserum to β-endorphin given intravenously had no effect on serum prolactin, but in subsequent work intracerebroventricular antiserum lowered serum prolactin to 56% of the control level (Ragavan and Frantz 1981). It therefore seems likely that the endogenous modulation of prolactin occurs via β-endorphin, most neurons of which are located in the hypothalamic arcuate nucleus. In the male rat β-endorphin might sensitize the lactotroph to TRH by a direct effect (Buydens et al. 1978), whereas central opioid receptors might be involved in the prolactin response to stress (Velkeniers et al. 1987). Little effect of naloxone is seen in female rats, and it appears that opioid regulation is contingent upon the neonatal presence of testosterone (Limonta et al. 1989).

The physiological significance of endogenous opioid modulation of prolactin release remains unclear. As has been mentioned, low doses of naloxone can reduce the prolactin response to a variety of stressful stimuli. In addition, suckling-induced prolactin release can be blocked by naloxone in the rat and the *sheep* (Knight et al. 1986). The prolactin-releasing effect of arginine vasotocin also appears to be opioid-mediated (Blask et al. 1984). However, there is no clear understanding of the actual physiological function of this modulation, nor of the situations in which it is subject to change.

3.3.3 Human Studies

An extensive series of human studies have confirmed that morphine (Tolis et al. 1978), β-endorphin, analogs of Met-enkephalin such as DAMME, dermorphin (Degli Uberti et al. 1983), and many opiate alkaloids uniformly stimulate prolactin release (reviews: Grossman 1983, 1987). This stimulation is readily antagonized by low doses of naloxone, such as 0.8 mg, suggesting that in man, as in the rat, receptors of high naloxone sensitivity are involved. Based on receptor binding studies, these are most likely to be of the μ-receptor subclass, although it should be noted that the probable endogenous ligand in the rat is β-endorphin, which may preferably activate the so-called ε-receptor. However, as morphine has only a very low affinity for ε-receptors, it seems unlikely that such receptors are particularly important in the human.

Prolactin responses to opioids and opiates can be blocked by dopamine antagonists such as metoclopramide and domperidone (Delitala et al. 1981a, b). However, the prolactin response to a maximal dose of metoclopramide cannot be further potentiated by morphine, suggesting that some dopamine tone is necessary for morphine to stimulate prolactin release (Delitala et al. 1983a). This in turn is compatible with the suggestion that opiates stimulate prolactin secretion by decreasing dopamine release from the hypothalamus. Certainly, blockade of α_1- or α_2-adrenoceptors or muscarinic cholinoceptors (Delitala et al. 1983b; Bouloux et al. 1986) does not seem to affect opioid-mediated prolactin release, although there is some (indirect) evidence that serotonin may be involved (Foresta et al. 1985). Thus, the present data suggest a direct interaction between an opiate receptor and the tuberoinfundibular dopaminergic system as the predominant mechanism for opiate-stimulated prolactin release in the human, with a possible lesser role for serotonin.

There is also generally uniform agreement that narcotic antagonists such as naloxone can rarely be shown to influence serum prolactin under a variety of conditions. While an early study reported that naloxone could lower basal resting serum prolactin in normal subjects, this has not been the experience in many other studies. Similarly, one report that 1.6 mg naloxone attenuated the prolactin response to hypoglycemia (Saltiel et al. 1982) has not been confirmed using either higher (16 mg: Grossman et al. 1981a) or lower (0.8 mg/2 h: Wakabayashi et al. 1980) doses of naloxone. Furthermore, no dose of naloxone had any effect on the elevation in serum prolactin seen during sleep (Martin et al. 1979), in the puerperium (Grossman et al. 1982a), following suckling (Lodico et al. 1983) or food ingestion (Ishizuka et al. 1983), or in renal failure (Smith et al. 1985). Acute severe exercise in normal fit subjects produces a rise in serum prolactin which is not reversible by naloxone (Meyer et al. 1980; Grossman et al. 1984). Corenblum and Taylor (1981) reported that apprehension before surgery produced a rise in prolactin which was blocked by 0.8 mg naloxone, while the stress of surgery per se produced a rise in prolactin which was blocked by 1.6 mg (but not 0.8 mg) naloxone. However, it is worth pointing out that a variety of drugs were found to suppress apprehension-induced prolactin release, and it is possible that in this study the 'control' level was artificially elevated. A suppressive effort of naloxone on surgically-induced prolactin release was apparently confirmed by Pontiroli et al. (1982) using 0.4 mg naloxone, but there still remains the possibility that naloxone modifies the effect of the anesthetic used rather than revealing any intrinsic opioid mechanism. More difficult to discount is the study of Moretti et al. (1983), who found that 16 mg naloxone totally abolished prolactin release in response to acute severe exercise in professional athletes. This was subsequently confirmed in a further randomized double-blind study (Grossman and Moretti 1986). As such an effect was absent in those subjects who were not professional athletes (see above), it would appear that some aspect of training per se increases the opioid domination of prolactin release. However, in preliminary

studies we have been unable to demonstrate a shift in the opioid control of exercise-induced prolactin release during a training program (Grossman and Bouloux, unpublished observations), so that full understanding of these fascinating data remains unattained.
Pathophysiologically, there is little to suggest that changes in opioid tone are responsible for disorders of prolactin secretion in the human. Pathological hyperprolactinemia does not respond to naloxone with a fall in serum prolactin (Grossman et al. 1982b), and there is no evidence that any state of prolactin deficiency is due to opioid excess.

3.3.4 Conclusions

Prolactin is under the stimulatory control of opioid peptides, principally via the endorphinergic modulation of hypothalamic dopamine. While this pathway may mediate stress-induced prolactin release in the rat, there is little evidence that, except under exceptional conditions, this potential mechanism of regulation is important in the human.

3.4 Gonadotrophins

3.4.1 Morphine and Related Substances, Opioid Peptides

Early studies in the *rat* indicated that morphine was able to block ovulation in females and cause testicular regression in males. It was subsequently apparent that opioid peptides such as β-endorphin had similar effects, that these changes were due to substantial inhibition of the pulsatile release of luteinizing hormone (LH) (Kinoshita et al. 1980) and that these responses could be blocked by naloxone. Changes in follicle-stimulating hormone (FSH) were more difficult to demonstrate, presumably due to the long circulatory half-life of FSH compared with LH. A direct effect of opiates on the pituitary release of LH and FSH has rarely been demonstrated, while an antagonist of gonadotropin-releasing hormone (GnRH) is able to block the release of LH induced by naloxone (Blank and Roberts 1982, see below). This suggests that the principal effect of opiates is at a suprapituitary level, leading directly or indirectly to inhibition of the GnRH pulse generator in the hypothalamus.

3.4.2 Opiate Antagonists

Naloxone clearly elevates LH, and usually FSH, in the male and female *rat* under many conditions. Initial studies concentrated on the administration of bolus doses of naloxone to intact rats, and emphasized the reproducibility of the phenomenon (van Vugt and Meites 1980), while more recent research has concentrated on the site, receptor and physiological milieu in which endogenous opioids act to modulate GnRH. Panerai et al. (1983) noted that quaternary derivatives of naloxone were still able to modify serum LH in the rat, and suggested a site of action at or near the median eminence. [Pechnick et al. (1987) claimed that the lack of effect of N-methylmorphine indicated a site *within* the blood-brain barrier.] Similarly, Wilkes and Yen (1981) reported that naloxone could directly enhance the release of GnRH from rat hypothalami in vitro, while Drouva et al. (1981) found that morphine inhibited GnRH release from hypothalamic slices. As further confirmation of a hypothalamic site of action, direct injection of antisera to β-endorphin and dynorphin (but not enkephalin) into the hypothalamus of immature female rats leads to an elevation of serum LH (Schulz et al. 1981). However, not all the evidence is in agreement. The depression of serum LH induced by morphine is attenuated by lesions of the cortical amygdaloid nucleus (Lakoski and Gebhart 1981), while other workers have found it difficult to demonstrate a clear effect of morphine on the rat hypothalamus in vitro (Lengyel et al. 1985). Furthermore, instillation of β-endorphin into the amygdala of the *pig* leads to a fall in LH, whereas no change is seen following intrahypothalamic injection (Parvizi and Ellendorff 1980).

There is also a lack of agreement as to the opiate receptor subtype involved in the regulation of GnRH. Panerai et al. (1985) found that LH was modified mainly by μ-receptor antagonists, with some effect of κ-antagonists but none of δ-blockers. Similar conclusions were reached by Pfeiffer et al. (1987), although these workers were less impressed with the results of κ-agonists. Marko and Romer (1983) were able to demonstrate a more consistent suppressive effect of a κ-agonist. Leadem and Kalra (1985) were unable to clearly distinguish the effects of different opiate-receptor agonists, but in view of the high potency of β-endorphin suggested that an endorphin-specific ε-receptor was involved; Wiesner et al. (1985) also found β-endorphin to be specially effective, but in their study the effect was blocked by a δ-antagonist. However, the δ-antagonist used is relativley weak, and may not be as specific an antagonist as is currently claimed. In general, in spite of the lack of all agreement, data in the rat tend to support the view that the opioid modulation of GnRH occurs at hypothalamic μ- or ε-receptors, with β-endorphin as the endogenous ligand. An additional action by dynorphin-related peptides at κ-receptors cannot be excluded.

The opioid-induced suppression of LH is highly dependent on the steroid milieu, and is clearly modified by gonadal steroids. The ovulation-associated rise in LH is highly sensitive to opioid inhibition (Koves et al. 1981), while naloxone is much more effective in elevating LH in the presence of testosterone or estradiol (Sylvester et al. 1982; van Vugt et al. 1982). It has therefore been suggested that opioids are responsible, at least in part, for the feedback effects of such steroids on GnRH release. While this may be the case, it has also been plausibly suggested that endogenous opioids may operate in parallel to steroidal feedback pathways, such pathways providing the background activity against which the opioid effects then become manifest. On the other hand, an increase in opioid inhibition does not appear to be responsible for the suppression of gonadotrophin release seen in hyperprolactinemia (Sweeney et al. 1985) or during starvation (Kuderling et al. 1984).

Several groups of workers have concluded that the modulation of GnRH by opioids involves an obligatory interaction with a noradrenergic pathway (Kalra and Simpkins 1981; van Vugt et al. 1981; Blank and Bohnet 1982; Adler and Crowley 1984), quite possibly involving epinephrine rather than norepinephrine (Kalra and Crowley 1982). This has been developed into a model explaining opioid involvement in pubertal development and the estrus cycle. In brief, it is thought that during sexual maturation in the rat there is an increasing noradrenergic drive to the GnRH neurons at α_1-adrenoceptors, which becomes increasingly resistant to presynaptic opioid inhibition and hence steroidal feedback (Wilkinson and Bhanot 1982). In the mature rat, the estrogen-progesterone-induced ovulatory surge in LH (and presumably GnRH) is envisaged as a consequence of a reduced opioid inhibition of noradrenergic input: as the opioid tone decreases, the release of noradrenergic activity leads to a massive surge in GnRH (Gabriel et al. 1986). This timed decrease in opioid control may in turn be a response either to a neural clock (Allen et al. 1988) or to a surge in progesterone (Jacobson and Kalra 1989). Thus, changes in opioid sensitivity and/or tone may be essential components in the rat's reproductive life cycle. In this respect it is worthwhile to mention that recently Field and Kuhn (1989) reported that opioid inhibition of LH secretion diminishes with age in cycling rats, suggesting a possible role of diminished opiate tone in reproductive senescence.

The secretion of β-endorphin by the arcuate neurons in the hypothalamus may also underlie the inhibition of LH release seen during acute stress. Central activation of corticotropin-releasing factor-41 (CRF-41) during stress stimulates hypothalamic β-endorphin neurons, which in turn inhibit GnRH release (Petraglia et al. 1986a; Almeida et al. 1988; see Fig. 1). There is substantial evidence for each stage of this hypothetical pathway, although it is currently unclear as to whether opiate receptors are actually on GnRH neurons. Based on data on the control of adreno-corticotropin, there may well be dual opioid pathways, controlling GnRH release both indirectly (via noradrenergic pathways, as in maturation and ovulation) and directly (during stress). There is also evidence that one or other of these pathways may show the phenomenon of tolerance (Cicero et al. 1984; Landymore and Wilkinson 1988).

Data in *other species* are more scanty than in the rat, but generally point to opioid inhibition of GnRH via β-endorphin, although the details of the control vary (see for review also: Brooks et al. 1986). In seasonally breeding animals, opioid inhibition appears to be lost during reproductive quiescence, and is therefore not causal in its onset; thus, introduction of a ram to a prepubertal ewe will lead to a rise in LH in the correct breeding month, and this may be inhibited by an opioid; at an earlier stage, no increase in LH is seen and the ewe is refractory to opioid-induced inhibition (Stansfield et al. 1987). Opioid tone therefore appears during reproductive activity, and is essentially absent during quiescence. It is also suggested that in the developing *sheep* endogenous opioid mechanisms are an important inhibitory mechanism controlling pulsatile LH secretion (Ebling et al. 1989). However, according to these authors, it is not likely that opioid inhibition underlies the decrease in sensitivity to steroid feedback and increase in pulsatile LH secretion that occur at puberty.

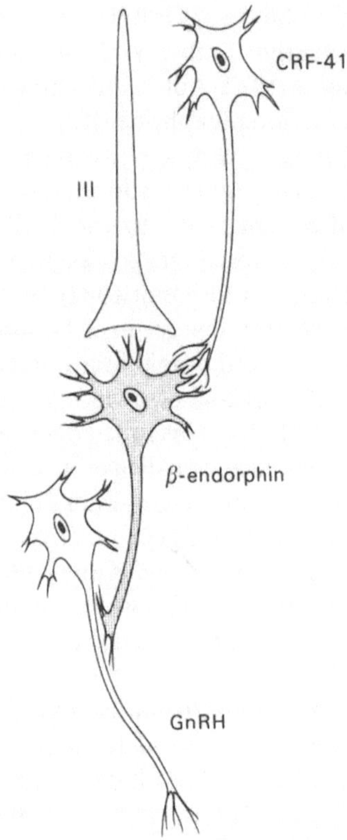

Fig. 1. Suggested schema for the modulation of GnRH by β-endorphin in response to stress. It is postulated that CRF-41 neurons in the paraventricular nucleus abutting the third ventricle (*III*) stimulate β-endorphin release, which in turn inhibited the pulsatile release of GnRH

3.4.3 Human Studies

In the human, as in the rat, morphine induces a profound suppression of serum LH, and similar results are seen with the Met-enkephalin analog DAMME (Stubbs et al. 1978) and β-endorphin (Reid et al. 1981b), which lead to an inhibition of the pulsatile release of LH (Grossman et al. 1981b). Unlike the rat, in the absence of gonadal steroids LH still remains sensitive to opioid inhibition, and levels fall in menopausal women infused with DAMME and probably dermorphin (Grossman et al. 1981b; Petraglia et al. 1985), but not β-endorphin (Reid 1983; Grossman 1987). As the gonadotropin responses to GnRH are unaltered by DAMME, it is assumed that this inhibition occurs above the level of the pituitary (Grossman et al. 1981b). This interaction does not seem to involve muscarinic cholinoceptors, histamine receptors (Chiodera et al. 1987), dopamine receptors or α_1- or α_2-adrenoceptors, and suggests that opioids directly inhibit the release of hypothalamic GnRH (Delitala et al. 1981a, b, 1983b; Veldhuis et al. 1984). Direct synaptic contact between GnRH and presumptive β-endorphin neurons have recently been demonstrated in juvenile *monkeys* (Thind and Goldsmith 1988). As the opioid responses are highly sensitive to naloxone (Grossman et al. 1986a), and κ-agonists are ineffective (Pfeiffer et al. 1986; Gilbeau et al. 1987), it is probable that μ- or ε-receptors are responsible for the process (Delitala et al. 1983b).

Naloxone infusion induces a prompt increase in serum LH and FSH in normal subjects: there is agreement that this is at least in part due to an increase in the pulse frequency, but some disagreement exists as to whether LH pulse amplitude also increases (Delitala et al. 1981c; 1983b; Grossman et al. 1981b; Ellingboe et al. 1982). In women, the degree of endogenous opioid tone varies during the menstrual cycle, being slight in the early follicular phase, later increasing considerably at the ovulatory surge of LH. There is also significant opioid inhibition of LH release during the progesterone-dominated luteal phase (Quigley and Yen 1980; Blankstein et al. 1981; Moult et al. 1981). In postmenopausal women this opioid tone is lost, but reappears with estrogen therapy (Melis et al. 1984) or progesterone therapy (Casper and Alapin-Rubillovitz 1985), and particularly with both (Casper et al. 1984; Shoupe et al. 1985). In men, the oral opiate antagonist naltrexone overcomes dihydrotestosterone-induced changes in LH pulse amplitude (Veldhuis et al. 1984), although there is some disagreement as to the effects of androgen replacement therapy (reviewed in Grossman 1987). However, the current data support the concept that an important determinant of the degree of opioid inhibition of the GnRH pulse generator is the level of gonadal steroids.

There is no convincing evidence that opioid inhibition of GnRH is increased during prepubertal life, and indeed, endogenous opioid tone appears to be acquired pari passu with the onset of pubertal maturation (Fraioli et al. 1984; Sauder et al. 1984; Mauras et al. 1986; Petraglia et al. 1986b). Similarly, opioid tone is absent in patients with anorexia nervosa (Grossman et al. 1982b), as well as in Kallman's syndrome. However, it is now clear that the predominant mechanism by which hyperprolactinemia causes hypogonadotropic amenorrhea in humans is by an increase in the opioid regulation of GnRH release (Quigley et al. 1980; Grossman et al. 1982b). As this modulation is paralleled by an increase in dopaminergic tone (Seki et al. 1986), and as dopamine increases β-endorphin release from the human hypothalamus in vitro (Rasmussen et al. 1987), a reasonable speculation is that hyperprolactinemia increases dopamine and hence β-endorphin, which in turn directly inhibits the release of GnRH. This obligatory mediation by dopamine would also account for the otherwise unexpected finding that hyperprolactinemia induced by dopamine antagonist drugs does not increase the opioid inhibition of LH (Petraglia et al. 1987), while bromocriptine does enhance the opioid inhibition of LH (Melis et al. 1988).

Other forms of amenorrhea are more variably associated with changes in opioid modulation. Hypothalamic amenorrhea may occasionally respond to opioid antagonism with a rise in LH (Khoury et al. 1987) or a return of menstruation (Wildt and Leyendecker 1987); this is most likely to happen when the amenorrhea is induced by stress. Similarly, recent studies have demonstrated that infusion of CRF-41 into normal women suppresses gonadotropin release (Barbarino et al. 1989). The absent LH pulsatility associated with either pure weight loss (Grossman et al.

1982b) or changes in body fat consequent on excercise (Dixon et al. 1987) is also not reversible by naloxone. However, it remains possible that pure exercise-induced amenorrhea, if such an entity exists, is an opioid-mediated disorder (Grossman 1988). Hyperandrogenemia may also induce amenorrhea or oligomenorrhea, as in women with the polycystic ovary syndrome: although acute infusion of dihydrotestosterone leads to opioid-mediated suppression of GnRH (Vermesh et al. 1987), women with this syndrome have an increase in LH pulse frequency. There is no evidence that this is due to any change in opioid inhibition (Barnes and Lobo 1985). Finally, perimenopausal hot flushes do not appear to be opioid-mediated (DeFazio et al. 1984).

3.4.4 Conclusions

In summary, both human and rat data suggest that an endogenous opioid, probably β-endorphin, directly or indirectly inhibits the release of hypothalamic GnRH. In the rat, diminishing sensitivity of β-endorphin occurs during sexual maturation, while a chronic increase in opioid tone may occur during stress and a transient decrease accompanies ovulation. In the human, the evidence in favor of stress-related changes is more circumstantial, but as in the rat there may be a short-lived decrement in opioid inhibition at the initiation of the mid-cycle LH surge (Rossmanith et al. 1988). What appears to be more certain is that hyper-prolactinemic amenorrhea is opioid-mediated, probably via dopaminergic enhancement of hypothalamic β-endorphin. Studies with long-acting orally active opiate antagonists such as nalmefene (van Vugt et al. 1989) will clearly be of great interest.

3.5 Adrenocorticotrophin

3.5.1 Morphine and Related Substances, Opioid Peptides, Opiate Antagonists

Morphine and related drugs have long been known to cause a profound suppression of the pituitary-adrenal axis when given chronically to *rats*, early studies demonstrating a hypothalamic site of action (Briggs and Munson 1955; George and Way 1955). However, all acute studies of morphine and opioid peptides have shown a rapid onset and marked rise in serum corticosterone and, when measured, plasma adrenocorticotropin (ACTH). Most such studies are also in agreement in demonstrating a suprapituitary site of action (Lotti et al. 1969), although Lamberts et al. (1983) reported that the Met-enkephalin analog DAMME inhibited basal and stimulated release of ACTH from rat anterior pituitary lobes in vitro. However, other workers have been unable to demonstrate a pituitary site of action for opioids (Buckingham 1982; Rittmaster et al. 1985; Tsagarakis et al. 1989; 1990) and have concluded that both stimulatory and inhibitory effects occur above the level of the pituitary.

Utilizing a bioassay for CRF, Buckingham (1982) found that morphine and several enkephalin analogs stimulated the hypothalamo-pituitary-adrenal axis both in vivo and in vtiro at a hypothalamic site, involving an interaction with μ-receptors and, to a lesser extent, κ-receptors (Buckingham and Cooper 1986). This was thought to occur via an interaction with an inhibitory noradrenergic pathway (Buckingham and Cooper 1986). Pfeiffer et al. (1985) also reported in vivo stimulation at both μ- and κ-receptors. These findings are compatible with those mentioned above on the stimulatory role of opioids on the pituitary-adrenal axis. Iyengar et al. (1987) also noted potent stimulation of pituitary-adrenal activity by all opioids tested, their cross-tolerance technique identifying roles for μ-, ε-, κ- and δ-receptors. However, there are discordant reports. The stimulation of ACTH by an enkephalin analog was followed rapidly by significant inhibition in one study, while β-endorphin can also inhibit bioactive CRF release in vitro at certain concentrations. Furthermore, naloxone alone at high doses will stimulate the release of ACTH and corticosterone (Eisenberg 1980), probably at a site outside the blood-brain barrier (Jezova 1985). Both naloxone and κ-agonist stimulation appear to occur via CRF-41. However, this does not seem to be true for morphine (Nikolarakis et al. 1987). Indeed, the in vitro release of CRF-41 is potently inhibited by morphine (Tsagarakis et al. 1989a; see Fig. 2). Both dynorphin and β-endorphin will also inhibit the release of CRF-41 into portal blood, while levels are increased by naloxone (Plotsky 1986). Taken together, these data indicate that endogenous opioids tonically inhibit the pituitary-adrenal axis at receptors directly located on CRF-41 neurons. It may be tentatively suggested that the stimulatory effects of morphine are exerted not on CRF-41 but

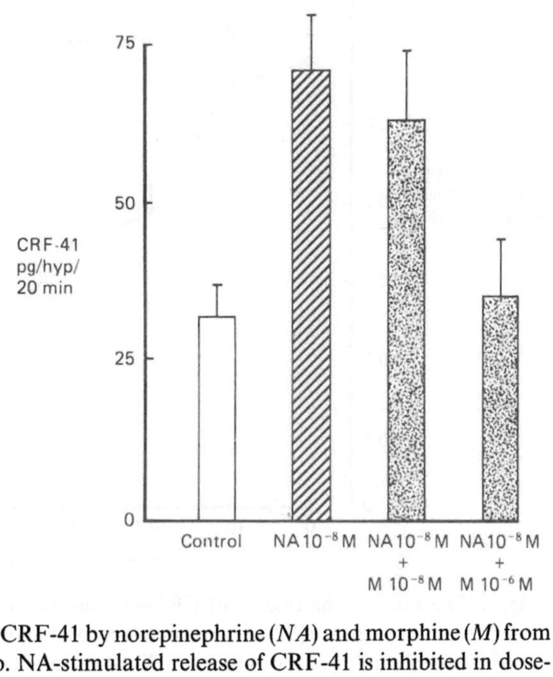

Fig. 2. Modulation of the release of CRF-41 by norepinephrine (*NA*) and morphine (*M*) from rat hypothalamic fragments in vitro. NA-stimulated release of CRF-41 is inhibited in dose-dependent fashion by morphine. (From Tsagarakis et al. 1988, with the permission of the authors and publishers)

via another regulatory substance, either increasing stimulatory factors such as catecholamines or decreasing a corticotropin-inhibiting factor. To further confuse the issue, κ-receptor antagonists appear to enhance the release of ACTH (Cover and Buckingham 1989). The δ-receptor does not appear to be involved in the opioid control of ACTH in vivo (Cover and Buckingham 1989) or CRF-41 in vitro (Tsagarakis et al. 1990). Unfortunately, this explanation cannot incorporate the actions of κ-receptor agonists, which appear to increase ACTH in vivo (Pfeiffer et al. 1985) and CRF bioactivity in vitro (Buckingham and Cooper 1986) via CRF-41 (Nikolarakis et al. 1987), but which inhibit the release of CRF-41 from portal blood (Plotsky 1986) and incubated hypothalami in vitro (Tsagarakis et al. 1990; see Fig. 3). It has also been suggested that morphine stimulates the release of ACTH by activating both CRF secretion and peripheral sympathetic neuronal pathways (Nikolarakis et al. 1989). At present, there is no unifying hypothesis to account for these findings.

Reports on the effects of opioid peptides on pituitary-adrenal responses in *sheep* are conflicting. Even recent reports present conflicting data, mentioning either that in this species opioid involvement in the (tonic) regulation of pituitary-adrenal function is absent (Brooks and Challis, 1989), or that endogenous and exogenous

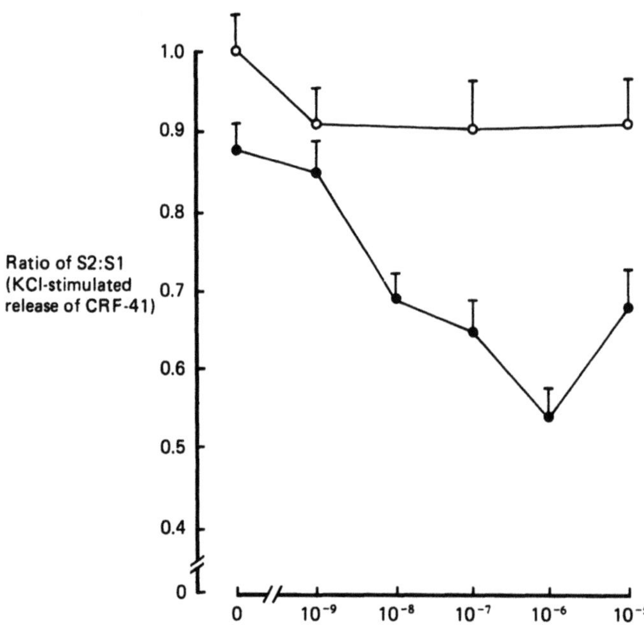

Fig. 3. The ratio of the release of CRF-41 from rat hypothalami incubated in vitro in the absence (*S1*) or presence (*S2*) of two opiate-receptor agonists. The δ-opiate receptor agonist D-Pen^{2,5}-enkephalin(○–○) produces an S2:S1 ratio not significantly differing from unity at all concentrations, while the κ-opiate receptor agonist U-50448 inhibits CRF-41 release. In dose-dependent fashion. (From Tsagarakis et al. 1990, with the permission of the authors)

opioids modulate cortisol release in non-stressed sheep as well as in sheep subjected to psychological stress (Parrott and Thornton 1989).

3.5.2 Human Studies

While there is considerable uncertainty regarding the actions of opiates in the rat, in man there is clear evidence that morphine (McDonald et al. 1959), β-endorphin (Taylor et al. 1983), DAMME (Gaillard et al. 1981), dermorphin (Degli Uberti et al. 1985) and loperamide (Ambrosi et al. 1986) all inhibit the release of ACTH. However, the pituitary-adrenal inhibition revealed by several opiate alkaloids, as well as DAMME, did not appear to be antagonized by 4 mg (del Pozo et al. 1980; Delitala et al. 1983b) or even 8 mg (Gaillard et al. 1981) of naloxone, although the effects of loperamide were readily reversed by 0.8 mg (Ambrosi et al. 1986). This suggests that the responses to the majority of opiate drugs are mediated by opiate receptors of low naloxone sensitivity. This conclusion has also been reached by investigating the role of endogenous opioids in modulating the pituitary-adrenal axis.

Few areas of research in opioid-neuroendocrine research are as confusing as the site of action of opioids in regulating ACTH release in the human. The induced release of ACTH by lysine vasopressin is inhibited by DAMME (del Pozo et al. 1980), while both DAMME (Allolio et al. 1986; Grossman et al. 1986b) and morphine (Rittmaster et al. 1985) inhibit pituitary-adrenal activation in response to CRF-41. This might suggest that exogenous opioids act directly on the pituitary to suppress ACTH release. Few animal studies are in agreement, but this may be due to interspecies differences. However, against this theory is the fact that endogenous opioids appear to inhibit ACTH at a suprapituitary site (see below), and it seems a priori unlikely that exogenous and endogenous opioids act at different loci. An alternative explanation is that opiate receptors inhibit the release of a series of CRFs, such that the response to any given factor in isolation is attenuated by the absence of its synergizing or potentiating factors. In the human, the most probably candidates as CRFs are CRF-41 and vasopressin, but excluding catecholamines (Al-Damluji et al. 1987). To date, there is no report of the effect of exogenous opiates on the pituitary-adrenal response to the combined administration of CRF-41 and vasopressin.

Naloxone alone in low doses has no effect on the pituitary-adrenal axis: at doses above 4 mg there is marked stimulation of ACTH and cortisol under basal conditions (Volavka et al. 1979; Morley et al. 1980; Blankstein et al. 1980; Deuss et al. 1985; Grossman et al. 1986a). Peak plasma ACTH levels of ca. 40–60 ng/l were attained with a naloxone dose of 8–10 mg; it seems probable that no further rise in response is seen above 15–20 mg naloxone (Volavka et al. 1979; Bouloux et al. 1989). The degree of opioid tone varies during the day, being greatest at around 18:00 and least around midnight (Grossman et al. 1982c); thus, the nocturnal fall in ACTH release is not a consequence of a rise in endogenous opioid tone. However, this circadian variation is very similar to that induced by *d*-amphetamine, and indeed the rise in cortisol induced by naloxone is blocked by the α_1-adrenoceptor antagonist thymoxamine (Grossman and Besser 1982). This suggests that opioids regulate ACTH release by inhibiting an excitatory noradrenergic drive, which previous studies have located at a suprapituitary site (Al-Damluji et al. 1987). Work in other systems has demonstrated inhibitory presynaptic α_2-adrenoceptors in noradrenergic pathways; if our theory is correct, then increasing noradrenergic drive by means of naloxone should also increase α_2-adrenoceptor inhibition. We have recently tested this hypothesis, and found that the α_2-antagonist idazoxan enhanced the rise in cortisol and ACTH induced by naloxone, while having no effect under basal conditions (Bouloux et al., unpublished observations). This therefore supports the idea that endogenous opioids modulate pituitary-adrenal activity by interacting with the noradrenergic input to CRF (see Fig. 4). However, it should again be emphasized that this may not wholly be CRF-41, as there is evidence that naloxone and CRF-41 are additive in their effects on ACTH, and the α_1-adrenoceptor control of ACTH is probably principally via vasopressin (Al-Damluji S., unpublished observations).

As the dose of naloxone required to elevate ACTH is some 10–20 times greater than that which will elevate the gonadotropins (Grossman et al. 1986a), and as the human hypothalamus is rich in κ-receptors but relatively deficient in δ-receptors (Pfeiffer

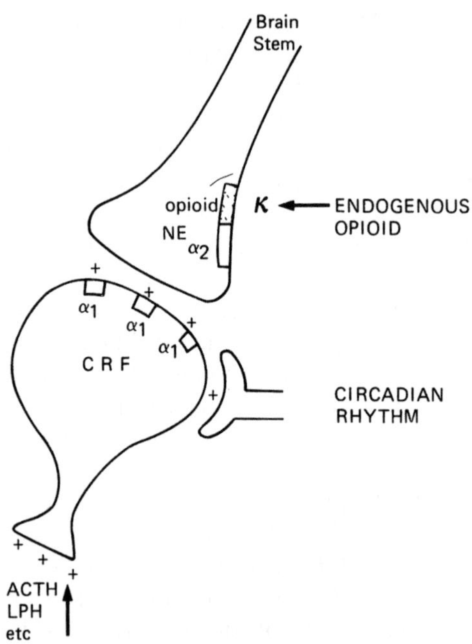

Fig. 4. Probable mechanism underlying the opioid control of CRF, and hence ACTH, in man. Ascending pathways containing epinephrine or norepinephrine (*NE*) stimulate CRF release at α_1-adrenoceptors, and are in turn subject to tonic opioid inhibition at κ-receptors. Such pathways may mediate certain changes during stress; the κ-opiate receptor may also be intimately related to the α_2-adrenoceptor. In the rat, part of the κ-opiate modulation may occur directly on CRF-41 neurons

et al. 1982), it appears probable that opioids act predominantly at κ-opiate receptors to influence CRF release. Nevertheless, the potent activity of drugs such as morphine, which have very little κ-receptor activity, suggests that μ-receptors may also be involved. It will be recalled that in the rat the evidence indicates that both μ- and κ-receptors (but not δ-receptors) are involved in the direct modulation of CRF-41 (Tsagarakis et al. 1990).

3.5.3 Conclusions

In summary, μ-receptor agonists stimulate the pituitary-adrenal axis in the rat, probably via pathways not including CRF-41; by contrast, there is an endogenous opioid pathway which inhibits the release of CRF-41 at κ-receptors. In man the situation is clearer, in that all opioids and opiates inhibit pituitary-adrenal activity at naloxone-insensitive sites, and it seems likely that the endogenous opioid activates κ-receptors to decrease noradrenergic stimulation. However, without more selective κ-receptor antagonists and without knowledge of the other factors involved in the CRF complex this area will remain one of uncertainty and contradictions.

4 Overall Conclusions

It is clear that the principal effect of opioids on vasopressin release is profoundly inhibitory in both rat and man, but might be stimulatory in dogs and rabbits. Similar inhibitory considerations apply to the release of oxytocin in the rat, while in man this also seems likely. This control is relatively insensitive to naloxone, and most probably occurs principally at κ-receptors.

With regard to the anterior pituitary, opioids stimulate GH in both rat and man, but the precise mechanism remains uncertain. There are data suggesting dual control via GHRH and somatostatin, but as opioids are generally inhibitory, dominant control via the latter peptide seems most likely. TSH is inhibited by opioids in the rat and stimulated in man; it seems likely that in the former this is due to inhibition of the hypothalamic release of TRH, while in the latter an inhibition of dopamine or somatostatin is most credible. Prolactin is stimulated in both species, the majority of the control occurring via dopamine. However, modulation of a prolactin-releasing factor may also occur.

Opioids inhibit the pulsatile release of GnRH in rat and man; the endogenous control of this peptide by opioids appears to be relatively important in both species. It seems likely that this occurs via β-endorphin, probably originating from the neurons of the arcuate nucleus.

Finally, opioids are profoundly inhibitory to the pituitary-adrenal axis in man, probably via an interaction with an noradrenergic pathway at κ-receptors. In the rat, opioids act in a dual fashion on ACTH and related peptides, but their chronic effect is suppressive, and again predominantly at κ-receptors.

In general, opioids can be said to be inhibitory to most hypothalamic peptides affecting pituitary function, and they may operate at the level of the hypothalamic regulatory peptide as well as inhibiting a catecholaminergic input.

Are there any general conclusions one can make that would include the totality of the opioid modulation of pituitary function? This is relatively difficult, as opiate receptors and the effects of opioid manipulation vary according to species, drug, dose, route of administration and physiological situation. Nevertheless, some broad schematic outlines may be dimly perceived beneath the mass of data now in existence. The predominant hypothalamic opioids are β-endorphin and dynorphin, while the principal receptors in man and rat are of the μ- and κ-subtypes. The endogenous ligand or ligands for the μ-receptors are not known, but may include β-endorphin, while there is firmer evidence that dynorphin activates endogenous κ-receptors. Vasopressin, oxytocin and CRF-41 all appear to be under important inhibitory regulation by dynorphin or a related peptide, while GnRH (and prolactin, in the rat) is modulated by β-endorphin. The dynorphinergic network seems to moderate activity in the vasopressin, oxytocin and CRF-41 neurons, probably in an autocrine or paracrine fashion: these peptides may in turn regulate GnRH release via β-endorphin. Thus, dynorphin acts mainly to *counterregulate* stress responsiveness, while β-endorphin *mediates* such responses. However, further speculation is not possible on the basis of current knowledge. All we can say is that, in light of the fascinating accumulation of information on the endogenous opioids in the first 15 years since their discovery, the next decade and a half seems likely to finally establish their precise pathophysiological significance.

Acknowledgements. The authors gratefully acknowledge the help of Miss P. Bakels and Miss E. Stawowska in preparing and typing the manuscript.

5 References

Adachi T, Hisano S, Daikoku S (1985) Intragranular colocalization of immunoreactive methionine-enkephalin and oxytocin within the nerve terminals in the posterior pituitary. J. Histochem Cytochem 33:891–899

Adler BA, Crowley WR (1984) Modulation of luteinizing hormone release and catecholamine activity by opiates in the female rat. Neuroendocrinology 38:248–253

Al-Damluji S, Cunnah C, Grossman AB, Perry L, Ross G, Coy D, Rees LH, Besser GM (1987) Effect of adrenaline on basal and ovine corticotrophin releasing factor stimulated ACTH secretion in man. J Endocrinol, 112:145–150

Algeri A, Consolazione A, Calderini G, Achillin G, Canas EP, Garattini S (1978) Effect of the administration of (d-ala)2-methionine-enkephalin on the serotonin metabolism in rat brain. Experientia 34:1488–1489

Allen LG, Hahn E, Caton D, Kalra SP (1988) Evidence that a decrease in opioid tone on proestrus changes the episodic pattern of luteinizing hormone (LH) secretion: implications in the preovulatory LH hypersecretion. Endocrinology 122:1004–1013

Allolio B, Winkelmann W, Hipp FX, Kaulen D, Mies R (1982) Effects of a Met-enkephalin analog on adrenocorticotropin (ACTH), growth hormone, and prolactin in patients with ACTH hypersecretion. J Clin Endocrinol Metab 55:1–7

Allolio B, Deuss U, Kaulen D, Leonhardt U, Kallabis D, Hamel E, Winkelmann W (1986) FK 33–824, a met-enkephalin analog, blocks corticotropin-releasing hormone-induced adrenocorticotropin secretion in normal subjects but not in patients with Cushing's disease. J Clin Endocrinol Metab 63:1427–1431

Almeida OFX, Nikolarakis KE, Herz A (1988) Evidence for the involvement of endogenous opioids in the inhibition of luteinizing hormone by corticotrophin-releasing factor. Endocrinology 122:1034–1041

Al Zein M, Lutz-Bucher B, Koch B (1984) Modulation by Leu-enkephalin of peptide release from perifused neurointermediate pituitary. Neuroendocrinology 39:392–396

Ambrosi B, Bochicchio D, Faglia G (1986) Loperamide, an opiate analogue, inhibits plasma ACTH levels in patients with Addison's disease. Clin Endocrinol (Oxf) 24:483–489

Arancibia S, Taopia-Arancibia L, Roussel JP, Assenmacher I, Astier H (1986) Effects of morphine on cold-induced TRH release from the median eminence of unanesthetized rats. Life Sci 38:59–66

Arita J, Porter JC (1984) Relationship between dopamine release into hypophysial portal blood and prolactin release after morphine treatment in rats. Neuroendocrinology 38:62–67

Arnauld E, Cirino M, Layton BS, Renaud LP (1983) Contrasting actions of amino acids, acetylcholine, noradrenaline and leucine enkephalin on the excitability of supraoptic vasopressin-secreting neurons. Neuroendocrinology 36:187–196

Ashton N, Balment RJ, Blackburn TP (1989) κ-Opioid-induced changes in renal water and electrolyte management and endocrine secretion. Br J Pharmacol 97:769–776

Auriac A, Azam J, Dumas JC, Roux G, Montrastruc JL (1985) Modifications du métabolisme hydrique du rat sous l'influence de la naloxone: role de la vasopressine. Biomed Pharmacother 39:89–99

Aziz LA, Forsling ML, Woolf CJ (1981a) The effect of intracerebroventricular injections of morphine on vasopressin release in the rat. J Physiol (Lond) 311:401–409

Aziz LA, Forsling ML, Woolf CJ (1981b) Does Met-enkephalin influence vasopressin release? J Physiol 315:9P

Barbarino A, de Marinis L, Tofani A, della Case S, d'Amico C, Mancini A, Corsello S, Sciuto R, Barini A (1989) Corticotropin-releasing hormone inhibition of gonadotropin release and the effect of opioid blockade. J Clin Endocrinol Metab 68:523–528

Barnes RB, Lobo RA (1985) Central opioid activity in polycystic ovary syndrome with and without dopaminergic modulation. J Clin Endocrinol Metab 61:779–782

Bentley AM, Wallis M (1986) Effects of two enkephalin analogues, morphine sulphate, dopamine and naloxone on prolactin secretion from rat anterior pituitary glands in vitro. J Endocrinol 109:313–320

Besser GM, Delitala G, Grossman AB, Yeo T (1980) Metergoline and cyproheptadine suppress prolactin release by a non-5-hydroxytryptaminergic non-dopaminergic mechanism. Br J Pharmacol 70:5–7

Bhanot R, Wilkinson M (1983) Opiatergic control of gonadotropin secretion during puberty in the rat: a neurochemical basis for the hypothalamic 'gonadostat'? Endocrinology 113:596–603

Bhargava HN, Das S (1986) Evidence for opiate action at the brain receptors for thyrotropin-releasing hormone. Brain Res 368:262–267

Bicknell RJ, Leng G (1982) Endogenous opiates regulate oxytocin but not vasopressin secretion from the neurohypophysis. Nature 298:161–162

Bicknell RJ, Dean ADP, Dyer RG, Ingram CD, Leng G, Mansfield S, Marsh MIC, Yates JO (1983a) Evidence that endogenous opiates regulate the speed of parturition in rats. Proceedings of the Physiological Society, Sept. 23–24, Edinburgh, 142P

Bicknell RJ, Ingram CD, Leng G (1983b) Oxytocin release is inhibited by opiates from the neural lobe, not those from the intermediate lobe. Neurosci Lett 43:227–230

Bicknell RJ, Chapman C, Leng G (1985a) Effects of opioid agonists and antagonists on oxytocin and vasopressin release in vitro. Neuroendocrinology 41:142:148

Bicknell RJ, Chapman C, Leng G (1985b) Neurohypophysial opioids and oxytocin secretion: source of inhibitory opioids. Exp Brain Res 60:192–196

Bicknell RJ, Chapman C, Leng G, Russell JA (1985c) Vasopressin release following naloxone administration to rats treated chronically with intracerebroventricular (i.c.v.) morphine. Proceedings of the Physiological Society, Febr 22–23, Bristol, p 73P

Bicknell RJ, Chapman C, Leng G, Russell JA (1985d) Chronic morphine exposure and oxytocin neurones in rats: lack of both morphine dependence and cross-tolerance to endogenous opioids in the neurohypophysis. J Physiol (Lond) 361:32P

Bicknell RJ, Sirinathsinghji DJS, Zhao B-G (1987) Opioid-catecholamine interactions and secretion in the neurohypophysis. Neurosci Lett [Suppl] 29:S39

Bicknell RJ, Leng G, Lincoln DW, Russell JA (1988a) Naloxone excites oxytocin neurones in the supraoptic nucleus of lactating rats after chronic morphine treatment. J Physiol (Lond) 396:297–317

Bicknell RJ, Coombes JE, Fink G, Russell JA, Sheward WJ (1988b) Effects of naloxone and long-term morphine administration on oxytocin release into hypophysial portal blood in urethane-anaesthetized rats. Proceedings of the Physiological Society, Febr 12–13, Bristol, 43 P

Bicknell RJ, Leng G, Russell JA, Dyer RG, Mansfield S, Zhao BG (1988c) Hypothalamic opioid mechanisms controlling oxytocin neurones during parturition. Brain Res Bull 20:743–749

Bisset GW, Chowdrey HS, Feldberg W (1978) Release of vasopressin by enkephalin. Br J Pharmacol 62:370–371

Blanco E, Carretero J, Sanchez F, Riesco JM, Vazquez R (1989) Sex-specific effects of Met-enkephalin treatment on vasopressin immunoreactivity in the rat supraoptic nucleus. Neuropeptides 13:115–120

Blank MS, Bohnet HG (1982) Alpha-adrenergic but not beta-adrenergic or serotonergic antagonists block naloxone-induced LH release. Neuroendocrinol Lett 5:3–8

Blank MS, Roberts DL (1982) Antagonists of gonadotropin-releasing hormone blocks naloxone-induced elevations in serum luteinizing hormone. Neuroendocrinology 35:309–312

Blankstein J, Reyes FI, Winter JDS, Faiman C (1980) Effects of naloxone upon prolactin and cortisol in normal women. Proc Soc Exp Biol Med 16:363–365

Blankstein J, Reyes FI, Winter JSD, Faiman C (1981) Endorphins and the regulation of the human menstrual cycle. Clin Endocrinol (Oxf) 14:287–294

Blask DE, Vaughan MK, Champney TH, Johnson LY, Vaughan GM, Becker RA, Reiter RJ (1984) Opioid and dopamine involvement in prolactin release induced by arginine vasopressin in the male rat. Neuroendocrinology 38:56–61

Bluet-Pajot MT, Schaub C (1977) Interaction of morphine and hypoglycaemia in stimulating the release of growth hormone in the rat. J Endocrinol 74:335–335

Bondy CA, Gainer H, Russell JT (1988) Dynorphin A inhibits and naloxone increases the electrically stimulated release of oxytocin but not vasopressin from the terminals of the neural lobe. Endocrinology 122:1321–1327

Bouloux PMG, Grossman AB, Allolio B, Delitala G, Besser GM (1986) Effects of adrenergic antagonism on the growth hormone, prolactin and cortisol response to a synthetic opioid agonist in man. Horm Res 23:83–90

Bouloux PMG, Newbould E, Causon R, Perry L, Rees LH, Besser GM, Grossman A (1989) The differential effect of high-dose naloxone on the plasma adrenaline response to the cold-pressor test. Clin Sci 76:625–630

Brady LS, Herkenham M (1987) Dehydration reduces κ-opiate receptor binding in the neurohypophysis of the rat. Brain Res 425:212–217
Brady LS, Rothman RB, Herkenham M (1988) Physiological regulation of neurohypophyseal κ-opiate receptors. Brain Res 443:398–402
Briggs FN, Munson PL (1955) Studies on the mechanism of stimulation of ACTH secretion with aid of morphine as blocking agent. Endocrinology 57:205–219
Briski K, Quigley K, Meites J (1984) Counteraction by naltrexone of stress-induced inhibition of TSH release: role of noradrenergic system. Proc Soc Exp Biol Med 177:354–359
Brooks AN, Lamming GE, Haynes NB (1986) Endogenous opioid peptides and the control of gonadotrophin secretion. Res Vet Sci 41:285–299
Brooks AN, Challis JRG (1989) Effects of CRF, AVP and opioid peptides on pituitary-adrenal responses in sheep. Peptides 10:1291–1293
Brownell J, del Pozo E, Donatsch P (1980) Inhibition of vasopressin secretion by a met-enkephalin (FK 33-824) in humans. Acta Endocrinol (Copenh) 94:304–308
Bruni JF, van Vugt D, Marshall S, Meites J (1977) Effects of naloxone, morphine and methionine enkephalin on serum prolactin, luteinizing hormone, follicle stimulating hormone, thyroid stimulating hormone and growth hormone. Life Sci 21:461–466
Buckingham JC (1982) Secretion of corticotrophin and its hypothalamic releasing factor in response to morphine and opioid peptides. Neuroendocrinology 35:111–116
Buckingham JC, Cooper TA (1984) Differences in hypothalamo-pituitary-adrenocortical activity in the rat after acute and prolonged treatment with morphine. Neuroendocrinology 38:411–417
Buckingham JC, Cooper TA (1986) Pharmacological characterization of opioid receptors influencing the secretion of corticotrophin releasing factor in the rat. Neuroendocrinology 44:36–40
Bunn SJ, Hanley MR, Wilkin GP (1986) Kappa opioid binding sites on pituitary astrocytes (Abstr P 29). 17th International Narcotic Research Conference, San Francisco
Buydens P, Velkeniers B, Golstein J, Finné E, Vanhaelst L (1987) Opioid modulation of thyrotropin releasing hormone induced prolactin secretion. Life Sci 40:1207–1214
Carter DA, Lightman SL (1984a) Inhibition of vasopressin secretion by a κ-opiate receptor agonist. Neuroendocrinol Lett 6:95–100
Carter DA, Lightman SL (1984b) Modulation of vasopressin secretion by opiates: Evidence for a κ-receptor mediated action in the nucleus tractus solitarius (Abstr). 1st Symposium of the European NeuroEndocrine Association (ENEA), Basel, p 148
Carter DA, Lightman SL (1987a) Temporal changes in the activity of endogenous opioid mechanisms regulating oxytocin secretion in saline-loaded rats. Neurosci Lett 82:191–195
Carter DA, Lightman SL (1987b) Opioid control of oxytocin secretion: Evidence of distinct regulatory actions of two opiate receptor types. Life Sci 40:2289–2296
Carter DA, Williams TDM, Lightman SL (1986) A sex difference in endogenous opioid regulation of the posterior pituitary response to stress in the rat. J Endocrinol 111:239–244
Casper RF, Alapin-Rubillovitz S (1985) Progestins increase endogenous opioid peptide activity in postmenopausal women. J Clin Endocrinol Metab 60:34–36
Casper RF, Bhanot R, Wilkinson M (1984) Prolonged elevation of hypothalamic opioid peptide activity in women taking oral contraceptives. J Clin Endocrinol Metab 58:582–584
Chavkin C, James IF, Goldstein A (1982) Dynorphin is a specific endogenous ligand of the κ-opioid receptor. Science 213:413–415
Cheung CY (1984) Does β-endorphin modulate basal and dopamine-inhibited prolactin release by an action at the anterior pituitary? Neuroendocrinology 39:489–495
Chihara K, Arimura A, Coy DH, Schally AV (1978) Studies on the interaction of endorphins,

substance P, and endogenous somatostatin in growth hormone and prolactin release in rats. Endocrinology 102:281–290
Chiodera P, Coiro V (1990) Inhibitory effect of ethanol on the arginine vasopressin response to insulin-induced hypoglycemia and the role of endogenous opioids. Neuroendocrinology 51:501–504
Chiodera P, Volpi R, Delsignore R, Ippolito L, Rossi G, Camellini L, Marchesi C, Gnudi A, Coiro V (1987) Muscarinic cholinergic and histaminergic H_1 and H_2 receptors are not involved in the LH response to naloxone in man. Horm Res 25:120–124
Cicero TJ, Owens DP, Schmoeker PF, Meyer ER (1984) Enhancement of opiate-induced depressions in serum luteinizing hormone levels by the prior administration of naloxone in the male rat. Life Sci 35:1321–1328
Clarke G, Patrick G (1983) Differential inhibitory action by morphine on the release of oxytocin and vasopressin from the isolated neural lobe. Neurosci Lett 39:175–180
Clarke G, Wright DM (1984) A comparison of analgesia and suppression of oxytocin release by opiates. Br J Pharmacol 83:799–806
Clarke G, Wood P, Merrick L, Lincoln DW (1979) Opiate inhibition of peptide release from the neurohumoral terminals of hypothalamic neurones. Nature 282:746–748
Clarke G, Lincoln DW, Wood P (1980) Inhibition of vasopressin neurones by intraventricular morphine. J Physiol (Lond) 303:60P
Cocchi D, Santagostino A, Gil-Ad I, Ferri S, Müller EE (1977) Leu-enkephalin-stimulated growth hormone and prolactin release in the rat: comparison with the effect of morphine. Life Sci 20:2041–2046
Collado-Escobar D, Rovati LC, Ganzetti I, Cocchi D, Panerai AE, Li CH (1986) The β-endorphin-induced secretion of growth hormone but not of prolactin is inhibited by an endogenous opioid antagonist. Eur J Pharmacol 129:385–387
Coombes JE, Russell JA (1987) Inhibition of oxytocin secretion via κ-opioid receptors in the neurohypophysis in vivo (Abstr p 17) Meeting. "Brain Opioid Systems in Reproduction", Babraham
Corenblum B, Taylor PJ (1981) Mechanisms of control of prolactin release in response to apprehension stress and anesthesia-surgery stress. Fertil Steril 36:712–715
Cover PO, Buckingham JC (1989) Effects of selective opioid-receptor blockade on the hypothalamo-pituitary-adrenocortical responses to surgical trauma in the rat. J Endocrinol 121:213–220
De Bodo RC (1944) The antidiuretic action of morphine, and its mechanism. J Pharmacol Exp Ther 82:74–85
DeFazio J, Verheugen C, Chetkowski R, Nass T, Judd HL, Meldrum DR (1984) The effects of naloxone on hot flushes and gonadotrophin secretion in postmenopausal women. J Clin Endocrinol Metab 58:578–581
Degli Uberti EC, Trasforini G, Salvadori S, Tomatis R, Margutti A, Bianconi M, Rotola C, Pansini R (1983) Prolactin-releasing activity of dermorphin, a new synthetic potent opiate-like peptide, in normal human subjects. J Clin Endocrinol Metab 56:1032–1034
Degli Uberti EC, Petraglia F, Transforini G, Salvadori S, Margutti A, Bianconi M, Teodori V, Facchinetti F, Tomatis R, Genazzani AR, Pansini R (1985) Dermorphin reduces the metyrapone-evoked release of adrenocorticotropin, β-endorphin, and β-lipotropin in man. J Clin Endocrinol Metab 61:1018–1022
Delitala G, Grossman A, Besser GM (1981a) The modulation of opiate-induced changes in anterior pituitary function by a dopamine agonist and antagonist. Neuroendocrinol Lett 3:147–152
Delitala G, Devilla L, Arata L (1981b) Opiate receptors and anterior pituitary hormone secretion in man. Effect of naloxone infusion. Acta Endocrinol (Copenh) 97:150–156

Delitala G, Grossman A, Besser GM (1981c) Changes in pituitary hormone levels induced by Met-enkephalin in man – the role of dopamine. Life Sci 29:1537–1544
Delitala G, Giusti M, Mazzocchi L, Granziera L, Tarditi W, Giordano G (1983a) Participation of endogenous opiates in regulation of the hypothalamic-pituitary-testicular axis in normal men. J Clin Endocrinol Metab 57:1277–1281
Delitala G, Grossman A, Besser M (1983b) Differential effects of opiate peptides and alkaloids on anterior pituitary hormone secretion. Neuroendocrinology 37:275–279
Delitala G, Mannelli M, Giusti M, Maggi M, de Feo ML, Giordano G, Giusti G, Serio M (1984) Relation of opiates to hormonal secretion in man. In: Delitala G Motta M, Serio M (eds): Opioid modulation of endocrine function. Raven, New York
Delitala G, Tomasi PA, Palermo M, Ross RJM, Grossman A, Besser GM (1989) Opioids stimulate growth hormone (GH) release in man independently of GH-releasing hormone. J Clin Endocrinol Metab 69:356–358
Del Pozo E, von Graffenried B, Derrer F, Marbach P (1980) endocrine effect of a methionine-enkephalin derivative (FK 33–824) in man. Horm Res 13:90–97
Dencker Christensen J, Fjalland B (1982) Lack of effect of opiates on release of vasopressin from isolated rat neurohypophysis. Acta Pharmacol Toxicol (Copenh) 50:113–116
Deuss U, Allolio B, Kaulen D, Fischer H, Winkelmann W (1985) Effects of high-dose and low-dose naloxone on plasma ACTH in patients with ACTH hypersecretion. Clin Endocrinol (Oxf) 22:273–279
De Wied D, van Ree JM, de Jong W (1974) Narcotic analgesics and the neuroendocrine control of anterior pituitary function. In: Zimmermann E, George R (eds) Narcotics and the hypothalamus. Raven, New York, pp 251–266
Dixon G, Eurman P, Stern BE, Schwartz B, Rebar RW (1987) Hypothalamic function in amenorrheic runners. Fertil Steril 42:377–383
Dores RM, Akil H (1987) Species-specific processing of prodynorphin in the posterior pituitary of mammals. Endocrinology 120:230–238
Duke HN, Pickford M, Watt JA (1951) The antidiuretic action of morphine: Its site and mode of action in the hypothalamus of the dog. Q J Exp Physiol 36:149–158
Drouva SV, Epelbaum J, Tapia-Arancibia L, Laplante E, Kordon C (1981). Opiate receptors modulate LHRH and SRIF release from mediobasal hypothalamic neurons. Neuroendocrinol 32:163–167
Dunne M, Haddock J, Lightman S, Seckl J (1987) Differential inhibition of vasopressin and oxytocin by opiates in man. Proceedings of the Physiological Society, Middlesex, March 1987: Abstr C180, p 205P
Dupont A, Cusan L, Garon M, Labrie F, Hao Li C (1977) β-Endorphin: Stimulation of growth hormone release in vivo. Proc Natl Acad Sci USA 74:358–359
Dupont A, Lépine J, Langelier P, Mérand Y, Rouleau D, Vaudry H, Gros C, Barden N (1980) Differential distribution of β-endorphin and enkephalins in rat and bovine brain. Regul Pept 1:43–52
Ebling FJP, Schwartz ML, Foster DL (1989) Endogenous opioid regulation of pulsatile luteinizing hormone secretion during sexual maturation in the female sheep. Endocrinology 125:369–383
Eisenberg RM (1980) Effects of naloxone on plasma corticosterone in the opiate-naive rat. Life Sci 26:935–943
Ellingboe J, Veldhuis JD, Mendelson JH, Kuehnle JC, Mello NK (1982) Effect of endogenous opioid blockade on the amplitude and frequency of pulsatile luteinizing hormone secretion in normal men. J Clin Endocrinol Metab 54:854–857
Enjalbert A, Ruberg M, Arancibia S, Priam M, Kordon C (1979) Endogenous opiates block dopamine inhibition of prolactin secretion in vitro. Nature 280:595–597

Evans RG, Olley JE, Rice GE, Abrahams JM (1989a) μ- and κ-Opiate receptor agonists reduce plasma neurohypophysial hormone concentrations in water-deprived and normally hydrated rats. Clin Exp Pharmacol Physiol 16:191–197

Evans RG, Olley JE, Rice GE, Abrahams JM (1989b) Effects of subacute opioid administration during late pregnancy in the rat on the initiation, duration and outcome of parturition and maternal levels of oxytocin and arginine vasopressin. Clin Exp Pharmacol Physiol 16:169–178

Falke N (1988) Dynorphin (1–8) inhibits stimulated release of oxytocin but not vasopressin from isolated neurosecretory endings of the rat neurohypophysis. Neuropeptides 11:163–167

Falke N, Martin R (1988) Opiate binding differentially associated with oxytocin and vasopressin nerve endings from porcine neurohypophyses. Exp Brain Res 70:145–154

Ferland L, Fuxe K, Eneroth P, Gustafsson J-A, Skett P (1977) Effects of methionine-enkephalin on prolactin release and catecholamine levels and turnover in the median eminence. Eur J Pharmacol 43:89–90

Feuerstein G, Molineaux CJ, Rosenberg JG, Zerbe RL, Cox BM, Faden AI (1985) Hemorrhagic shock and the central vasopressin and opioid peptide system of rats. Am J Physiol 249:E244–E250

Field E, Kuhn CM (1988) Opioid inhibition of luteinizing hormone release declines with age and acyclicity in female rats. Endocrinology 123:2626–2631

Firemark HM, Weitzman RE (1979) Effects of β-endorphin, morphine and naloxone on arginine vasopressin secretion and the electroencephalogram. Neuroscience 4:1895–1902

Flanagan LM, Verbalis JG, Stricker EM (1988) Naloxone potentiation of effects of cholecystokinin and lithium chloride on oxytocin secretion, gastric motility and feeding. Neuroendocrinology 48:668–673

Foley KM, Kouride IA, Inturrisi CE, Kaiko RF, Zaroulis CG, Posner JB, Houde RW, Li CH (1979) β-Endorphin: analgesic and hormonal effects in humans. Proc Natl Acad Sci USA 76:5377–5381

Foresta C, Scanelli G, Indino M, Federspil G, Scandellari C (1985) Naloxone reduces the fenfluramine-induced prolactin release in man. Clin Endocrinol (Oxf) 22:539–543

Forsling ML, Aziz LA (1983) Release of vasopressin in response to hypoxia and the effect of aminergic and opioid antagonists. J Endocrinol 99:77–86

Forsling ML, Matziari C, Aziz L (1988) A comparison of the vasopressin response of rats to intraperitoneal and intravenous administration of hypertonic saline, and the effect of opioid and aminergic antagonists. J Endocrinol 116:217–224

Fraioli F, Cappa M, Fabbri A, Gnessi L, Moretti C, Borrelli P, Isidori A (1984) Lack of endogenous opioid inhibitory tone on LH secretion in early puberty. Clin Endocrinol (Oxf) 20:299–305

Fujimoto JM (1971) The kidney. In: Clouet DH (ed) Narcotic Drugs: biochemical pharmacology. Plenum, New York, pp 366–393

Fujimoto S, Hisada S (1978) Effects of intracerebroventricular morphine and enkephalins on the urine outflow in ethanol-anaesthetized rats. Nagoya Med J 23:1–10

Gabriel SM, Berglund LA, Simpkins JW (1986) A decline in endogenous opioid influence during the steroid-induced hypersecretion of luteinizing hormone in the rat. Endocrinology 118:558–561

Gaillard RC, Grossman AB, Smith R, Rees LH, Besser GM (1981) The effects of a met-enkephalin in patients with adrenocortical disease. Clin Endocrinol (Oxf) 14:471–478

Garthwaite TL, Hagen TC (1979) Evidence that serotonin stimulates a prolactin-releasing factor in the rat. Neuroendocrinology 29:215–220

Gaymann W, Martin R (1987) A re-examination of the localization of immunoreactive

dynorphin (1–8), [Leu]enkephalin and [Met]-enkephalin in the rat neurohypophysis. Neuroscience 20:1069–1080
George R (1971) Anterior pituitary gland. In: Clouet DH (ed) Narcotic drugs: biochemical pharmacology. Plenum, New York, pp 283–299
George R, Lomax P (1972) Hormones. In: Mule SJ Brill H (eds) Chemical and biological aspects of drug dependence. CRC, Cleveland, pp 523–543
George R, Way EL (1959) The role of the hypothalamus in pituitary-adrenal activation and anti-diuresis by morphine. J Pharmacol Exp Ther 125:111–115
Gerstberger R, Barden N (1986) Dynorphin 1–8 binds to opiate kappa receptors in the neurohypophysis. Neuroendocrinology 42:376–382
Giarman NJ, Condouris GA (1954) The antidiuretic action of morphine and some of its analogs. Arch Int Pharmacodyn 97:28–33
Gilbeau PM, Hosobuchi Y, Lee NM (1986) Dynorphin effects on plasma concentrations of anterior pituitary hormones in the nonhuman primate. J Pharmacol Exp Ther 238:974–977
Gilbeau PM, Hosobuchi Y, Lee NM (1987) Consequence of dynorphin-A administration on anterior pituitary hormone concentrations in the adult male rhesus monkey. Neuroendocrinology 45:284–289
Gosden RG, Humphreys EM, Johnston V, Liddle S, Russell JA (1985) Morphine acts centrally to interrupt established parturition in rats. Proceedings of the Physiological Society, Febr 22–23, Bristol, p 70P
Grandison L, Fratta W, Guidotti A (1980) Location and characterization of opiate receptors regulating pituitary secretion. Life Sci 26:1633–1642
Grell S, Dencker Christensen J, Fjalland B (1985) Morphine antidiuresis in conscious rats: contribution of vasopressin and blood pressure. Acta Pharmacol Toxicol (Copenh) 56: 38–43
Grell S, Fjalland B, Christensen JD (1989) Biphasic effect of a κ-opioid receptor agonist on plasma oxytocin levels in rats. Eur J Pharmacol 166:41–48
Grossman AB (1983) Brain opiates and neuroendocrine function. In: Clin Endocrinol Metab Neuroendocrinology 12:725–746
Grossman AB (1987) Opioid peptides and reproductive function. Semin Reprod Endocrinol 5:115–124
Grossman AB (1988) Exercise and the GnRH pulse generation. In: Jumnra M, Shizume K, Yoshida S (eds) Progress in endocrinology. Elsevier, Amsterdam, pp 441–448
Grossman AB, Besser GM (1982) Opiates control ACTH through a noradrenergic mechanism. Clin Endocrinol (Oxf) 17:287–290
Grossman AB, Rees LH (1983) The neuroendocrinology of opioid peptides. Br Med Bull 39:83–88
Grossman A, Clement-Jones V (1983) Enkephalins and endorphins. Clin Endocrinol Metab 12:31–56
Grossman A, Moretti C (1986) Opioid peptides and their relationship to hormonal changes during acute exercise. In: Benzi G, Packer L, Siliprandi N (eds) Problems in the Biochemistry of Physical Exercise and Training. Elsevier, Amsterdam, pp 375–386
Grossman AB, Milles J, Baylis PS, Besser GM (1980) Inhibition of vasopressin release by an opiate peptide in man. Lancet 2:1108–1109
Grossman AB, Stubbs WA, Gaillard R, Delitala G, Rees LH, Besser GM (1981a) Studies on the opiate control of prolactin, GH and TSH. Clin Endocrinol (Oxf) 14:321–324
Grossman AB, Moult P, Gaillard RC, Delitala G, Toff WD, Rees LH, Besser GM (1981b) The opioid control of LH and FSH release: effects of a met-enkephalin analogue and naloxone. Clin Endocrinol (Oxf) 14:41–47

Grossman AB, West S, Evans JM, Williams S, Rees LH, Besser GM (1982a) The role of opiate peptides in the control of prolactin in the puerperium and TSH in primary hypothyroidism. Clin Endocrinol (Oxf) 16:317–320

Grossman AB, Moult PJA, McIntyre H, Evans J, Silverstone T, Rees LH, Besser GM (1982b) Opiate mediation of amenorrhoea in hyperprolactinaemia and in weight-loss related amenorrhoea. Clin Endocrinol (Oxf) 17:379–388

Grossman AB, Gaillard RC, McCarthy P, Rees LH, Besser GM (1982c) Opiate modulation of the pituitary-adrenal axis: effects of stress and circadian rhythm. Clin Endocrinol (Oxf) 17:279–286

Grossman AB, Bouloux P, Price P, Drury P, Lam K, Turner T, Besser GM, Sutton J (1984) The role of opioid peptides in the hormonal responses to acute exercise in man. Clin Sci 67:483–491

Grossman AB, Moult PJA, Cunnah D, Besser GM (1986a) Different opioid mechanisms are involved in the modulation of gonadotrophin and ACTH release in man. Neuroendocrinology 42:357–360

Grossman AB, Delitala G, Coy DH, Besser GM (1986b) An analogue of Met-enkephalin attenuates the cortisol response to ovine corticotrophin releasing factor. Clin Endocrinol (Oxf) 25:421–426

Gudelsky GA, Porter JC (1979) Morphine and opioid peptide-induced inhibition of the release of dopamine from tuberoinfundibular neurons. Life Sci 25:1697–1702

Guiol C, Montastruc JL, Montastruc P (1984) Naloxone reduces diuretic responses induced by water, alcohol or congenital lack of vasopressin in rats. Br J Pharmacol 81:85:90

Haldar J (1982) Release of antidiuretic hormone by morphine in rats: and in vivo and in vitro study. Proc Soc Exp Biol Med 169:113–120

Haldar J, Bade V (1981) Involvement of opioid peptides in the inhibition of oxytocin release by heat stress in lactating mice. Proc Soc Exp Biol Med 168:10–14

Haldar J, Sawyer WH (1978) Inhibition of oxytocin release by morphine and its analogs. Proc Soc Exp Biol Med 157:476–480

Haldar J, Hoffman DL, Zimmerman EA (1982) Morphine, β-endorphin and [D-Ala^2]Met-enkephalin inhibit oxytocin release by acetylcholine and suckling. Peptides 3:663–668

Hammer RP, Bridges RS (1987) Preoptic area opioids and opiate receptors increase during pregnancy and decrease during lactation. Brain Res 420:48–56

Handley CA, Keller AD (1950) Changes in renal function produced by morphine in normal dogs and dogs with diabetes insipidus. J Pharmacol Exp Ther 99:33–37

Hart IC, Cowie AT (1978) Effect of morphine, naloxone and an enkephaline analogue on plasma prolactin, growth hormone, insulin and thyroxine in goats. J Endocrinol 77: 16P–17P

Hartman R, Rosella-Dampman LM, Emmert SE, Summy-Long JY (1986a) Inhibition of release of neurohypophysial hormones by endogenous opioid peptides in pregnant and parturient rats. Brain Res 382:352–359

Hartman R, Rosella-Dampman LM, Emmert SE, Summy-Long JY (1986b) Ontogeny of opioid inhibition of vasopressin and oxytocin release in response to osmotic stimulation. Endocrinology 119:1–11

Hartman R, Rosella-Dampman LM, Summy-Long JY (1987) Endogenous opioid peptides inhibit oxytocin release in the lactating rat after dehydration and urethane. Endocrinology 121:536–543

Haskins JT, Gudelsky GA, Moss RL, Porter JC (1981) Iontophoresis of morphine into the arcuate nucleus: effects on dopamine concentrations in hypophysial portal plasma and serum prolactin concentrations. Endocrinology 108:767–771

Hellebrekers LJ, Mol JA, van den Brom WE, van Wimersma Greidanus TjB (1987) Effect of

methadone on plasma arginine vasopressin level and urine production in conscious dogs. Eur J Pharmacol 136:279–286
Herkenham M, Rice KC, Jacobson AE, Rothman RB (1986) Opiate receptors in rat pituitary are confined to the neural lobe and are exclusively kappa. Brain Res 382:365–371
Herz A, Millan MJ (1984) Opioid peptides in the hypothalamic-pituitary axis. In: Hughes J, et al. (eds) Opioids: past, present and future, Taylor and Francis, London, pp 79–102
Hoehe M, Duka T, Doenicke A (1988) Human studies on the μ opiate receptor agonists fentanyl: neuroendocrine and behavioral responses. Psychoneuroendocrinology 13:397–408
Höllt V, Haarmann I, Seizinger BR, Herz A (1981) Levels of dynorphin-(1–3) immunoreactivity in rat neurointermediate pituitaries are concomitantly altered with those of leucine enkephalin and vasopressin in response to various endocrine manipulations. Neuroendocrinology 33:333–339
Honer WG, Thompson C, Lightman SL, Williams TDM, Checkley SA (1986) No effect of naloxone on plasma oxytocin in normal men. Psychoneuroendocrinology 11:245–248
Huidobro F, Huidobro-Toro P (1979) Antidiuretic effect of morphine and vasopressin in morphine tolerant and non-tolerant rats, differential effects on urine composition. Eur J Pharmacol 59:55–64
Huidobro-Toro JP (1980) Antidiuretic effect of β-endorphin and morphine in Brattleboro rats: development of tolerance and physical dependence after chronic morphine treatment. Br J Pharmacol 71:51–56
Huidobro-Toro JP, Huidobro F, Croxatto R (1979) Effects of β-endorphin and D-Alanine$_2$ enkephalinamide on urine production and urinary electrolytes in the rat. Life Sci 24:697–704
Iovino M, Rouhani S, Peonaru S (1983) Dual effect of Met-enkephalin on plasma ADH concentration in the rat. Neuroendocrinol Lett 5:239–243
Ishizuka B, Quigley ME, Yen SSC (1983) Pituitary hormone release in response to food ingestion: evidence for neuroendocrine signals from gut to brain. J Clin Endocrinol Metab 57:1111–1116
Ito S, Iwanaga T, Yui R, Yamaguchi K, Hama H, Kamoi K, Shibata A (1981) Presence of α-neo-endorphin-like immunoreactivity in the posterior lobe of the pituitary gland. Life Sci 29:1457–1461
Iversen LL, Iversen SD, Bloom FE (1980) Opiate receptors influence vasopressin release from nerve terminals in rat neurohypophysis. Nature 284:350–351
Iyengar S, Kim HS, Wood PL (1987) μ-, δ-, κ- and ε-opioid receptor modulation of the hypothalamic-pituitary-adrenocortical (HPA) axis: subchronic tolerance studies of endogenous opioid peptides. Brain Res 435:220–226
Jacobson W, Kalra SP (1989) Decreases in mediobasal hypothalamic and preoptic area [^{3}H]naloxone binding are associated with the progesterone-induced luteinizing hormone surge. Endocrinology 124:199–206
Jezova D (1985) Stimulation of ACTH release by naloxone: central or peripheral action? Life Sci 37:1007–1013
Johnson JH (1982) Release of prolactin in response to microinjection of morphine into mesencephalic dorsal raphe nucleus. Neuroendocrinology 35:169–172
Johnson MR, Bower M, Seckl JR, Lightman SL (1990) Neurohypophysial secretion to insulin-induced hypoglycemia and its regulation by endogenous opioids in women. Acta Endocrinol (Copenh) 122:467–471
Jones SA, Summerlee AJS (1986) Relaxin acts centrally to inhibit oxytocin release during parturition: an effect that is reversed by naloxone. J Endocrinol 111:99–102
Jordan J, Veisseire M, Borson-Chazot F, Mornex R (1986) In vitro effects of endogenous

opiate peptides on thyrotropin function: inhibition of thyrotropin-releasing hormone release and absence of effect on thyrotropin release. Neurosci Lett 67:289–294
Judd AM, Hedge GA (1982) The roles of opioid peptides in controlling thyroid stimulating hormone release. Life Sci 31:2529–2536
Judd AM, Hedge GA (1983) Direct pituitary stimulation of thyrotropin secretion by opioid peptides. Endocrinology 113:706–710
Kalra SP, Crowley WR (1982) Epinephrine synthesis inhibitors block naloxone-induced LH release. Endocrinology 111:1403–1405
Kalra SP, Simpkins JW (1981) Evidence for noradrenergic mediation of opioid effects on luteinizing hormone secretion. Endocrinology 109:776–782
Kamoi K, White K, Robertson GL (1979) Opiates elevate the osmotic threshold for vasopressin (VP) release in rats. Clin Res 27:254A
Kato Y, Matsushita N, Katakami H, Shimatsu A, Imura H (1981) Stimulation by dynorphin of prolactin and growth hormone secretion in the rat. Eur J Pharmacol 73:353–355
Keil LC, Rosella-Dampman LM, Emmert S, Chee O, Summy-Long JY (1984) Enkephalin inhibition of angiotensin-stimulated release of oxytocin and vasopressin. Brain Res 297:329–336
Khachaturian HK, Lewis ME, Schäfer MK-H, Watson SJ (1985) Anatomy of the CNS opioid systems. Trends Neurosci 8:111–119
Khorram O, McCann SM (1986) Interaction of α-melanocyte-stimulating hormone with β-endorphin to influence anterior pituitary hormone secretion in the female rat. Endocrinology 119:1071–1075
Khoury SA, Reame NE, Kelch RP, Marshall JC (1987) Diurnal patterns of pulsatile luteinizing hormone secretion in hypothalamic amenorrhea: reproducibility and responses to opiate blockade and an α_2-adrenergic agonist. J Clin Endocrinol Metab 64:755–762
Kinoshita F, Nakai Y, Katakami H, Kato Y, Yajima H, Imura H (1980) Effect of β-endorphin on pulsatile luteinizing hormone release in conscious castrated rats. Life Sci 27:843–846
Knepel W, Reimann W (1982) Inhibition by morphine and β-endorphin of vasopressin release evoked by electrical stimulation of the rat medial basal hypothalamus in vitro. Brain Res 238:484–488
Knepel W, Nutto D, Hertting G (1982) Evidence for inhibition by β-endorphin of vasopressin release during foot shock-induced stress in the rat. Neuroendocrinology 34:353–356
Knepel W, Nutto D, Meyer DK (1983) Naloxone increases vasopressin secretion from the neurointermediate lobe of the hypophysis of the rat: search for the endogenous agonist. Life Sci 33:499–502
Knight PG, Howles CM, Cunningham FJ (1986) Evidence that opioid peptides and dopamine participate in the suckling-induced release of prolactin in the ewe. Neuroendocrinology 44:29–35
Koenig JI, Mayfield MA, McCann SM, Krulich L (1984) Differential role of the opioid μ and δ receptors in the activation of prolactin (PRL) and growth hormone (GH) secretion by morphine in the male rat. Life Sci 34:1829–1837
Koenig JI, Meltzer HY, Devane GD, Gudelsky GA (1986) The concentration of arginine vasopressin in pituitary stalk plasma of the rat after adrenalectomy or morphine. Endocrinology 118:2534–2539
Kokka N, Garcia JF, Elliott HW (1973) Effects of acute and chronic administration of narcotic analgesics on growth hormone and corticotrophin (ACTH) secretion in rats. Prog Brain Res 39:347–360
Koves K, Marton J, Molnar J, Halasz B (1981) (D-Met2,Pro5)-Enkephalinamide-induced blockade of ovulation and its reversal by naloxone in the rat. Neuroendocrinology 32: 82–86

Krulich L, Koenig JI, Conway S, McCann SM, Mayfield MA (1986a) Opioid κ receptors and the secretion of prolactin (PRL) and growth hormone (GH) in the rat. Neuroendocrinology 42:75–81
Krulich L, Koenig JI, Conway S, McCann SM, Mayfield MA (1986b) Opioid κ receptors and the secretion of prolactin (PRL) and growth hormone (GH) in the rat. Neuroendocrinology 42:82–87
Kuderling I, Dorsch G, Warnhoff M, Pirke KM (1984) The actions of prostaglandin E_2, naloxone and testosterone on starvation-induced suppression of luteinizing hormone-releasing hormone and luteinizing-hormone secretion. Neuroendocrinology 39:530–537
Kuhn CM, Bartolome MB (1985) Effect of chronic methadone administration on neuroendocrine function in young adult rats. J Pharmacol Exp Ther 234:204–210
Lacoumenta S, Young VTF, Burrin JM (1988) Opioid control of hypothalamic AVP release in vitro. J Endocrinol [Suppl] 117:181
Lakoski JM, Gebhart GF (1981) Attenuation of morphine's depression of serum luteinizing hormone by lesions in the amygdala. Neuroendocrinology 33:105–111
Lamberts SWJ, Janssens ENW, Bons EG, Uitterlinden P, Zuiderwijk JM, Del Pozo E (1983) The met-enkephalin analog FK 33-824 directly inhibits ACTH release by the rat pituitary gland in vitro. Life Sci 32:1167–1173
Landymore KM, Wilkinson M (1988) Influence of neonatal opioid blockade or injections of gonadotrophin-releasing hormone on the timing of puberty in female rats: correlation of opioid effects with occupation of hypothalamic μ-opioid receptors. J Endocrinol 119: 447–452
Leadem CA, Kalra SP (1985) Effects of endogenous opioid peptides and opiates on luteinizing hormone and prolactin secretion in ovariectomized rats. Neuroendocrinology 41:342–352
Leander JD (1983a) A kappa opioid effect: increased urination in the rat. J Pharmacol Exp Ther 224:89–94
Leander JD (1983b) Further study of kappa opioids on increased urination. J Pharmacol Exp Ther 227:35–41
Leander JD, Zerbe RL, Hart JC (1985) Diuresis and suppression of vasopressin by kappa opioids: comparison with mu and delta opioids and clionidine. J Pharmacol Exp Ther 234:463–469
Leander JD, Hart JC, Zerbe RL (1987) Kappa agonist-induced diuresis: evidence for stereoselectivity, strain differences, independence of hydration variables and a result of decreased plasma vasopressin levels. J Pharmacol Exp Ther 242:33–39
Lehtinen A-M, Fyhrquist F, Kivalo I (1984) The effect of fentanyl on arginine vasopressin and cortisol secretion during anesthesia. Anesth Analg 63:25–30
Leng G, Russell JA (1987) Opioids, oxytocin and parturition (Abstr C21). Meeting "Brain Opioid Systems in Reproduction", Babraham
Leng G, Mansfield S, Bicknell RJ, Dean ADP, Ingram CD, Marsh MIC, Yates JO, Dyer RG (1985) Central opioids: a possible role in parturition? J Endocrinol 106:219–224
Leng G, Russell JA, Bicknell RJ, Hunter NJ (1987) Opiate withdrawal kindles oxytocin hypersecretion (Abstr p 19). Meeting "Brain Opioid Systems in Reproduction", Babraham
Lengyel AMJ, Grossman AB, Bouloux PMG, Rees LH, Besser GM (1985) Effects of dopamine and morphine on immunoreactive somatostatin and LH-releasing hormone secretion from hypothalamic fragments in vitro. J Endocrinol 106:317–322
Lien EL, Morrison A, Dvonch W (1979) The effects of partial opiate agonists on plasma prolactin and growth hormone levels in the rat. Life Sci 25:1709–1716
Lightman SL, Forsling ML (1980a) The effect of the methionine enkephalin analogue DAMME on the vasopressin response to tilt in man. Clin Sci 59:501–503

Lightman SL, Forsling ML (1980b) Evidence for endogenous opioid control of vasopressin release in man. J Clin Endocrinol Metab 50:569–571
Lightman S, Seckl J (1987) Opioid-mediated inhibition of oxytocin in man (Abstr C26). Meeting "Brain Opioid Systems in Reproduction", Babraham
Lightman SL, Langdon N, Forsling ML (1980) Effects of the opiate antagonist naloxone and the enkephalin analog DAMME on the vasopressin response to a hypertonic stimulus in man. J Clin Endocrinol Metab 51:1447–1449
Lightman SL, Iversen LL, Forsling ML (1982) Dopamine and [D-Ala^2, D-Leu^5]enkephalin inhibit the electrically stimulated neurohypophyseal release of vasopressin in vitro: evidence for calcium-dependent opiate action. J Neurosci 2:78–81
Lightman S, Forsling M, Todd K (1983a) Hypothalamic integration of dopaminergic and opiate pathways controlling vasopressin secretion. Endocrinology 113:77–80
Lightman SL, Ninkovic M, Hunt SP, Iversen LL (1983b) Evidence for opiate receptors on pituicytes. Nature 305:235–237
Limonta P, Dondi D, Maggi R, Martini L, Piva F (1989) Neonatal organization of the brain opioid systems controlling prolactin and luteinizing hormone secretion. Endocrinology 124:681–686
Lodico G, Stoppelli I, Delitala G, Maioli M (1983) Effects of naloxone infusion on basal and breast-stimulation-induced prolactin secretion in puerperal women. Fertil Steril 40: 600–603
Lomax P, George R (1966) Thyroid activity following administration of morphine in rats with hypothalamic lesions. Brain Res 2:361–367
Lomax P, Kokka N, George R (1970) Thyroid activity following intracerebral injection of morphine in the rat. Neuroendocrinology 6:146–152
Lotti, VJ, Kokka N, George R (1969) Pituitary-adrenal activation following intrahypothalamic microinjection of morphine. Neuroendocrinology 4:326–332
Lutz-Bucher B, Koch B (1980) Evidence for a direct inhibitory effect of morphine on the secretion of posterior pituitary hormones. Eur J Pharmacol 66:375–378
Lutz-Bucher B, Koch B (1981) Specific inhibitory effect of morphine on experimentally-induced release of posterior pituitary hormones. Neuroendocrinol Lett 3:261–266
Männistö PT (1983) Central regulation of thyrotropin secretion in rats: methodological aspects, problems and some progress. Med Biol 61:92–100
Männistö PT, Rauhala P, Tuominen R, Mattila J (1984) Dual action of morphine on cold-stimulated thyrotropin secretion in male rats. Life Sci 35:1101–1107
Mansour A, Lewis ME, Khachaturian H, Watson SJ (1986) Multiple opioid receptor subtypes in the pituitary adrenal axis. A cross-species study (Abstr P3) 17th International Narcotic Research Conference, Sán Francisco
Marchand C (1970) The diuretic and antidiuretic effect of morphine sulphate in rats. Proc Soc Exp Biol Med 133:1303–1306
Marko M, Romer D (1983) Inhibitory effect of a new opioid agonist on reproductive endocrine activity in rats of both sexes. Life Sci 33:233–240
Martin JB, Tolis G, Wood I, Guyda H (1979) Failure of naloxone to influence physiological growth hormone and prolactin secretion. Brain Res 168:210–215
Martin R, Voigt KH (1981) Enkephalins co-exist with oxytocin and vasopressin in nerve terminals of rat neurohypophysis. Nature 289:502–504
Martin R, Geis R, Holl R, Schäfer M, Voigt KH (1983) Co-existence of unrelated peptides in oxytocin and vasopressin terminals of rat neurohypophyses: immunoreactive methionine^5-enkephalin-, leucine^5-enkephalin- and cholecystokinin-like substances. Neuroscience 8:213–227

Mata MM, Gainer H, Klee WA (1977) Effect of dehydration on the endogenous opiate content of the rat neuro-intermediate lobe. Life Sci 21:1159–1162
Mauras N, Veldhuis D, Rogol AD (1986) Role of endogenous opiates in pubertal maturation: opposing actions of naltrexone in prepubertal and late pubertal boys. J Clin Endocrinol Metab 62:1256–1263
May P, Mittler J, Manougian A, Ertel N (1979) TSH release-inhibiting activity of leucine-enkephalin. Horm Metab Res 11:30–33
Mayer DJ, Wessel J, Kobberling J (1980) Failure of naloxone to alter exercise-induced growth hormone and prolactin release in normal men. Clin Endocrinol (Oxf) 13:413–416
Maysinger D (1985) Opioids and hypophysis. Acta Pharm Jogosl 35:1–14
Maysinger D, Vermes I, Tilders F, Seizinger BR, Gramsch C, Höllt V, Herz A (1984) Differential effects of various opioid peptides on vasopressin and oxytocin release from the rat pituitary in vitro. Naunyn Schmiedebergs Arch Pharmacol 328:191–195
McDonald RK, Evans FT, Weise VK, Patrick RW (1959) Effect of morphine and nalorphine on plasma hydrocortisone levels in man. J Pharmacol Exp Ther 125:241–247
Meites J, Bruni JF, van Vugt DA, Smith AF (1979) Relation of endogenous opioid peptides and morphine to neuroendocrine functions. Life Sci 24:1325–1336
Melis GB, Paoletti AM, Gambacciani M, Mais V, Fioretti P (1984) Evidence that estrogens inhibit LH secretion through opioids in postmenopausal women using naloxone. Neuroendocrinology 39:60–63
Melis GB, Gambacciani M, Paoletti AM, Mais V, Cagnacci A, Petacchi FD, Fioretti P (1985) Sex steroids modulate prolactin response to naloxone in postmenopausal women. Neuroendocrinology 41:138–141
Melis GB, Cagnacci A, Gambacciani M, Paoletti AM, Caffi T, Fioretti P (1988) Chronic bromocriptine administration restores luteinizing hormone response to naloxone in postmenopausal women. Neuroendocrinology 47:159–163
Micevych P, Elde R (1980) Relationship between enkephalinergic neurons and the vasopressin-oxytocin neuroendocrine system of the cat: an immunohistochemical study. J Comp Neurol 190:135–146
Michne WF, Lewis TR, Michalec SJ (1978) Novel developments of N-methylbenzomorphan narcotic antagonists. In: van Ree JM and Terenius L (eds) Characteristics and function of opioids. Elsevier/North-Holland Amsterdam, pp 197–206
Millan MJ, Herz A (1985) The endocrinology of the opioids. Int Rev Neurobiol 26:1–83
Miller M (1975) Inhibition of ADH-release in the rat by narcotic antagonists. Neuroendocrinol 19:241–251
Miller M (1980) Role of endogenous opioids in neurohypophysial function of man. J Clin Endocrinol Metab 50:1016–1020
Miller M (1981) Effect of opioids on antidiuretic hormone (ADH) release from the rat neurohypophysial system (Abstr 417). 63rd Annual Meeting of the Endocrine Society
Mitsuma T, Nogimori T (1983a) Effects of leucine-enkephalin on hypothalamic-pituitary-thyroid axis in rats. Life Sci 32:241–248
Mitsuma T, Nogimori T (1983b) β-Neoendorphin inhibits thyrotropin secretion in rats. Acta Endocrinol (Copenh) 104:437–442
Montastruc JL, Morales-Olivas F, Montastruc P (1980) Effect of morphine on diuresis in normal and diabetes insipidus dogs. Arch pharmacol Toxicol 6:287–290
Moos F, Richard P (1979) Effects of dopaminergic antagonist and agonist on oxytocin release induced by various stimuli. Neuroendocrinology 28:138–144
Moos F, Richard P (1982) Excitatory effect of dopamine on oxytocin and vasopressin reflex releases in the rat. Brain Res 241:249–260

Moretti C, Fabbri A, Gnessi L, Cappa M, Calzolari A, Fraiolo F, Grossman AB, Besser GM (1983) Naloxone inhibits exercise-induced release of PRL and GH in athletes. Clin Endocrinol 18:135–138

Morley JE (1983) Neuroendocrine effects of endogenous opioid peptides in human subjects: a review. Psychoneuroendocrinology 8:361–379

Morley JE, Baranetski NG, Wingert TD, Carlson HE, Hershman JM, Melmed S, Levin SR, Jamison KR, Weitzman R, Chang RJ, Varner AA (1980) Endocrine effects of naloxone-induced opiate receptor blockade. J Clin Endocrinol Metab 50:251–257

Morley JE, Willenbring ML, Krahn DD, Carlson GA, Briggs JE, Levine AS, Shafer RB (1984) Opioid control of thyroid function. In: Delitala G et al. (eds) Opioid modulation of endocrine function. Raven, New York, 267–275

Moult PJ, Grossman A, Evans J, Rees LH, Besser GM (1981) The effect of naloxone on pulsatile gonadotrophin release in normal subjects. Clin Endocrinol (Oxf) 14:321–324

Muraki T, Nakadate T, Tokunaga Y, Kato R (1980) Effect of morphine on the release of thyroid-stimulating hormone stimulated by exposure to cold, thyroidectomy and the administration of thyrotropin releasing hormone in male rats. J Endocrinol 86:357–362

Nikolarakis KE, Pfeiffer A, Stalla GK, Herz A (1987) The role of CRF in the release of ACTH by opiate agonists and antagonists in rats. Brain Res 421:373–376

Nikolarakis KE, Pfeiffer A, Stalla GK, Herz A (1989) Facilitation of ACTH secretion by morphine is mediated by activation of CRF releasing neurons and sympathetic neuronal pathways. Brain Res 498:385–388

Nilaver G, Hoffman DL, Zimmerman EA (1979) Multiple projections of the magnocellular hypothalamus to other brain regions including cerebral cortex. Trans Am Neurol Assoc 104:205–208

Nordmann JJ, Dayanithi G, Cazalis M (1986) Do opioid peptides modulated, at the level of the nerve endings, the release of neurohypophysial hormones? Exp Brain Res 61:560–566

Nutt JG, Jasinski DR (1974) Diuretic action of the narcotic antagonist oxilorphan. Clin Pharmacol Ther 15:361–367

Oiso Y, Iwasaki Y, Kondo K, Takatsuki K, Tomita A (1988) Effect of the opioid kappa-receptor agonist U50488H on the secretion of arginine vasopressin. Neuroendocrinology 48:658–662

Oka T, Negishi K, Suda M, Sawa A, Fujimo M, Wakimasu M (1982) Evidence that dynorphin (1–13) acts as an agonist on opioid κ-receptors. Eur J Pharmacol 77:137–141

Palkovits M, Brownstein MJ, Zamir N (1983) Immunoreactive dynorphin and α-neo-endorphin in rat hypothalamo-neurohypophyseal system. Brain Res 278:258–261

Pan J-T, Teo K-L (1989) Fentanyl stimulates prolactin release through μ-opiate receptors, but not the serotonergic system. Endocrinology 125:1863–1869

Panerai AE, Casanueva F, Martini A, Mantegazza P, de Giulio AM (1981) Opiates act centrally on GH and PRL release. Endocrinology 108:2400–2402

Panerai AE, Martini A, Casanueva F, Petraglia F, di Giulio AM, Mantegazza P (1983) Opiates and their antagonists modulate luteinizing hormone acting outside the blood brain barrier. Life Sci 32:1751–1756

Panerai AE, Petraglia F, Scerdote P, Genazzani AR (1985) Mainly μ-opiate receptors are involved in luteinizing hormone and prolactin secretion. Endocrinology 117:1096–1099

Parrott RF, Thornton SN (1989) Opioid influences on pituitary function in sheep under basal conditions and during psychological stress. Psychoneuroendocrinology 14:451–459

Parvizi N, Ellendorff F (1980) β-Endorphin alters luteinizing hormone secretion via the amygdala but not the hypothalamus. Nature 286:812–813

Pechnick RN, George R, Poland RE (1985) Identification of multiple opiate receptors through neuroendocrine responses. I Effects of agonists. J Pharmacol Exp Ther 232: 163–169

Pechnick RN, George R, Poland RE (1987) The effects of the systemic administration of N-methylmorphine chloride, a quaternary analogue of morphine that does not cross the blood-brain barrier, on the release of anterior pituitary hormones in the rat. Psychoneuroendocrinology 12:67–71

Peters GR, Ward NJ, Antal EG, Lai PY, DeMaar EW (1987) Diuretic actions in man of a selective kappa opioid agonist: U-62, 066E. J Pharmacol Exp Ther 240:128–131

Petraglia F, Degli Uberti EC, Trasforini G, Facchinetti F, Margutti A, Volpe A, Salvadori S, Tomatis R, Genazzani AR (1985) Dermorphin decreases plasma LH Levels in human: evidence for a modulatory role of gonadal steroids. Peptides 6:689–872

Petraglia F, Vale W, River C (1986a) Opioids act centrally to modulate stress-induced decrease in luteinizing hormone in the rat. Endocrinology 119:2445–2450

Petraglia F, Bernasconi S, Iughetti L, Loche S, Romanini F, Facchinetti F, Marcellini C, Genazzani AR (1986b) Naloxone-induced luteinizing hormone secretion in normal, precocious, and delayed puberty. J Clin Endocrinol Metab 63:1112–1116

Petraglia F, de Leo V, Nappi C, Facchinetti F, Monetmagno U, Brambilla F, Genazzani AR (1987) Differences in the opioid control of luteinizing hormone secretion between pathological and iatrogenic hyperprolactinemic states. J Clin Endocrinol Metab 64:508–510

Pfeiffer A, Pasi A, Mehraien P, Herz A (1982) Opiate receptor binding sites in human brain. Brain Res 348:87–96

Pfeiffer A, Herz A, Lorriaux DL, Pfeiffer DG (1985) Central kappa- and mu-opiate receptors mediate ACTH-release in rats. Endocrinology 116:2688–2690

Pfeiffer A, Braun S, Mann K, Meyer H-D, Brantil V (1986) Anterior pituitary hormone responses to a κ-opioid agonist in man. J Clin Endocrinol Metab 62:181–185

Pfeiffer DG, Pfeiffer A, Shimohigashi Y, Merriam GR, Lorriaux DL (1983) Predominant involvement of mu-rather than delta- or kappa-opiate receptors in LH secretion. Peptides 4:647–649

Pfeiffer DG, Pfeiffer A, Almeida PFX, Herz A (1987) Opiate suppression of LH secretion involves central receptors different from those mediating opiate effects on prolactin secretion. J Endocrinol 114:469–476

Pittman QJ, Hatton JD, Bloom FE (1980) Morphine and opioid peptides reduce paraventricular neuronal activity: studies on the rat hypothalamic slice preparation. Proc Natl Acad Sci USA 77:5527–5531

Pitzel L, König A (1984) Lack of response in the release of oxytocin and vasopressin from isolated neurohypophyses to dopamine, met-enkephalin and leu-enkephalin. Exp Brain Res 56:221–226

Plotsky PM (1986) Opioid inhibition of immunoreactive corticotropin-releasing factor secretion into the hypophysial-portal circulation of rats. Regul Pept 16:235–242

Pontiroli AE, Baio G, Stella L, Crescenti A, Giraldi AM (1982) Effects of naloxone on prolactin, luteinizing hormone and cortisol responses to surgical stress in humans. J Clin Endocrinol Metab 55:378–380

Price J, Grossman AB, Besser GM, Rees LH (1983) Dopaminergic control of the rat thyrotroph. Neuroendocrinology 36:125–129

Quigley ME, Sheehan KL, Casper RF, Yen SSC (1980) Evidence for an increased opioid inhibition of luteinizing hormone secretion in hyperprolactinemic patients with pituitary microadenoma. J Clin Endocrinol Metab 50:427–430

Quigley ME, Yen SSC (1980) The role of endogenous opiates on LH secretion during the menstrual cycle. J Clin Endocrinol Metab 51:179–181

Quirion R, Pilapil C (1984) Opioid receptor binding profile of [^{3}H]-dynorphin A-(1–8) in rat and guinea pig brain. Eur J Pharmacol 99:361–363
Ragavan VV, Frantz AG (1981) Opioid regulation of prolactin secretion: evidence for a specific role of β-endorphins. Endocrinology 109:1769–1771
Rasmussen DD, Liu JH, Wolf PL, Yen SC (1987) Neurosecretion of human hypothalamic immunoreactive β-endorphin: in vitro regulation by dopamine. Neuroendocrinology 45:197–200
Rathbun RC, Kattau RW, Leander JD (1983) Effects of mu- and kappa-opioid receptor agonists on urinary outout in mice. Pharmacol Biochem Behav 19:863–866
Rauhala P (1988) Regulation of thyrotropin and prolactin secretion by opioids in the rat. Thesis, University of Helsinki
Rauhala P, Tuominen RK, Männistö PT (1987) Opioid peptides in the regulation of TSH and prolactin secretion in the rat. Acta Endocrinol (Copenh) 114:383–388
Rauhala P, Tuominen RK, Männistö PT (1988) Opiate receptor subtypes in the regulation of thyrotropin and prolactin secretion in the rat. Horm Res 29:218–222
Rayner VC, Robinson ICAF, Russell JA (1988) Chronic intracerebroventricular morphine and lactation in rats: dependence and tolerance in relation to oxytocin neurones. J Physiol (Lond) 396:319–347
Redding TW, Bowers CY, Schally AV (1966) The effects of morphine and other narcotics on thyroid function in mice. Acta Endocrinol (Copenh) 51:391–399
Reid RL (1983) Endogenous opioid activity and the premenstrual system. Lancet 2:786
Reid RL, Yen SSC, Artman H, Fisher DA (1981a) Effects of synthetic β-endorphin on release of neurohypophyseal hormones. Lancet 2:1169–1170
Reid RL, Hoff JD, Yen SSC, Li CH (1981b) Effects of exogenous βh-endorphin on pituitary hormone secretion and its disappearance rate in normal human subjects. J Clin Endocrinol Metab 52:1179–1184
Riskind PN, Millard WJ, Martin JB (1984) Opiate modulation of the anterior pituitary hormone response during suckling in the rat. Endocrinology 114:1232–1237
Rittmaster RS, Cutler GB, Sobel DO, Goldstein DS, Koppelman MCS, Loriaux DL, Chrousos GP (1985) Morphine inhibits the pituitary-adrenal response to ovine corticotrophin-releasing hormone in normal subjects. J Clin Endocrinol Metab 60:891–895
Rivier C, Vale W, Ling N, Brown M, Guillemin R (1977) Stimulation in vivo of the secretion of prolactin and growth hormone by β-endorphin. Endocrinology 100:238–241
Rockhold RW, Crofton JT, Share L (1981) Vasopressin release does not contribute to pressor action of enkephalin in SHR. Hypertension 3:410–415
Rockhold RW, Crofton JT, Wang BC, Share L (1983) Effect of intracarotid administration of morphine and naloxone on plasma vasopressin levels and blood pressure in the dog. J Pharmacol Exp Ther 224:386–390
Rockhold RW, Crofton JT, Brooks DP, Share L (1986) Naloxone does not improve cardiovascular or blunt vasopressin responses in spontaneously hypertensive rats following graded hemorrhage. Neuroendocrinology 43:657–663
Rosella-Dampman LM, Keil LC, Chee O, Summy-Long JY (1983) Naltrexone effects on plasma vasopressin concentration elevated and lowered by various stimuli. J Pharmacol Exp Ther 226:373–381
Rosella-Dampman LM, Emmert SE, Keil LC, Summy-Long JY (1985) Differential effects of naloxone on the release of neurohypophysial hormones in normotensive and spontaneously hypertensive rats. Brain Res 325:205–214
Rossier J (1982) Functions of β-endorphin and enkephalins in the pituitary. Front Neuroendocrinol 7:191–209

Rossier J, Battenberg E, Pittman Q, Bayon A, Koda L, Miller R, Guillemin R, Bloom F (1979) Hypothalamic enkephalin neurones may regulate the neurohypophysis. Nature 277:653–655

Rossmanith WG, Mortola JF, Yen SSC (1988) Role of endogenous opioid peptides in the initiation of the midcycle luteinizing hormone surge in normal cycling women. J Clin Endocrinol Metab 67:695–700

Roti E, degli Uberti E, Salvadori S, Bianconi M, Emanuele R, Rotola C, Trasforini Robuschi G, Tomatis R, Gnudi A, Pansini R, Braverman LE (1984) Dermorphin, a new opioid peptide, stimulates thyrotropin secretion in normal subjects. J Endocrinol Invest 7: 211–214

Russell JA, Spears N (1983) Morphine inhibits suckling-induced oxytocin secretion in conscious lactating rats but also disrupts maternal behavior. Proceedings of the Physiological society, Edinburgh, September 1983: Abstr. C114 p 143P

Russell JA, Antoni FA, Coombes JE, Robinson ICAF (1988) Naloxone provokes secretion of oxytocin and not other pituitary hormones altered by stress in urethane-anaesthetized morphine-treated rats. J Endocrinol [Suppl] 117:95

Russell JA, Gosden RG, Humphreys EM, Cutting R, Fitzsimons N, Johnston V, Liddle S, Scott S, Stirland JA (1989) Interruption of parturition in rats by morphine; a result of inhibition of oxytocin secretion. J Endocrinol 121:521–536

Saltiel H, Passa PH, Kuhn JM, Canivet J (1982) Mediation par les peptides opioides de la reponse antehypophysaire à l'hypoglycemie insulinique. Nouv Presse Med 11:847–850

Samel M (1958) Blocking effect of morphine on the secretion of thyroid-stimulating hormone in rats. Nature 181:845–846

Samson WK, McDonald JK, Lumpkin MD (1985) Naloxone-induced dissociation of oxytocin and prolactin releases. Neuroendocrinology 40:68–71

Sauder SE, Case GD, Hopwood NJ, Kelch RP, Marshall JC (1984) The effects of opiate antagonism on gonadotropin secretion in children and in women with hypothalamic amenorrhea. Pediatr Res 18:322–328

Schnieden H, Blackmore EK (1955) The effect of nalorphine on the antidiuretic action of morphine in rats and men. Br J Pharmacol 10:45–50

Schriefer JA (1985) Effect of co-localized peptides on oxytocin and vasopressin release, in vitro. Abstr Soc Neurosci 11:171.4

Schulz R, Wilhelm A, Prike KM, Gramsch C, Herz A (1981) β-Endorphin and dynorphin control serum luteinizing hormone level in immature rats. Nature 294:757–759

Seckl JR, Lightman SL (1987) Effect of naloxone on oxytocin and vasopressin release during vaginocervical stimulation in the goat. J Endocrinol 115:317–322

Seckl JR, Haddock JA, Dunne MJ, Lightman SL (1988) Opioid-mediated inhibition of oxytocin during insulin-induced hypoglycemic stimulation of vasopressin in man. Acta Endocrinol (Copenh) 118:77–81

Seki K, Kato K, Shima K (1986) Parallelism in the luteinizing hormone responses to opioid and dopamine antagonists in hyperprolactinemic women with pituitary microadenoma. J Clin Endocrinol Metab 63:1225–1228

Seybold VS, Miller JW, Lewis PR (1978) Investigation of a dopaminergic mechanism for regulating oxytocin release. J Pharmacol Exp Ther 207:605–610

Shaar CJ, Clemens JA (1980) The effects of opiate agonists on growth hormone and prolactin release in rats. Fed Proc 39:2539–2543

Shaar CJ, Frederickson CA, Dininger NB, Jackson L (1977) Enkephalin analogues and naloxone modulate the release of growth hormone and prolactin – evidence for regulation by an endogenous opioid peptide in brain. Life Sci 21:853–860

Sharp B, Morley JE, Carlson HE, Gordon J, Briggs J, Melmed S, Hershman JM (1981) The

role of opiates and endogenous opioid peptides in the regulation of rat TSH secretion. Brain Res 219:335–344
Sherman TG, Civelli O, Douglas J, Herbert E, Watson SJ (1986) Coordinate expression of hypothalamic pro-dynorphin and pro-vasopressin mRNAs with osmotic stimulation. Neuroendocrinology 44:222–228
Shibuki K, Leng G, Way S (1988) Effects of naloxone and of intraperitoneal hypertonic saline upon oxytocin release and upon supraoptic neuronal activity. Neurosic Lett 88:75–80
Shoupe D, Montz FJ, Lobo RA (1985) The effects of estrogen and progestin on endogenous opioid activity in oophorectomized women. J Clin Endorinol Metab 60:178–183
Slizgi GR, Ludens JH (1982) Studies on the nature and mechanism of the diuretic activity of the opioid analgesic ethylketocyclazocine. J Pharmacol Exp Ther 220:585–591
Slizgi GR, Ludens JH (1986) Role of ADH in ethylketocyclazocine-induced diuresis: studies in the Brattleboro rat. Life Sci 38:2437–2440
Slizgi GR, Taylor CJ, Ludens JH (1984) Effects of the highly selective kappa opioid, U-50,488, on renal function in the anesthetized dog. J Pharmacol Exp Ther 230:641–645
Smith CB (1972) Neurotransmitters and the narcotic analgesics. In: Mule SJ, Brill H (eds) Chemical and biological aspects of drug dependence. CRC, Cleveland, pp 495–504
Smith R, Grossman AB, Gimson AES, Besser GM, Rees LH (1985) Effect of liver and renal dysfunction on circulating metenkephalin immunoreactivity. Neurosci Lett 60:301–305
Spampinato S, Locatelli V, Cocchi D, Vicentini L, Bajusz S, Ferri S, Muller EE (1979) Involvement of brain serotonin in the prolactin-releasing effect of opioid peptides. Endocrinology 105:163–170
Spiegel K, Kourides IA, Pasternak GW (1982) Prolactin and growth hormone release by morphine in the rat; Different receptor mechanisms. Science 217:745–747
Stansfield SC, Knight PG, Al-Mauly NZ, Bryant MJ (1987) Involvement of endogenous opioid peptides in the neuroendocrine response of ewe lambs to the introduction of a ram. J Endocrinol 115:333–339
Stubbs WA, Jones A, Edwards CRW, Delitala G, Jeffcoate WJ, Ratter SJ (1978) Hormonal and metabolic responses to an enkephalin analogue in normal man. Lancet 2:1225–1227
Summy-Long JY, Rosella LM, Keil LC (1981a) Effects of centrally administered endogenous opioid peptides on drinking behavior, increased plasma vasopressin concentration and pressor response to hypertonic sodium chloride. Brain Res 221:343–357
Summy-Long JY, Keil LC, Deen K, Severs WB (1981b) Opiate regulation of angiotensin-induced drinking and vasopressin release. J Pharmacol Exp Ther 217:630–637
Summy-Long JY, Miller DS, Rosella-Dampman LM, Hartman RD, Emmert SE (1984) A functional role for opioid peptides in the differential secretion of vasopressin and oxytocin. Brain Res 309:362–366
Summy-Long JY, Denlinger C, Palm D, Hartman RD, Rosella-Dampman LM (1986) Naloxone effects on plasma vasopressin and oxytocin concentrations elevated by histamine, nicotine, isoproterenol and an acute increase in [NACl] in cerebrospinal fluid. Neuroendocrinology 44:157–162
Summy-Long JY, Rosella-Dampman LM, McLemore GL, Koehler E (1990) Kappa opiate receptors inhibit release of oxytocin from the magnocellular system during dehydration. Neuroendocrinology 51:376–384
Sweeney CA, Morgan WW, Smith MS, Bartke A (1985) Altered sensitivity to an opiate antagonist, naloxone, in hyperprolactinemic male rats. Neuroendocrinology 34:269–273
Sylvester PW, van Vugt DA, Aylsworth CA, Hanson EA, Meites J (1982) Effects of morphine and naloxone on inhibition by ovarian hormones of pulsatile release of LH in ovariectomized rats. Neuroendocrinology 34:269–273
Székely JI (1983) Opioid peptides and the neuroendocrine functions. In: Székely JI, Ronai AZ

(eds) Pharmacology. CRC, Boca Raton Opioid peptides, Vol 2
Taylor T, Dluhy RG, Williams GH (1983) β-Endorphin suppresses adrenocorticotropin and cortisol levels in normal human subjects. J Clin Endocrinol Metab 57:592–596
Ten Haaf JA, van Wimersma Greidanus TjB, Maigret C, de Wied D (1986) Effect of the opioid peptide beta-endorphin on the in vivo release of vasopressin in rats under various conditions. Neuroendocrinology 44:102–107
Ten Haaf JA, Maigret C, Andringa-Bakker EAD, van Wimersma Greidanus TjB (1987) Dynorphin-(1–13) is a potent in vivo suppressor of vasopressin levels in the rat. Acta Endocrinol (Copenh) 114:96–101
Thind KK, Goldsmith PC (1988) Infundibular gonadotropin-releasing hormone neurons are inhibited by direct opioid and autoregulatory synapses in juvenile monkeys. Neuroendocrinology 47:203–216
Thornton SN, Parrott RF (1989) Naloxone affects the release of cortisol, but not of vasopressin or oxytocin, in dehydrated sheep. Acta Endocrinol (Copenh) 120:50–54
Todd K, Lightman SL (1986) Oxytocin release during coitus in male and female rabbits: effect of opiate receptor blockade with naloxone. Psychoneuroendocrinology 11:367–371
Tolis G, Hickey J, Guyda H (1975) Effects of morphine on serum growth hormone, cortisol, prolactin and thyroid stimulating hormone in man. J Clin Endocrinol Metab 41:797–800
Tolis G, Dent R, Guyda H (1978) Opiates, prolactin and the dopamine receptor. J Clin Endocrinol Metab 47:200–203
Trudeau VL, Meijer JC, van de Wiel DFM, Bevers MM, Erkens JHF (1988) Effects of morphine and naloxone on plasma levels of LH, FSH, prolactin and growth hormone in the immature male pig. J Endocrinol 119:501–508
Tsagarakis S, Holly JMP, Rees LH, Besser GM, Grossman AB (1988) Acetylcholine and norepinephrine stimulate CRF-41 release from rat hypothalamus in vitro. Endocrinology 123:1962–1969
Tsagarakis S, Navara P, Rees LH, Besser GM, Grossman AB (1989) Morphine directly modulates the release of corticotropin releasing factor-41 from rat hypothalamus in vitro. Endocrinology 124:2330–2335
Tsagarakis S, Rees LM, Bresser GM, Grossman A (1990) Opiate receptor sub-type regulation of CRF-41 release from rat hypthalamus in vitro. Neuroendocrinology 51:599–605
Tseng H-F, Loh HH, Li CH (1978) β_h-Endorphin: antidiuretic effects in rats. Int J Pept Protein Res 12:173–176
van de Heijning BJM, Smelik T, van Wimersma Greidanus TjB (1989) Effects of dynorphin A fragments on the release of vasopressin (AVP) and oxytocin (OXT) in vivo and in vitro. International Narcotics Research Conference, Ste.-Adèle, Canada, July 9–14 (1989), Abstr. p 128, P 159
van de Heijning BJM, Maigret C, Koekkoek-Van den Herik I, van Wimersma Greidanus TjB (1990) A tonic suppressive action of opioids on oxytocin but not vasopressin release in vivo and in vitro in the rat. Neurosci Res Commun 7:43–52
van Leeuwen FW, Pool C, Sluiter A (1983) Enkephalin immunoreactivity in synaptoid elements on glial cells in the rat neural lobe. Neuroscience 8:229–241
van Loon GR, de Souza EB, Ho D, Shin SH (1979) β-Endorphin-induced prolactin secretion is mediated by suppression of release of newly synthesized hypothalamic dopamine. Can J Physiol Pharmacol 58:436–439
van Loon GR, Ho D, Kim C (1980) β-Endorphin-induced decrease in hypothalamic dopamine turnover. Endocrinology 6:76–80
van Vugt DA, Meites J (1980) Influence of endogenous opiates on anterior pituitary function. Fed Proc 39:2533–2538
van Vugt DA, Aylesworth CF, Sylvester PW, Leung FC, Meites J (1981) Evidence for

hypothalamic noradrenergic involvement in naloxone-induced stimulation of luteinizing hormone release. Neuroendocrinology 33:261–264
van Vugt DA, Sylvester PW, Aylesworth CF, Meites J (1982) Counteraction of gonadal steroid inhibition of luteinizing hormone release by naloxone. Neuroendocrinology 34:274–278
van Vugt DA, Webb MY, Reid RL (1989) Comparison of the duration of action of nalmefene and naloxone on the hypothalamic-pituitary axis of the rhesus monkey. Neuroendocrinology 49:275–280
van Wimersma Greidanus TjB (1987) Opiates, opioid peptides and the release of vasopressin. Front Horm Res 17:112–125
van Wimersma Greidanus TjB, ten Haaf JA (1984) Opioids and the posterior pituitary. In: Delitala G, et al. (eds) Opioid modulation of endocrine function. Raven New York, pp 125–136
van Wimersma Greidanus TjB, ten Haaf JA (1985) The effects of opiates and opioid peptides on oxytocin release. In: Amico JA, Robinson AG (eds) Oxytocin: clinical and laboratory studies. Elsevier, Amsterdam, pp 145–153
van Wimersma Greidanus TjB, Thody AJ, Verspaget H, de Rotte AA, Goedemans JH, Croiset G, van Ree JM (1979) Effects of morphine and β-endorphin on basal and elevated plasma levels of α-MSH and vasopressin. Life Sci 24:579–586
van Wimersma Greidanus TjB, van Ree JM, Goedemans HJH, van Dam AF, Andringa-Bakker AD, de Wied D (1981) Effects of β-endorphin fragments on plasma levels of vasopressin. Life Sci 29:783–788
van Wimersma Greidanus TjB, Andringa-Bakker EAD, ten Haaf JA (1983) Endorphins and the posterior pituitary (Abstr 239). 14th Acta Endocrinologica Congress, Stockholm
van Wimersma Greidanus TjB, ten Haaf JA, Mens WBJ (1985) The regulation of the function of the posterior pituitary with special reference to the role of opioid peptides. Front Horm Res 14:197–212
van Wimersma Greidanus TjB, Hellebrekers LJ, van de Heijning B (1988) Opioids and posterior pituitary function in vitro (Abstr). 8th International Congress of Endocrinology, Kyoto, p 251
Veldhuis JD, Rogol AD, Samojik E, Ertel NH (1984) Role of endogenous opiates in the expression of negative feedback actions of androgen and estrogen on pulsatile properties of luteinizing hormone secretion in man. J Clin Invest 74:47–55
Velkeniers B, Vanhaelst L, Golstein J (1983) Dissociation between the effects of β- and γ-endorphin on PRL and TSH secretion in rat. Horm Metab Res 15:521–522
Velkeniers B, Buydens P, Baldys A, De Boel S, Finné E, Golstein J, Vanhaelst L (1987) In vivo and in vitro effect of naloxone on prolactin response to TRH in rat. Life Sci 40:2415–2420
Vermesh M, Silva PD, Lobo RA (1987) Endogenous opioids modulate the inhibitory effects of androgen on the hypothalamic-pituitary axis of normal women. J Clin Endocrinol Metab 65:1183–1186
Vizi ES, Volbekas V (1978) Modulatory role of dopamine and β-endorphin in the release of oxytocin from neural lobe of pituitary (Abstr 1439). 7th International Congress of Pharmacology, Paris
Vizi ES, Volbekas V (1980) Inhibition by dopamine of oxytocin release from isolated posterior lobe of the hypophysis of the rat; disinhibitory effect of β-endorphin/enkephalin. Neuroendocrinology 31:46–52
Volavka J, Cho D, Mallya A, Bauman J (1979) Naloxone increases ACTH and cortisol levels in man. New Eng J Med 300:1056–1057
Volavka J, Bauman J, Pevnick J, Reker D, James B, Cho D (1980) Short-term hormonal effects of naloxone in man. Psychoneutroendocrinology 5:225–234

Von Bormann B, Weidler B, Dennhardt R, Sturm G, Scheld HH, Hempelmann G (1983) Influence of epidural fentanyl on stress-induced elevation of plasma vasopressin (ADH) after surgery. Anesth Analg 62:727–732

Von Voigtlander PF, Lahti RA, Luden JH (1983) U-50, 488. A selective and structurally novel non-mu (kappa) opioid agonist. J Pharmacol Exp Ther 224:7–12

Wakabayashi I, Demura R, Miki N, Phmura E, Miyoshi H, Shizume K (1980) Failure of naloxone to influence plasma growth hormone, prolactin, and cortisol secretions induced by insulin hypoglycemia. J Clin Endorinol Metab 50:597–599

Wakerley JB, Noble R, Clarke G (1983) Effects of morphine and D-Ala-D-Leu-enkephalin on the electrical activity of supraoptic neurosecretory cells in vitro. Neuroscience 10:73–81

Wammack R, Racké K (1988) Endogenous opioid inhibition of the release of oxytocin from the isolated rat neurohypophysis during high-frequency stimulation of the pituitary stalk. Neurosci Lett 92:114–118

Wang X, Tresham JJ, Congiu M, Scoggins BA, Coghlan JP (1989) Central effect of the enkephalin analogue FK-33824 on vasopressin secretion in conscious sheep. Acta Endocrinol (Copenh) 120:369–373

Watson SJ, Akil H, Ghazarossian VE, Goldstein A (1981) Dynorphin immunocytochemical localization in brain and peripheral nervous system: preliminary studies. Proc Natl Acad Sci USA 78:1260–1263

Watson SJ, Akil H, Fischli W, Goldstein A, Zimmerman E, Nilaver G, van Wimersma Greidanus TjB (1982) Dynorphin and vasopressin. Common localization in magnocellular neurons. Science 216:85–87

Way SA, Shibuki K, Leng G (1987) Systemic osmotic stimulation activates neurophpophysial opioid systems (Abstr p 18). Meeting "Brain Opioid Systems in Reproduction", Babraham

Weber E, Roth KA, Barchas JD (1982) Immunohistochemical distribution of α-neo-endorphin/dynorphin neuronal systems in rat brain: Evidence for colocalization. Proc Natl Acad Sci USA 79:3062–3066

Wehrenberg WB, McNicol D, Wardlaw SL, Frantz AG, Ferin M (1981) Dopaminergic and serotonergic involvement in opiate-induced prolactin release in monkeys. Endocrinology 109:544–547

Weitzman RE, Fisher DA, Minick S, Ling N, Guillemin R (1977) β-Endorphin stimulates secretion of arginine vasopressin in vivo. Endocrinology 101:1643–1646

Whitnall MH, Gainer H, Cox BM, Molineaux CJ (1983) Dynorphin-A-(1-8) is contained within vasopressin neurosecretory vesicles in rat pituitary. Science 222:1137–1139

Wiesner JB, Koening JI, Krulich L, Moss RL (1985) Posssible delta receptor mediation of the effect of β-endorphin on luteinizing hormone (LH) release, but not on prolactin (PRL) release, in the ovariectomized rat. Endocrinology 116:475–478

Wildt L, Leyendecker G (1987) Induction of ovulation by the chronic administration of naltrexone in hypothalamic amenorrhoea. J Clin Endocrinol Metab 64:1134–1335

Wilkes MM, Yen SSC (1980) Reduction by β-endorphin of efflux of dopamine and DOPAC from superfused medial basal hypothalamus. Life Sci 27:1397–13191

Wilkes MM, Yen SSC (1981) Augmentation by naloxone of efflux of LRF from superfused medial basal hypothalamus. Life Sci 28:2355–2359

Wilkinson M, Bhanot R (1982) A puberty-related attenuation of opiate peptide inhibition of LH secretion. Endocrinology 110:1046–1048

Wilson N, Ngsee J (1982) Antidiuretic effect of acute morphine administration in the conscious rat. Can J Physiol Pharmacol 60:201–204

Wright DM, Clarke G (1983) Opiates, neuropophyseal hormones and water balance. Alcohol Alcohol 18:337–342

Wright DM, Clarke G (1984) Inhibition of oxytocin secretion by μ and δ receptor selective enkephalin analogues. Neuropeptides 5:273–276
Wright DM, Clarke G, Pill CEJ (1982) Dynorphin suppresses the release of oxytocin. Regul Pept 4:384
Wright DM, Pill CEJ, Clarke G (1983) Effect of ACTH on opiate inhibition of oxytocin release. Life Sci 33:495–498
Yamada T, Nakao K, Itoh H, Morii N, Shiono S, Sakamoto M, Sugawara A, Saito Y, Mukoyama M, Arai H, Eigyo M, Matsushita A, Imura H (1988) Inhibitory action of leumorphin on vasopressin secretion in conscious rats. Endocrinology 122:985–990
Yamada T, Nakao K, Itoh H, Shirakami G, Sugawara A, Saito Y, Mukoyama M, Arai H, Hosoda K, Shiono S, Eigyo M, Matsushita A, Imura H (1989) Effects of naloxone on vasopressin secretion in conscious rats: Evidence for inhibitory role of endogenous opioid peptides in vasopressin secretion. Endocrinology 125:785–790
Yamada K, Nakano M, Yoshida S (1990) Inhibition of elevated arginine vasopressin secretion in response to osmotic stimulation and acute haemorrhage by U-62066E, a κ-opioid receptor agonist. Br J Pharmacol 99:384–388
Yamashita H, Inenaga K, Dyball EJ (1988) Chemical control of neurosecretory neurons in the SON. In: Yoshida S and Share L (eds) Recent progress in posterior pituitary hormones. Excerpta Medica, Amsterdam pp 3–10
Zamir N (1985) On the origin of Leu-enkephalin and Met-enkephalin in the rat neurohypophysis. Endocrinology 117:1687–1692
Zerbe RL, Henry DP, Robertson GL (1982) A new Met-enkephalin analogue suppresses plasma vasopressin in man. Peptides 3:199–201
Zhao B, Chapman C, Bicknell RJ (1988) Opioid-noradrenergic interactions in the neurohypophysis. Neuroendocrinology 48:16–24

The Development of Projections from Cerebral Cortex

G.M. Innocenti

Institut d'Anatomie, Université de Lausanne, rue du Bugnon 9, 1005 Lausanne, Switzerland

1 Introduction

The intriguing and fascinating complexity of biological structures probably reflects constraints imposed by three processes: phylogenesis, development and function. The latter provides the strongest justification for the existence of development neurobiology as a distinct discipline, within the broader field of developmental biology.

The specific function of nerve cells is communication with other nerve cells and/or the physical world around them. Nerve cells communicate with each other mainly, although not necessarily uniquely, through connections established by, in general long, arborescent processes. The network established by these connections is essential to what the nervous system does: an "adaptive structuration" of the surrounding "world". Indeed, within limits compatible with survival and reproduction, the structure of the world an animal perceives probably reflects that of its brain networks.

In some fundamental way these networks must be genetically constrained in a precise manner; species differences in brain structure, and hence, according to the above assumption, in the structure of the perceived world, provide convincing evidence for this. However, there must be limits to genetic determination. One could be that the amount of information in the genome may be insufficient to specify all the details of nervous system organization, including individual synapses (Changeux and Danchin 1976). Another is that each individual is an unpredictable genetic mosaic – the consequence both of sexual reproduction and evolution. Although evolution probably eliminated disastrous incompatibilities between genes, no inbuilt harmony among genes can be expected either. The product of one gene, say any protein crucial for cell generation, migration, differentiation, recognition etc., may not match qualitatively, quantitatively or temporally that of other genes. One would expect coordination of gene actions to be a requisite for the construction of a functional brain. How is coordination achieved? Genes capable of regulating other genes' action do exist. However, coordination seems also to be achieved through adjusting interaction between the products of genetic expression, i.e. neurons or parts of neurons.

Two of these adjusting mechanisms are in the developing nervous system: (a) neurons, dendrites, axons and synapses are first produced in excess and then some are maintained while others are eliminated out of interactions with other neuronal and/or non-neuronal elements; (b) the activity of a neural network can back-regulate the structure of the network which generates it.

This paper will try to track some of these concepts in the analysis of the development of cortical projections. Cortical projections were studied light-microscopically with the anterograde and retrograde anatomical tracer techniques which have become available during the last 10–15 years. Extracellular injections of axonally transported tracers do not visualize single axonal arbors or neurons but populations of retrogradely labeled neurons or anterogradely labeled axons. The light-microscopic analysis of these preparations does not reveal synaptic connections. Instead it delineates the gross pattern of interrelations among brain structures, which underlies connectivity. How these interrelations emerge in development is relevant for the old and difficult problem of the development of neural connections.

Although the phenomenon of massive neuronal death occurring at around the time of the establishment of connections was well known (Hamburger and Levi-Montalcini 1949; Cowan 1973), it was only in the second half of the 1970s that doubts began to emerge as to whether the model of a precisely and "ab initio" correctly target-oriented axonal growth, which appeared well grounded for the development of retinotectal projections in the fish (Sperry 1963), could be generalized to the whole nervous system, and in particular to the cerebral cortex of mammals. Evidence against such a generalization stemmed from the work of Hubel and Wiesel (Hubel et al. 1977) and of Rakic (1976) indicating initial overlap of geniculocortical projections in the developing area 17 and lateral geniculate body of the monkey. In the same years, Changeux and Danchin (1976), mostly on the evidence of transient multiple innervation in the development of the neuromuscular synapse, proposed the selective stabilization of synapses as a general mechanism in the development of neural connections.In their model, selection for stabilization was based on activity. This was compatible with the well-demonstrated role of visual experience in the segregation of geniculocortical projections as well as with the role of visual experience in the development of other response properties of cortical neurons (reviewed by Blakemore 1978; Wiesel 1982; Frégnac and Imbert 1984). The studies on the visual system, particularly those on ocular dominance were a gargantuan step beyond the classic embryology of the nervous system, derived from development biology. By focussing on the development of connections of neuronal ensembles, characterized by common functional properties, these studies validated the notion that building blocks larger than the neuron exist in the basic organization of the central nervous system (Mountcastle 1978). On the other hand they demonstrated how detailed knowledge about the function of a neuronal ensemble provides the crucial theoretical framework for the analysis of its development.

Nevertheless, in those years, the implications of the above studies, for a general theory of the formation of neural connections were probably not fully appreciated. The chemoaffinity hypothesis was phrased in sufficiently general and occasionally unclear terms (see discussion in Innocenti 1988) to accommodate the existence of a few exceptions, some of which could be interpreted as development errors (reviewed in Clarke 1981). Developmental "errors" were known at least since Cajal but did not prevent that author from proposing one of the first theories of neural development based on chemotropism (Cajal 1909).

However, the studies mentioned above were but the tip of the iceberg. It was found that the emergence of neural networks, at least at the cortical level – although the same turned out to apply to several other systems – had little to do with the precisely target-aimed growth of one axon to its postsynaptic partner which would have been expected on the basis of a literal interpretation of chemoaffinity theory. Long corticocortical projections are generated in excess in the course of neural development and reach structures from which they will be subsequently eliminated (Innocenti et al. 1977; Distel and Holländer 1980; Ivy and Killackey 1981). The magnitude of the projections eliminated in development and their reliable occurrence in different species and individuals precludes them from being considered as developmental errors. The very concept of developmental error suffers from intrinsic semantic difficulties (see Innocenti 1988 for discussion).

As previously discussed (Innocenti and Clarke 1984b; Innocenti 1987) no single, monotonic process seems adequate to explain the development of cortical connections[1]. Rather, an architectonic analogy seems to apply. Cortical connectivity is

[1] Throughout this paper the term "connection" will imply that synapses have been formed between neurons; "connectivity" will not. The second term will often designate juvenile conditions in which the existence of synapses was not proven. "Projection" was operationally defined above.

originally organized according to a juvenile Bauplan[2] as different from the adult Bauplan as romanesque architecture is to gothic. A number of well-characterized events mark the transition from the juvenile to the adult Bauplan. How these events are regulated is the object of active investigations, since this knowledge may open the way to controlling a crucial step in cortical development and possibly the evolution of neocortex (for discussions on development and evolution see Katz and Lasek 1978; Innocenti 1988).

2 The Adult Organization of Cortical Connectivity

While the existence of structural anisotropies in the cortex along a dimension perpendicular to its surface has been recognized at least since Gennari (1782; quoted by Glickstein and Rizzolatti 1984), the identification of boundaries between layers and sublayers has been notoriously difficult even in areas such as visual 17 in the monkey, where the lamination is especially clear cut (Billings-Gagliardi et al. 1974). The fact that neocortex is commonly divided into six layers reflects the acceptance of a conventional frame of reference rather than an objective feature of cortical structure.

Nevertheless, the analysis of cortical connectivity with neuroanatomical tracers demonstrated that, in general different cortical afferents reach different cortical laminae (Lund et al. 1979; Gilbert and Wiesel 1981; inter alios). Moreover, the different corticofugal projections have different and characteristic laminar origins, although within a given layer distinct neuronal populations can give rise to different projections and the projections to a given target can originate from more than one layer (Gilbert and Kelly 1975; Catsman-Berrevoets and Kuypers 1978; Innocenti 1980; Swadlow 1983; Innocenti et al. 1986; inter alios). As a rule of thumb, the supragranular layers (II and III) are the main source of projections to other cortical sites. The infragranular layers (V and VI) project to subcortical structures. Layer IV, the main thalamic recipient layer, sends short radial connections. In addition, probably all neurons, with corticofugal axons, also establish local connections through axon collaterals and these connections have both radial and tangential specificity (Gilbert and Wiesel 1981). Each cortical layer also contains neurons with exclusively local axons (Golgi type II neurons: Cajal 1911; Peters and Jones 1984).

A tangential trajectory across the cortex reveals different types of anisotropy, all in some way related to the radial anisotropy. Neocortex can be parcellated in distinct cytoarchitectonic areas basically according to thickness of layers and their cell composition, but the details of this parcellation evoked different degrees of

[2] The term Bauplan (construction plan) has obvious anthropological undertones. This may be justified by the view that evolution could be considered the engineer of the brain (for a similar belief see Glassman 1987). The term was chosen because it implies architectural purposefulness. It is not necessarily related to the usage by 19th century comparative anatomists (see Alberch 1984).

consensus. Some areas are easily identified by objective criteria, while the identification of others requires art and skill. Other criteria for parcellation can be used; for example, an area can contain a sensory representation distinct from those of surrounding areas. But these criteria do not always match the cytoarchitectonic parcellation (for a discussion, see Van Essen, 1985).

In the adult brain, probably each area is the preferential site of termination of a characteristic set of thalamic nuclei (Caviness and Frost 1980; Graybiel and Berson 1981; inter alios). Each area establishes connections with a characteristic subset of other cortical areas and these connections define lines of processing within the cortex (Mishkin et al. 1983; Livingstone and Hubel 1987a, b; Hubel and Livingstone 1987). In general, reciprocal connections exist between two areas but the radial locations of the neurons of origin both of the connections and of their terminations define the reciprocal hierarchical position of two areas along a processing line (Van Essen and Maunsell 1983; Van Essen 1985). Connections between areas obey largely unexplored mapping rules. At least in one case, that of the interhemispheric connections of visual and to some extent, primary somatosensory areas, the rules seem clear. These connections link in an exclusive manner parts of the primary sensory representations corresponding to sensory territories near the midline (Berlucchi 1972; Innocenti 1986), i.e. the line which divides the sensory worlds of the two hemispheres. In another case, that of the connections between areas 17 and 18 of the monkey, the mapping seems to achieve separation of processing lines with different retinal origins and thalamic relays (Hubel and Livingstone 1987; Livingstone and Hubel 1987a, b).

In addition, each area receives from and projects to a characteristic set of subcortical structures.

A different type of tangential cortical anisotropy derives from the fact that most, probably all cortical areas appear to be organized in radially oriented "modules" or "columns". Columns have been recognized both on functional (Mountcastle 1978; Hubel 1982) and anatomical grounds (Heimer et al. 1967; Hubel and Wiesel 1969; Jones et al. 1975; Goldman-Rakic and Schwartz 1982). The anatomical columns consist of discrete accumulations of thalamocortical terminals (Hubel and Wiesel 1969), sometimes associated with cytoarchitectonic specializations (Woolsey and Van der Loos 1970), or of discrete accumulations of corticocortical terminals (Heimer et al. 1967; Jones et al. 1975; Imig and Brugge 1978; Goldman-Rakic and Schwartz 1982; inter alios). Occasionally, corticocortically projecting neurons are organized in radial columns (Jones et al. 1975; but see Caminiti et al. 1985; see Innocenti 1986; for discussion and references). Cortical columns can appear as tangentially oriented bands or "blobs". The functional meaning of geniculo-cortical columns is known (Hubel 1982), that of corticocortical projection columns usually not, with two notable exceptions: (1) callosal termination columns in A1 are related to "ear dominance" bands (Imig and Brugge 1978); (2) origin and termination of projections from 17 to 18 in the monkey are related to afferent connectivity and functional properties of visual cortical neurons (Hubel and Livingstone 1987; Livingstone and Hubel 1987a, b).

Thus it appears that all the main principles of cortical organization recognized hitherto have their counterpart in some aspect of cortical connectivity.

3 The Juvenile Organization of Cortical Connectivity

Four features characterize juvenile cortical connectivity: (1) the existence of numerous transient projections; (2) radial specificity in the origin of the various projections; (3) some degree of topographical order in the tangential organization of the projections; (4) axons take specific routes in the white matter and cortex and exhibit a characteristic behavior at their site of termination.
A recent review by Payne et al. (1988b) complements this section in many respects.

3.1 The Transient Projections

Hubel et al. (1977) and Rakic (1976) provided the first evidence of transient overlap, in the visual cortex, of geniculocortical axons carrying information from the two eyes. This overlap is probably due to transient short branches of thalamocortical axons (Le Vay and Stryker 1979; Wiesel 1982) and therefore resembles the transient hypercollateralization of Purkinje cell axons and of climbing fibers (Cajal 1911) and the multiple innervation of the muscles (for references see Purves and Lichtman 1985).
A different type of exuberance, characterized by the production of long transient projections, was discovered in area 17 of the cat (Innocenti et al. 1977). It was found that while in the adult only the region near the border between areas 17 and 18 projects into the corpus callosum, the whole of areas 17 and 18 do so in the young kitten. Similarly, exuberant callosal projections were later found in somatosensory areas of the cat, where most of SI is devoid of callosal projections in the adult but not in the neonate (Innocenti and Caminiti 1980), as well as in the visual areas of the hamster, rat (Mooney et al. 1984; Olavarria and Van Sluyters, 1985) and rabbit (Chow et al. 1981), and in area 18 (Dehay et al. 1988b) and the somatosensory areas (Killackey and Chalupa 1986) of the monkey. Work in the somatosensory cortex of the rat (Ivy et al. 1979; Ivy and Killackey 1981) and in the auditory cortex of the cat (Feng and Brugge 1983) showed that in areas where callosal projections originate from more or less discrete cortical "columns", a continuous distribution of callosally projecting neurons is found in development. Apparently, in all areas and species where the development of callosal connections was studied, transient projections could be demonstrated. One interesting exception to this rule may be area 17 of the monkey, which either never projects into the corpus callosum (Dehay et al. 1988b) or does so but not as much as do the other areas (Chalupa et al. 1989), although the callosal projections from area 18 undergo the usual phase of exuberance. Unfortunately, at the moment, this negative finding is supported by the study of only three age points, embryonic days (E) 112, 114 and 133. The development of callosal connections has not yet been studied between E 133 and E 165, when exuberant callosal connections can be expected. Projections from area 17 seems to be missing at the time when the adjacent area 18 has already established them, but nearby cortical areas can establish callosal projections with remarkable time lags (Floeter and Jones 1985). In addition, the possibility that the transient callosal projections from area 17

may either fail to extend far beyond the midline or reach areas other than 17 and/or 18 was not explored. Neither possibility is unlikely, since the same transient projections can grow different distances in different species (O'Leary and Stanfield 1986), and exuberant callosal projections to heterotopic areas have been demonstrated (Innocenti and Caminiti 1980; Innocenti and Clarke 1984a, b).

Transient intrahemispheric projections were also discovered in the cat (Innocenti and Clarke 1984a; Dehay et al. 1984; Price and Blakemore 1985a, b; Clarke and Innocenti 1986; Price and Zumbroich 1989). Two of these projections have been particularly well characterized. One runs from the auditory to the visual cortex (Innocenti and Clarke 1984a, b; Innocenti et al. 1988; Dehay et al. 1988a), i.e. two areas between which only weak projections (Innocenti et al. 1988) exist in the adult. The other one is from area 17 to area 18 (Price and Blakemore 1985a, b). Like the auditory callosal projections mentioned above, this projection originates from discrete, vertically oriented modules in the adult, and from a tangentially continuous neuronal zone in the neonate. Shorter-range intra-areal or inter-areal transient projections have also been described in areas 17 (Luhmann et al. 1986; Katz and Wiesel 1987) and 18 (Price 1986) of the cat.

Transient corticosubcortical, and not only transient corticocortical projections, are formed in development (Distel and Hollander 1980). They include projections to the spinal cord (D'Amato and Hicks 1978; Distel and Hollander 1980; Stanfield et al. 1982; Bates and Killackey 1984; O'Leary and Stanfield 1986; Joosten and Van Eden 1989), the superior colliculus (Tsumoto et al. 1983; Thong and Dreher 1986), the cerebellum (Distel and Hollander 1980; Tolbert and Panneton 1983), the pons (Adams et al. 1983; Mihailoff et al. 1984), the nucleus ruber (Leonard and Goldberg 1987), the trigeminal nuclei (Tolbert et al. 1984) and other structures in the medulla (Iriki et al. 1988).

In summary, the selective connectivity between one area, or part of one area, and other cortical or subcortical centers, as well as the mapping rules of the projection, including their columnar origin, differ in the juvenile and adult brain.

Till recently, the extravagance of exuberant projections established by cortical neurons contrasted with the seemingly parsimonious development of thalamocortical connectivity (Payne et al. 1988a, b). In addition to the lack of segregation of geniculocortical afferents, mentioned above, only a moderate degree of laminar exuberance, a transient projection to layer I, was described (Kato et al. 1984; 1986). Lately, however, long transient thalamocortical projections have also been found (Bruce and Stein 1988; Minciacchi and Granato 1989). Indeed, the number of transient projections discovered in the developing brain increases at a fast rate and their comprehensive list is bound soon to become obsolete.

One may ask if the fact that a projection is visualized by pathway tracing experiments in the young animal, but not in the adult, provides sufficient evidence that the projection is eliminated in development. Given that a tracer may not have the same uptake and transport efficiency in the young and in the adult, the elimination should preferentially be corroborated by: (a) multiple tracers or experimental approaches other than axonal transport, for example, estimates of number of axons and (b) the stabilization of the transient projection into adulthood by experimental manipulation, but under conditions in which "de novo" growth of projections induced by the manipulation can be excluded. Only a few of the transient projections reported thus far pass both criteria, namely the callosal and intra-hemispheric projections to areas 17 and 18, the corticospinal projection from the visual cortex and the corticocerebellar projection from the frontal and somato-sensory cortex (see Sects. 5 and 6), although other projections pass the first criterion.

Conversely, the failure to demonstrate transient projections in the development of a given structure does not necessarily preclude its existence. Some of the factors which can lead to false negative results were discussed elsewhere (Berbel and Innocenti 1988).

3.2 The Radial Origin of Juvenile Cortical Projections

By and large, juvenile cortical projections originate from layers which will also give rise to the same kind of projection in the adult. Thus, in kittens, as in adult cats, the bulk of callosal projections from areas 17 and 18 originates from layers III and upper IV, with a few from layer VI. Layers II and V do not contribute substantial amounts of callosal projections, neither in the adult nor in kittens (Innocenti 1980; Caminiti and Innocenti 1980; Segraves and Innocenti 1985). Even more striking is the differential radial distribution of supragranular neurons with different targets in the auditory areas, including neurons with different transitory targets (Innocenti and Clarke 1984a; Dehay et al. 1988a). Similarly, the transient projections to the cerebellum (Tolbert and Panneton 1983), spinal cord (Stanfield and O'Leary 1985b) and tectum (Tsumoto et al. 1983) originate from layer V, which is also the main source of cortico-mesencephalic and cortico-pontine projections in the adult.

This is not to say that no changes occur in the relative contributions of different layers to a given projection in development. On the contrary, the interhemispheric projection from the posteromedial lateral suprasylvian cortex to areas 17 and 18 originates nearly exclusively from layer VI in adults (Segraves and Innocenti 1985) but also from layer III in the kitten (Innocenti and Clarke 1984b). Similarly, in the rat parietal cortex, the earliest callosal neurons are in layers Va and Vc-VIa. Later, the projection from layer III appears and the density of callosally projecting neurons greatly decreases in Vc and VIa (Ivy and Killackey 1981). Selective loss of projections from supragranular layers, but maintenance of the projection from infragranular layers was described in area 17 of the rat (Olavarria and Van Sluyters 1985).

The radial changes described above could just be another facet of the type of exuberance described in the preceding section. In the cat, subsets of layer III and layer VI neurons distributed throughout the cortex appear to be determined to send an axon through the corpus callosum. At a later stage, in certain areas, for example in most of area 17, neurons of both layers eliminate their callosal axon; in other areas and projections, e.g. the projection from suprasylvian areas to area 17, only neurons in layer III do so, and this shifts the origin of the projection to infragranular layers (Innocenti and Clarke 1984b). In addition, changes in the radial origin of a projection may be due to the fact that some neurons send an axon through the corpus callosum before they have reached their final radial position (Schwartz and Goldman-Rakic 1986), and that different layers establish projections at different times (Ivy and Killackey 1981).

As we discussed elsewhere (Innocenti and Clarke 1984b; for similar concepts see Rakic 1988) it appears that where a neuron sends its axon is determined rather early in development and is related to a neuron's radial position in the cortex.

This initial determination may condition pathway choice rather than target choice; the fact that axons reach a given target may be the consequence of having chosen a given pathway. The cortical white matter is organized in compartments related to the origin and termination of cortical projections, and individual tracts are topographically organized (Innocenti et al. 1983b; Nelson and Le Vay 1985; Nieuwenhuys et al. 1988; inter alios). There is also evidence that juvenile axons, including those which will be eliminated, travel in specific compartments in the white or gray matter (Distel and Holländer 1980; Katz andWiesel 1987; Innocenti et al. 1988).

It is well known to general embryologists that a cell position within the embryo can determine aspects of its differentiation, which may indicate that a cell's fate depends on "positional information" (Wolpert 1969). It appears improbable that positional information determines a neuron in a certain layer to send an axon to a given target/pathway. Firstly, neurons with identical radial position in the cortex are found in the adult to project to different targets, and neurons in different layers can project to the same target (Segraves and Innocenti 1985; Innocenti et al. 1986; inter alios). Secondly, neurons in the frontal cortex of the monkey appear to send an axon into a given pathway, namely the corpus callosum, before they have acquired their final radial position (Schwartz and Goldman-Rakic 1986). Thirdly, in the reeler mouse (Caviness 1977), and after X-irradiation or treatment with cytotoxic drugs (Jones et al. 1982; Jensen and Killackey 1984; Innocenti and Berbel 1989a, b) a neuron's radial position, but not its target/pathway choice, can be profoundly altered, although the usual relation seems to exist between the latter and the neuron's birthdate. Experiments aimed at testing directly the hypothesis that a neuron's birthdate also determines its pathway/target choice by etherochronic transplantation have given results compatible with this notion (McConnell 1988). Nevertheless, birthdate does not fully predict where a neuron sends its axon. The projection to a given structure can originate from several layers and the same layers also give rise to projections to other structures. For example, callosal neurons in the cat originate from layers VI, II and IV (Innocenti 1986) and these layers are generated at different times (Luskin and Shatz 1985). The callosally projecting neurons are a minority among neurons in each layer (Innocenti 1986).

Possibly, a neuron's clonal origin or serial position in a clone, rather than its birthdate is the crucial element in the initial pathway choice. Indeed, work in invertebrates has demonstrated that the serial position of a neuron in a clone determines or restricts its phenotype (Sulston and Horvitz 1977). On the other hand, neurons which are near each other in a cortical layer do not necessarily derive from the same germinal cell (Walsh and Cepko 1988; but see Luskin et al. 1988) but may owe their proximity to sideways displacement during migration along the fasciculating and defasciculating adjacent glial channels shown by Gadisseux

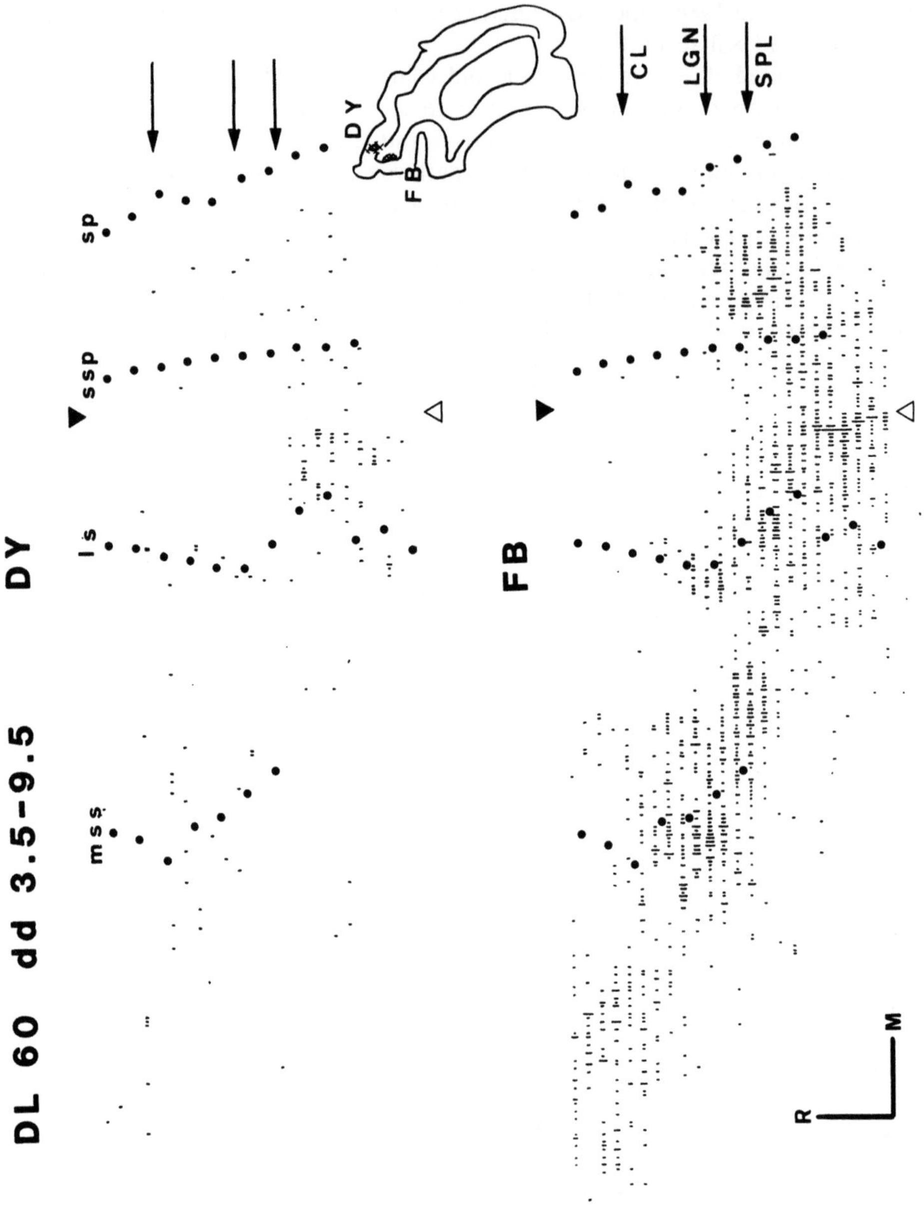
DL 60 dd 3.5–9.5
DY
FB
mss
ls
ssp
sp
D Y
F B
CL
LGN
SPL
R
M

et al. (1987). It is possible that clonal heterogeneity and intermixing plays a major role in determining the differential axonal growth of adjacent neurons in the cortex.

3.3 The Tangential Organization of the Juvenile Projections

A third characteristic feature of the juvenile projections is that they are neither totally diffuse nor random. One aspect of this is that although each area projects to and receives from a broader territory than in the adult, certain projections fail to form. A surprising absence is that of a projection from visual to auditory areas in kittens at ages when the auditory to visual projection and other transient projections are well developed (Innocenti and Clarke 1984a, b). Equally surprising is the apparent absence of transient projections from area 17 in the newborn monkey (Dehay et al. 1988b) already discussed above.

A second aspect of the tangential, juvenile organization is that in young kittens, an injection in areas 17 and 18 and involving the subcortical white matter in order to reach transient corticocortical axons, labels a characteristically shaped territory in the contralateral hemisphere (Innocenti and Clarke 1984a, b; Figs. 1, 2). This territory includes regions which project to areas 17 and 18 in the adult, as well as regions which lose this projection during maturation. The labeled territory is elongated, its greater extension running roughly mediolaterally. Its anteroposterior location changes systematically with the position of the injection site. The shape and location of this territory may reflect principles of axonal order in the callosal pathway and the direction of axonal growth. The anteroposterior cortical axis is

Fig. 1. Flat reconstructions of the occipital portion of the left hemisphere of a kitten injected on postnatal day 3.5 (P 3.5) and killed on P 9.5, showing distribution of layers II–VI callosal neurons retrogradely labeled by diamidino yellow (*DY, top*) or by fast blue (*FB, bottom*). Injection sites are shown on the section drawing. In each coronal section the labeled neurons were plotted with a computer microscope and projected onto a line running 400 μm below the pial surface. The line was divided into 100-μm segments and the number of neurons per segment represented by vertical lines whose lengths are proportional to the number of neurons (shortest line, one neuron). *Filled* and *empty triangles* denote the axis along which coronal sections were aligned. *Arrows* point to coronal levels through the caudal end of the lateral geniculate nucleus (*LGN*), splenium of the corpus callosum (*SPL*) and coronal end of the claustrum (*CL*). *Dots*, sulci: *mss*, middle suprasylvian sulcus; *ls*, lateral sulcus; *ssp*, suprasplenial sulcus; *sp*, splenial sulcus. Calibrations are 2 mm. *R*, rostral; *M*, medial; *DL60*, code number; *dd*, age in days. Note that the *DY* injection was restricted to the gray matter near the 17/18 border and labeled neurons almost exclusively in a cluster near *ls*, corresponding to the contralateral 17/18 border which projects into the corpus callosum in the adult; only a few neurons were labeled in medial area 17 (between *ssp* and *sp*) or in the auditory cortex (lateral to *mss*), which do not project to contralateral areas 17 and 18 in the adult. In contrast, the *FB* injection extended into the white matter under area 17 and labeled neurons at the 17/18 border, but also in the two regions which do not project to contralateral areas 17 and 18 in the adult. (From Innocenti and Clarke 1984b)

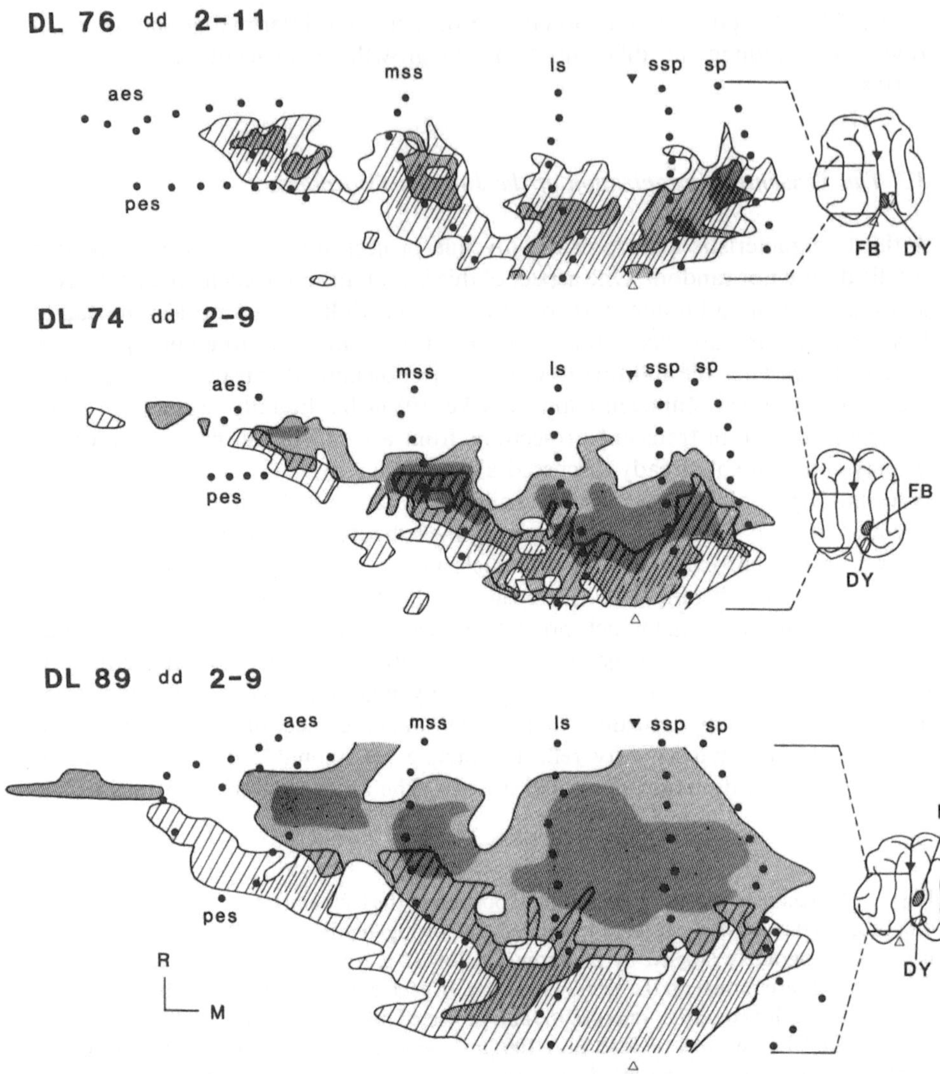

Fig. 2. Flat reconstructions of the occipital portion of the left hemisphere of three kittens (identified by code number and by age at injection and at death), showing regions containing layers II–VI callosal neurons retrogradely labeled by DY (*hatching*) or by FB (*shading*). The region reconstructed is that between *bars* on brain insets. Unlike in Fig. 1, these reconstructions are based on hand-drawn sketches of cell distributions. This procedure is less precise. Therefore, only two arbitrary levels in density of labeled neurons are indicated, i.e. high density (*dark* shading or hatching) and low density (*light* shading or hatching). Locations of injection sites are marked on brain insets. *Dots*, sulci: *aes*, anterior ectosylvian sulcus; *pes*, posterior ectosylvian sulcus; *R*, rostral; *M*, medial. Calibration bars are 2 mm. Notice that the labeled territories progressively separate, proportionally to the distance between injection sites. (From Innocenti and Clarke 1984b)

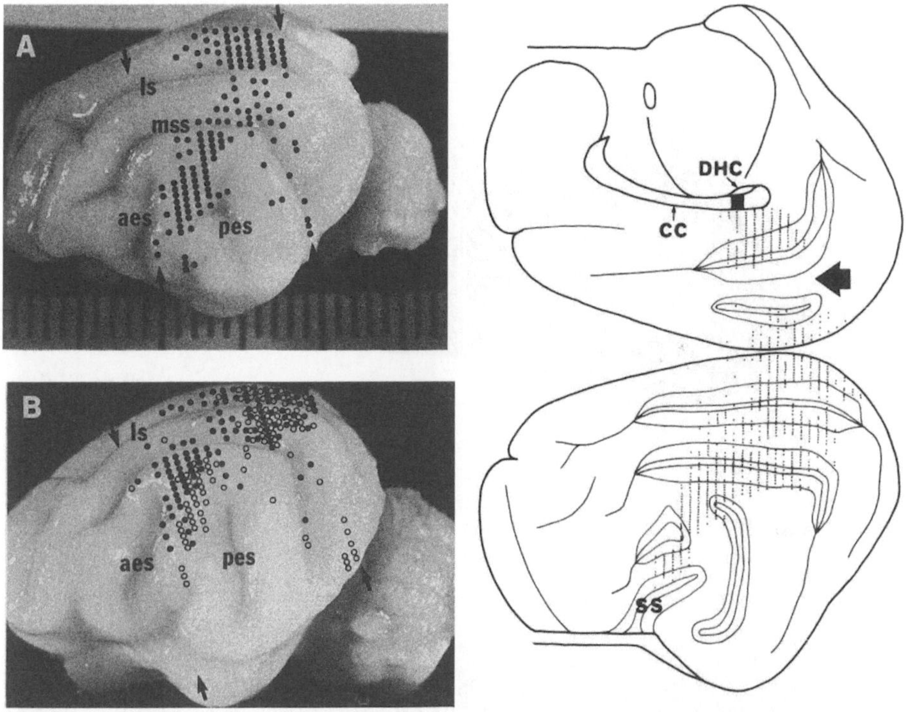

Fig. 3 A, B. *Left*, lateral view photographs of two kitten brains showing the distributions of neurons retrogradely labeled by *FB* (*filled dots*) or *DY* (*open dots*), from injections in contralateral areas 16 and 18 and the underlying white matter. *Above*, kitten DL 39, injected on P 2.5, killed on P 9.5; *below*, kitten DL 74, also shown in Fig. 2, injected on P2 killed on P9. In each photograph, *arrows* mark the rostral and caudal limits of the reconstructed regions. *Dots* are arrayed parallel to the plane of sectioning; their density along an array gives a rough indication of the density of labeled neurons. In the lower example, symbols for *FB* and *DY* labeled neurons from the same section are staggered. Sulci are abbreviated as in Figs. 1 and 2. (From Innocenti and Clarke 1984a). *Right*, reconstructions showing the distributions of neurons retrogradely labeled from applications (*in black*) of horseradish peroxidase to the presplenial part of the body of the corpus callosum (*CC*) of an adult cat. The bottom drawing shows a lateral view of the hemisphere, the top drawing an inverted, medial view; some sulci are open. *DHC*, dorsal hippocampal commissure; *SS*, sylvian sulcus. (From Nakamura and Kanaseki 1989). Notice the similarity in the shape and location of the labeled territories in the two types of experiments. However, no projections from medial area 17 (*arrow*) are labeled in the adult

mapped anteroposteriorly in the adult corpus callosum (Innocenti 1980; Pandya and Seltzer 1986), and such a topographical order seems to appear early in development (Olavarria and Van Sluyters 1986). In a recent study in which the topographic origin of callosal axons was investigated by discrete applications of HRP to the corpus callosum in adult cats (Nakamura and Kanaseki 1989), applications of the tracer to the presplenium retrogradely labeled a territory whose

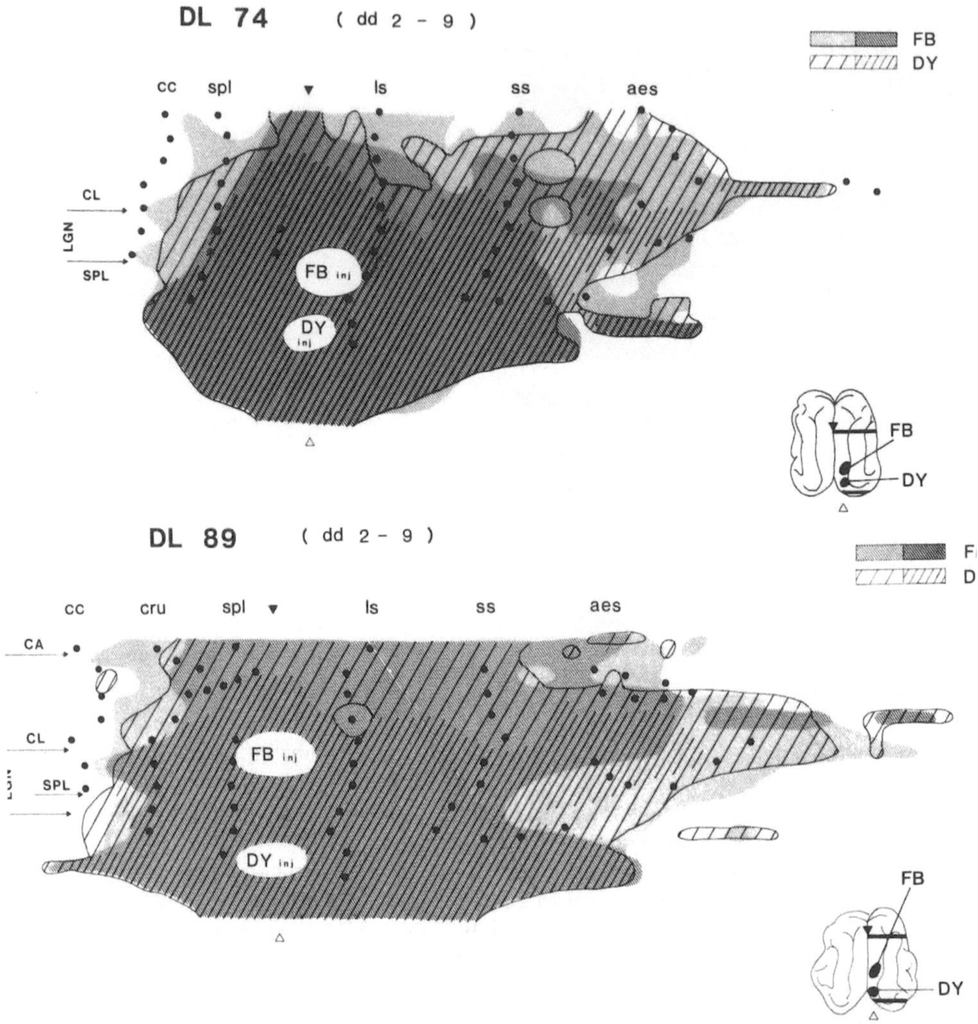

Fig. 4. Flat reconstructions of the occipital portions of the right hemispheres of two kittens, showing regions containing layers II–VI association neurons retrogradely labeled by DY (*hatching*) or FB (*shading*). The region reconstructed is that between *bars* in brain insets. All conventions as in Fig. 2, where the distribution of retrogradely labeled callosal neurons in these two kittens are also shown. Note (compare with Fig. 2) the broader rostrocaudal distribution of labeled association neurons than of callosal neurons. (From Clarke and Innocenti 1986)

location and extent were amazingly similar to that labeled by area 17 and 18 injections in the kitten (Fig. 3). As discussed earlier (Clarke and Innocenti 1986), the fact that intrahemispheric juvenile projections are less clearly topographically organized (Fig. 4) could reflect the lack of a preferential axis of growth of intrahemispheric projections.

To what extent order in the pathway is critical for the establishment of the projection is unclear. As in the case of the retinotectal projection, errors due to disordered growth may be corrected to some extent by redirection of axonal growth near the target (Thanos and Bonhoeffer 1986). Evidence that in partially "acallosal" mice the topography of callosal connections is close to normal but the trajectory of the axons may be abnormal and the projection is reduced (Olavarria et al. 1988b) does not clarify the issue in the absence of more detailed knowledge of the development of the agenesis, and the nature of the postulated disorder in axonal growth.

Some degree of topographical order, corresponding to the mediolateral cortical direction, was documented for the transient corticocerebellar projection (Tolbert 1989). The interpretation given was that this order matches the somatotopy of SI, from where part of the projection originates, to that of the vermis and paramedian lobule where part of the projection terminates. Nevertheless, the order found in this transient projection may also reflect the influence of factors ordering corticocerebellar axons along their pathway.

What may be responsible for the order along the pathway is unknown. In principle, axoaxonal interaction may suffice, although the growth of the first axons in the pathway and the subsequent addition of axons may not be unconstrained. Cell adhesion molecules may be involved (Edelman 1987), since in other systems the antibodies against these molecules cause loss of axoaxonal contacts and disorderly growth (Thanos et al. 1984).

3.4 Relations Between Juvenile Cortical Axons and Their Targets

Studies on several systems including neocortex suggest that "in vivo" axonal growth is not a continuous phenomenon but that it rather proceeds in spurts separated by pauses.

In the development of corticopetal afferents, one of the last pauses occurs at the bottom of the gray matter, where ingrowing axons appear to "rest" in a special compartment, usually referred to as the "subplate" (Rakic 1977; Lund and Mustari 1977; Wise et al. 1977; Wise and Jones 1978; Killackey and Bedford 1979; Shatz and Luskin 1986). This compartment contains transient neurons (Marin-Padilla 1978; Kostovic and Rakic 1980; Chun et al. 1987), descending dendrites of layer VI neurons (Marin-Padilla 1971, 1978), and astrocytes (Innocenti and Berbel 1990a) and is traversed by radial glia and migrating neurons. Whether any of these components causes or allows ingrowing axons to stop is presently unknown, although the transient neurons appear to receive transient synapses (Chun et al.1987). The pause of the geniculocortical axons in the subplate may be terminated by the positioning and possibly maturation of layer IV neurons, the main recipient of thalamocortical projections (Shatz and Luskin 1986). Thus, the waiting may be caused by the fact that the arrival of axons and the maturation of their target are not synchronous. In truth, what axons do during this waiting phase is unknown and one can think of additional or alternative roles for it.

Corticocortical axons, in particular callosal axons, were also found to pause in the subplate (Wise and Jones 1976; Goldman-Rakic 1982; Price and Zumbroich 1989),

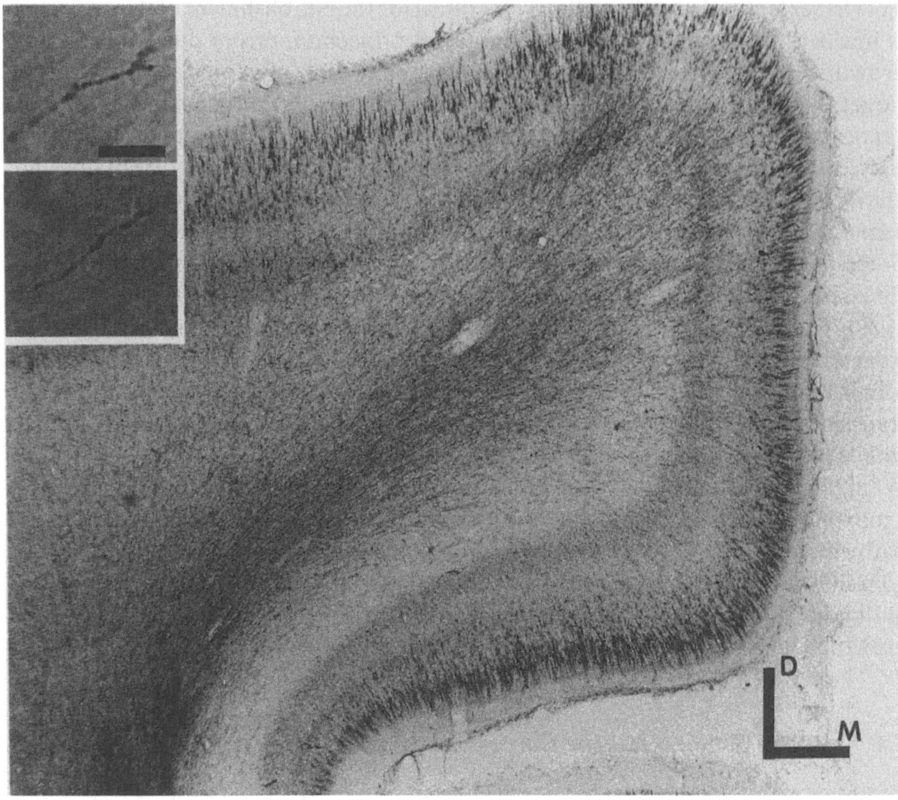

Fig. 5. Photomicrograph of a tetramethylbenzidine reacted section from the hemisphere contralateral to a wheat germ–horseradish peroxidase injection in a kitten injected on P1.5, killed on P2.5. Notice the continuous distribution of labeled neurons and the bundle of axons (from the corpus callosum), fanning out in the white matter. In the *upper right* portion of the section, faintly labeled axons have entered the gray matter in the region of the area 17/18 border. Other axons are directed to the gray matter of area 17 but without entering it. High-power photomicrographs of two of these axons from an adjacent section reacted with a modification of Hanker–Yates' technique are shown in the *inset*, oriented as in the full section. Calibration bars are 0.5 mm for the low-power and 10 mm for the high-power photomicrographs. *D*, dorsal; *M*, medial. (From Innocenti and Clarke 1984b)

although the significance of this pause is even less clear. Characteristically, the waiting phase of corticocortical axons is terminated by growth into specific "columnar" patterns resembling those of the adult (Wise and Jones 1976; Ivy et al. 1979; Innocenti 1981a; Goldman-Rakic 1982; Feng and Brugge 1983; Miller and Vogt 1984; Innocenti and Clarke 1984b; Olavarria and Van Sluyters 1985). Axons appear to enter at specific tangential locations, while they avoid others (Figs. 5, 6). This is clear in the case of callosal projections to the visual areas whose termination bears characteristic, fixed relations to the cytoarchitectonic borders (Innocenti 1981a; Innocenti and Clarke 1984b; Olavarria and Van Sluyters 1985).

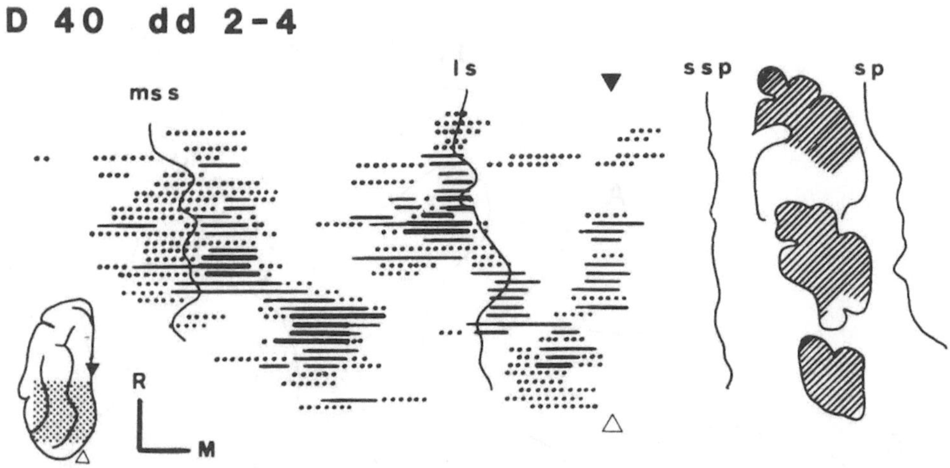

Fig. 6. Flat reconstruction of the occipital portion of the brain of a kitten injected on P2, killed on P4, showing the tangential distribution within the gray matter of terminating callosal axons. The region reconstructed corresponds to the *shaded portion* of the brain inset but it also comprises the cortex within the interhemispheric fissure. *Heavy lines*, heavy labeling; *light lines*, medium labeling; *dots*, light labeling. Outlines of coronal sections show the spread of horseradish peroxidase (*hatched*) at three rostrocaudal levels, in the injected hemisphere. Calibrations are 1 mm. *R*, rostral; *M*, medial. Note, in comparison with Figs. 1 and 2, the selective tangential distribution of terminating callosal axons. (From Innocenti and Clarke 1984b)

Juvenile callosal axons also reach other parts of the visual areas, notably most of area 17 and 18, where they do not terminate in the adult (Innocenti 1981a). At these locations, however, the overwhelming majority of callosal axons remains confined to the subplate or to the bottom of the gray matter (Innocenti 1981a; Innocenti and Clarke 1984b; Olavarria and Van Sluyters 1985).

Most of the callosal axons which enter the gray matter at the sites of adult termination originate from tangential sites destined to maintain the callosal projection (Innocenti and Clarke 1984b; Fig. 1). Most of the axons which remain confined to the white matter come from tangential sites from where callosal projections will be eliminated (Innocenti and Clarke 1984b). A few axons, however, possibly those whose fate depends on visual experience (Lund et al. 1978; Innocenti and Frost 1978, 1979; Cynader et al. 1981; see also Sect. 6.1) may enter the visual cortex and establish synapses.

Thus, callosal projections appear to acquire their adult pattern at their site of termination first and this may involve selection based on mutual recognition between the juvenile axons and some aspects of their targets.

Although the model above is based on the development of callosal projections, its general features may apply to the development of other corticocortical and corticosubcortical projections. Transient projections from auditory to visual areas enter only occasionally the gray matter and when they do, they do not appear to develop terminal arbors in it (Clarke and Innocenti 1986; Innocenti et al. 1988;

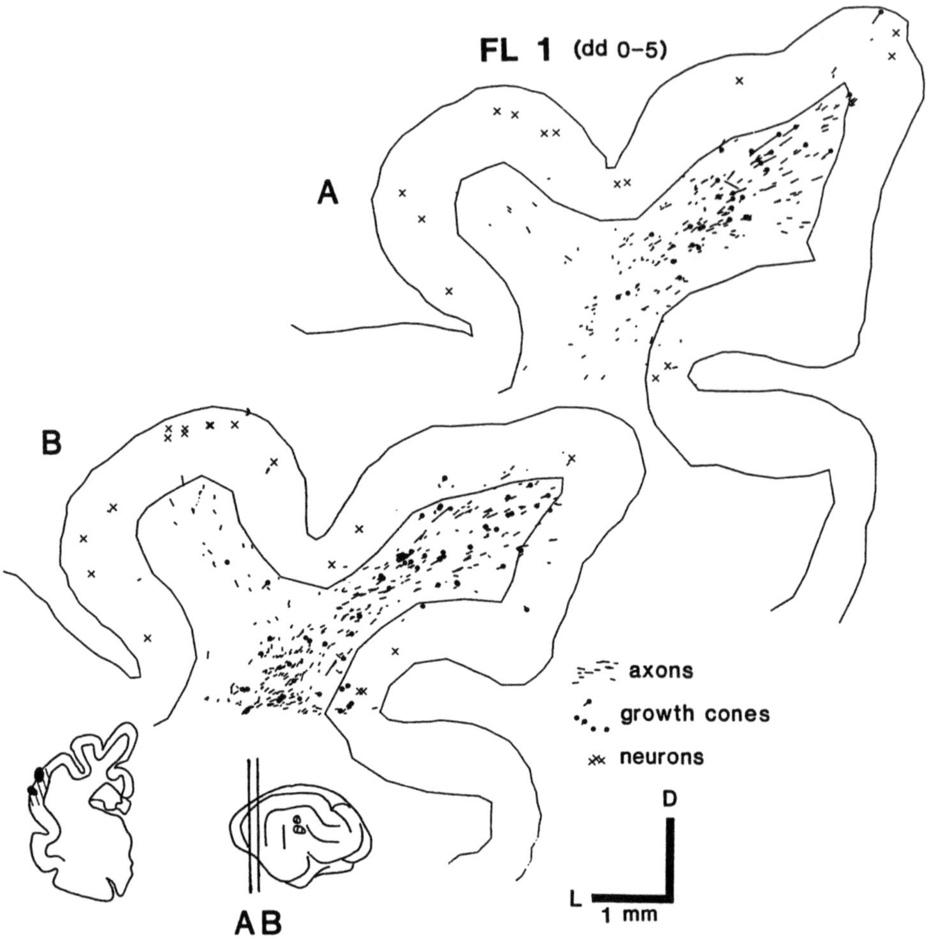

Fig. 7A, B. Computer-aided charts of coronal sections through areas 17 and 18 of the right hemisphere of a kitten, identified by code number and age at injection and sacrifice, showing the distribution of axons (*lines*) and neurons (*crosses*) labeled by a deposit of rhodamine-B-isothiocyanate (RITC) in ipsilateral AI and AII. Growth-cone-like endings are represented by *dots* on one end of the corresponding axon segment. The levels of the charted sections are indicated in the brain inset, together with the locations of the RITC deposits, which are also shown in the section drawing. Note that only a few of the transient axons from auditory cortex enter the visual cortex. (From Clarke and Innocenti 1986)

Fig. 7). Transient intrahemispheric projections between visual areas also fail to grow into the cortex (Price and Zumbroich 1989), and the analysis of axons from intracellularly filled single neurons in area 17 indicated that although these axons may extend tangentially over longer distances than in the adult, their radial growth takes, from the beginning, adult-like patterns (Katz and Wiesel 1987). Thus, transient intrahemispheric projections seem to lack some feature which allows target colonization. The fact that some transient axons do enter the cortex was stressed by Dehay et al. (1988a). Unfortunately, in their study of the transient auditory to visual projection, large injections of wheat germ agglutinin – horseradish peroxidase involved not only the auditory cortex but also the lateral suprasylvian cortex which is known to send projections to areas 17 and 18 in the adult (Symonds and Rosenquist 1984). The transient corticosubcortical projections do also not appear to enter regions of neurons to any significant extent and/or develop terminal fields in them (Distel and Hollander 1980; Stanfield and O'Leary 1985b). A few transient corticocerebellar axons may enter the deep cerebellar nuclei (Tolbert and Panneton 1983).

It should be stressed that at least in the case of the corticocortical projections, the fact that transient axons do not enter the cortex does not exclude the possibility that they may form synapses, since the subplate contains dendrites and interstitial neurons and the latter bear synapses (Chun et al. 1987).

4 Hypotheses on the Genesis of the Juvenile Organization: Cellular Specificities, Temporal Mismatches, Target-Directed Vs Pathway-Directed Axonal Growth

The evidence discussed above suggests that three factors (or groups of factors) may be crucial in organizing the juvenile projections.

The first factor appears to direct axonal growth into one of the several pathways available to the neuron. This pathway "choice" is probably conditioned by some kind of cellular specificity whose most probable determinant, by exclusion of others, seems to be either the clonal origin of a neuron or its serial position whithin a clone. The nature of the pathways and how they differ from each other cannot be specified. The hypothesis is compatible with the well-demonstrated selective nature of axonal growth, under the control of local cues (reviewed by Weiss 1955; Letourneau 1983; Purves and Lichtman 1985).

The second factor appears to be the topographical order of axons within the pathway. This has been long suspected to be a crucial factor in the establishment of orderly retinotectal projections (Cook and Horder 1977; Rager 1980), although it may not reflect with absolute precision the topological relations between the parent neurons, since individual axons change neighborhood relations along their course (Williams and Rakic 1985).

The third factor seems to be a mismatch between the time axons begin to grow or even arrive at the target and the maturation of either the target itself or of the competence of axons in recognizing the target. This can have two consequences: (a) axons pause near the target before entering it or (b) axons grow past the target and then either retract or grow an axon collateral into it. The first may occur when axons grow perpendicularly with respect to the cortex, the second when they grow tangentially. The second condition may be the cause of much of the exuberance observed in the developing nervous system (exuberance in the "elongation mode" according to Schneider et al. 1987).

This model has a certain number of implications, raises several questions and needs qualifications.

One implication is that that the juvenile connectivity is organized by axon-pathway and possibly ax-axonal interactions, while the adult connectivity is determined by axon-target interactions. As will be discussed below (see Sect. 7), the latter may include specific mechanisms for axon-target recognition as well as axo-axonal interactions and neural activity. These mechanisms do not prevent the occurrence of exuberance of axonal branches and synapses near the sites of axonal termination (exuberance in the "arborization mode" according to Schneider et al. 1987).

Although they may appear superfluous, arguments exist which refuse the possibility that the organization of the juvenile connectivity is directed from distant (i.e. several mm) signals released from the target, as earlier theories suggested (Cajal 1909). The first argument is development exuberance itself, i.e. the fact that axons grow towards brain sites which they will later abandon. Second, as mentioned above, several corticofugal or corticopetal axons begin to grow towards a target often before the target has been generated, has matured or reached its adult position.

The strongest arguments suggesting that axon-target interactions become decisive only after axons have reached their cortical targets, however, come from the fact that the adult organization appears first at the target and seems determined by the selective growth of axons into it. These interactions could be mediated by target-released factors (Heffner et al. 1990).

The main difficulty with this "dualistic" interpretation of the formation of connections is that one needs to postulate an additional mechanism relating pathways to targets in order to explain why axons, by choosing a certain pathway, reach an appropriate target, and, on the other hand, why many do not. Perhaps, an explanation could be sought in phylogenesis. Possibly, up to a certain level of complexity, target structures and pathways leading to them appeared in a one-to-one ratio. They may have carried similar markers, as Sperry proposed (1963). In such an "ideally primitive" brain, axons which chose a certain pathway reached one target only. In evolution, new structures may have appeared in excess to the available pathways. Consequently, a pathway gave access to several targets. In other words, in more evoluted brains, pathway-directed axonal growth uses more general cues than target-directed axonal growth. In evolution, pathway, pathway identification rules and rules for axonal ordering along the pathway may be relatively conserved while the neuronal groups which send into a pathway or receive from it are modified.

The model predicts that both exuberance and the waiting of axons near the target may become more conspicuous in structures which have undergone more drastic evolution. Such a model supports the relative importance that both phenomena seem to have acquired in the development of cerebral cortex, a structure which has undoubtedly undergone massive evolution.

The model neither implies nor excludes the formation of transient synapses. Furthermore, the fact that axonal "waiting" near the target and exuberance may have the same cause does not imply that they also have the same function. In addition to compensating for a time gap between developmental events which probably proceed independent of each other, axonal exuberance seems to have other "adjusting" functions in neural development (see Sect. 6) and is possibly involved in transient behavior (Iriki et al. 1988). The possibility that both axonal "waiting" and exuberance may have yet other functions must remain an open question.

5 Transition from the Juvenile to the Adult Organization

Two different mechanisms could be responsible for the partial elimination of juvenile projections characteristic of the transition from the juvenile to the adult corticocortical connectivity: neuronal death and selective axonal elimination. The first mechanism cannot be ruled out and indeed cell death seems to occur in the development of cerebral cortex, (Heumann and Leuba 1983; Finlay and Slattery 1983; Ferrer et al. 1989), including death of well-identified neuronal populations (Kostovic and Rakic 1980; Price and Blakemore 1985a; Chun et al. 1987). Neuronal death may be responsible for the elimination of the transient projection from deep layers in area 17 to area 18 (Price and Blakemore 1985a).

Nevertheless, the majority of the transient cortical projections appear to be lost by axonal elimination, in the absence of neuronal death. The experiment on which this concept is based involved labeling the neurons of origin of a juvenile projection which would be partially eliminated with a long-lasting retrograde tracer. The animal was allowed to survive until the projection was partially eliminated and then a different tracer was injected at the same location as the first tracer, or elsewhere. In the first experiment, differential distribution of the neurons labeled by the early and late injection indicated that the juvenile projection had been eliminated but the neurons from which it originated were still alive. After injection of the second tracer some of the neurons which had eliminated the juvenile projection could be double-labeled and this identified the location of their final target (Innocenti 1981a; O'Leary et al. 1981; Ivy and Killackey 1982; Tolbert and Panneton 1984; Price and Blakemore 1985a; see also below).

A second experiment was used to investigate whether a systematic relation exists between the site to which a neuron sends its transient axon and the target of its permanent axon. In this respect, supragranular and infragranular neurons seem to behave differently.

In the cat, supragranular neurons in visual area 17, which send a transient axon to contralateral areas 17 and 18 (Innocenti et al. 1986), and neurons in the auditory cortex which send a transient axon to areas 17 and/or 18 in the same or in the contralateral hemisphere (Clarke and Innocenti 1990) were investigated. In each of these populations, neurons were found with late, presumably permanent projections within the area where their cell body is located or to areas nearby, but none appeared to establish permanent projections to more distal areas (Fig. 8). Neurons with transient callosal axon in area SI of the rat and monkey also establish short permanent projections (Ivy and Killackey 1982; Chalupa and Killackey 1989).

In contrast, neurons with transient corticospinal axons from the visual cortex of the rat to the spinal cord, a population of layer V neurons, establish permanent

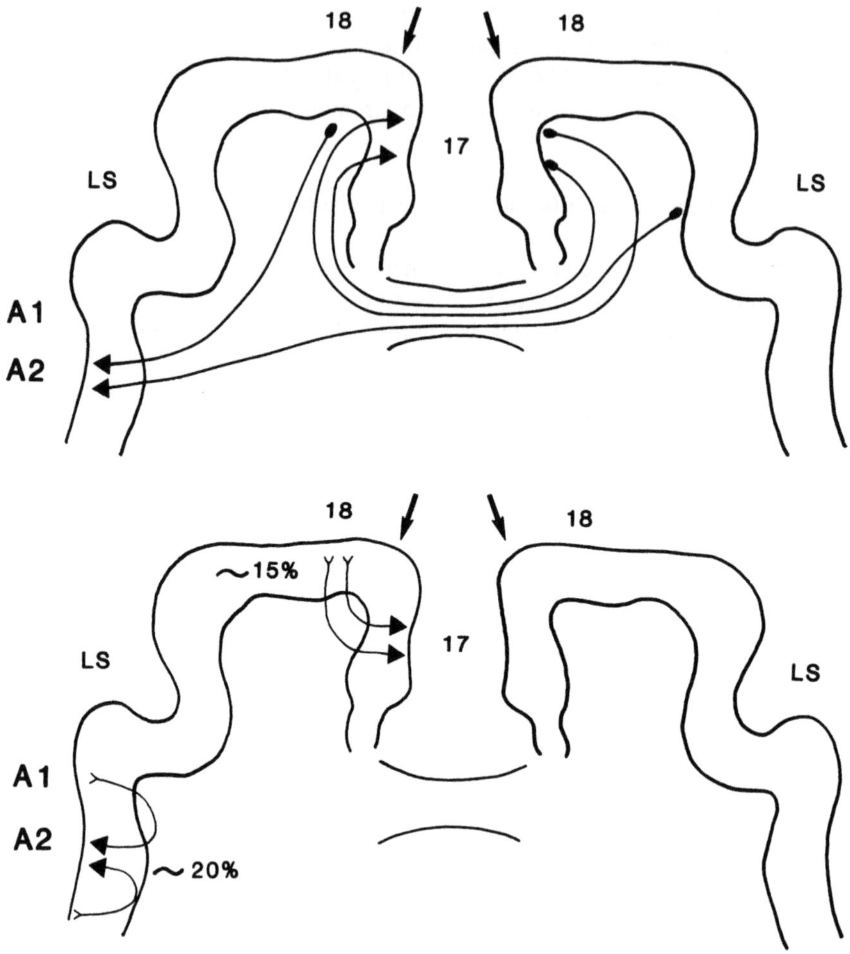

Fig. 8. *Top*, schematic representation of four types of neurons (*filled triangles*), characterized by different locations and site of transient projection. Two of these neurons are in the medial part of area 17 and their axons respectively reach the white matter under contralateral area 17 and the lateral suprasylvian areas (*LS*). Two other neurons are in areas A1 and A2 and their axons respectively reach the white matter under ipsi- and contralateral areas 17–18. *Bottom*, schematic representation of the presumed final projection of the same neurons shown above, after elimination of the transient projections. Notice that they all form relatively short projections. Percent values refer to the fraction of areas 17 and A1, A2 neurons with transient projections for which the local, final projection could be documented

projections to the superior colliculus or to the pons (O'Leary and Stanfield 1985). In the cat, neurons in the frontoparietal cortex with transient projection to the cerebellum, also layer V neurons, establish permanent projections to the spinal cord or brainstem (Tolbert and Panneton 1984).

Thus, the initial overproduction of axoplasm which will subsequently be eliminated seems to be greater for supragranular than infragranular neurons. Interestingly, though, the total axoplasmic production is probably similar in the two cases, and, in both, the connection to the final target may be through a modest collateral of the transient axon. More important, in all cases the initial transient projection seems to overshoot the final target, although to different degrees for different neuronal populations. Studies with retrogradely transported tracers suggested that at some stage, neurons in supra- and infragranular layers have axonal branches directed to their permanent target, in addition to their transient axon (Ivy and Killackey 1982; O'Leary and Stanfield 1985; Innocenti et al. 1986). Anterograde filling of the axons of both supragranular and infragranular neurons showed that, in both, the permanent axon forms as a side branch of the transient axon and the part distal to it will be deleted (Katz and Wiesel 1987; O'Leary and Terashima 1988). Therefore, the apparent lack of unique relations between sites reached by transient and permanent projections of supragranular and infragranular layer neurons may however be dictated by identical constraints on axonal growth and maintenance.

The emergence of the adult Bauplan of cortical connectivity, outlined above, seems to imply axonal loss in the pathways involved. Indeed, in the cat, during the first postnatal month, precisely at the time when transient callosal projections are being eliminated, the corpus callosum loses at least 70% of its axons (Berbel and Innocenti 1988). Similarly, the pyramidal tract of the rat eliminates in development at least 50% of its axons (Reh and Kalil 1982); this elimination coincides with that of corticospinal projections as shown with retrograde transport (O'Leary and Stanfield 1986). The decrease in the number of axons in the corpus callosum and in the pyramidal tract directly reflects the elimination of the transient projections and may be caused by the latter. In the monkey, however, the number of callosal axons decreases postnatally (La Mantia and Rakic 1984), while transient callosal projections from somatosensory areas and visual area 18, the only two transient projections thus far studied, are eliminated prenatally (Killackey and Chalupa 1986; Dehay et al. 1988b). The reason for this cross-species difference is not known (for discussion see Berbel and Innocenti 1988). Basically more complete and detailed information on the development of cortical connections in the monkey is needed. In the cat, the axonal loss in the corpus callosum is so massive that it provokes a decrease in its cross-sectional callosal area (Berbel and Innocenti 1988), already noticed (Innocenti and Caminiti 1980) in a study by Fleischhauer and Schlüter (1970) (Fig. 9). Maximal axonal elimination and the associated decrease in the cross-section of the callosum occur before the onset of myelination and during the fast phase of synaptogenesis in the visual cortex (Fig. 9). These temporal relations between morphogenetic events are interesting because they suggest that the fate of a juvenile callosal axon is decided before its myelination and that it may actually be influenced by the synaptogenesis in its target territory or by the signal(s) which trigger the latter.

The cross-sectional callosal area decreases in man during the last 2 gestational months and the first 2 postnatal months (Clarke et al. 1989). This decrease bears temporal relations to myelination of the corpus callosum (as determined by Yakovlev and Lecours 1967) and to the fast synaptogenesis in the cortex (as determined by Huttenlocher et al. 1982) similar to those observed in the cat (Fig. 10). It is therefore probably also an indicator of the occurrence of axonal elimination.

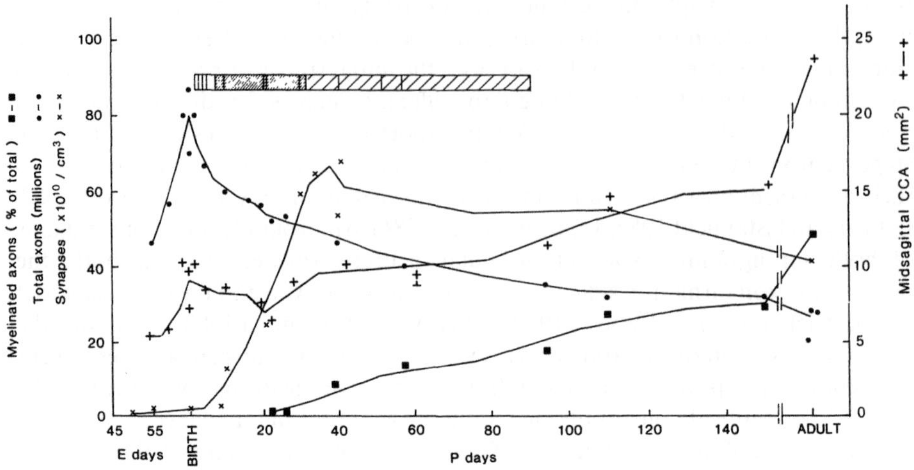

Fig. 9. Temporal relations of different aspects of cortical and callosal development in the cat, i.e., number of axons in the corpus callosum (*dots*), midsagittal cross-sectional callosal area (*CCA*, *crosses*), myelination of callosal axons (*squares*), density of synapses in the visual cortex (*Xs*, Cragg 1975). The *horizontal rectangle* represents the period of elimination of callosal projections from area 17, demonstrated with retrograde transport by Innocenti and Caminiti (1980). Each *vertical line* in the rectangle indicates the age of one kitten at the time of horseradish peroxidase injection. *Stippling* indicates the period of elimination of the projection, considered to extend from the age of the oldest kitten in which the projection was still fully exuberant to that of the youngest kitten in which the projection seemed to be as in the adult. Two phases are distinguished: the first (*heavy stippling*) corresponds to the bulk of the elimination, from most of area 17; the second (*light stippling*), corresponds to the elimination of projections close to the 17/18 border. Notice that most of the callosal axons and projections are eliminated before the onset of myelination. This massive elimination coincides with a pause in the growth of CCA and occurs during the fast increase in synaptic density. The peak in synaptic density is probably reached around P70, not around P37 as indicated here (Winfield 1981). Thus, the elimination of callosal axons and projections is most conspicuous during the initial 1 third to 1 fourth of synaptogenesis in area 17. (Modified from Berbel and Innocenti 1988)

Biochemical and immunohistochemical analysis of callosal axons during the postnatal period indicated that the three subunits of neurofilaments progressively increase during the first month (Figlewicz et al. 1988). In particular, a monoclonal antibody which recognizes a phosphorylated epitope of the heavy (approx. 200K) subunit shows that only between P11 and P18 have significant numbers of callosal axons acquired this protein, although possibly, until P29, in smaller amounts than in the adult (Figlewicz et al. 1988; Guadano-Ferraz 1990; Fig. 11). Even more interestingly, a second monoclonal antibody, which in the rat recognizes an unphosphorylated epitope of the heavy subunit of neurofilaments, begins to show significant axonal labeling only around P39 (Guadano-Ferraz et al. 1990; Fig. 11). The transition between the juvenile and the adult forms of the microtubule-

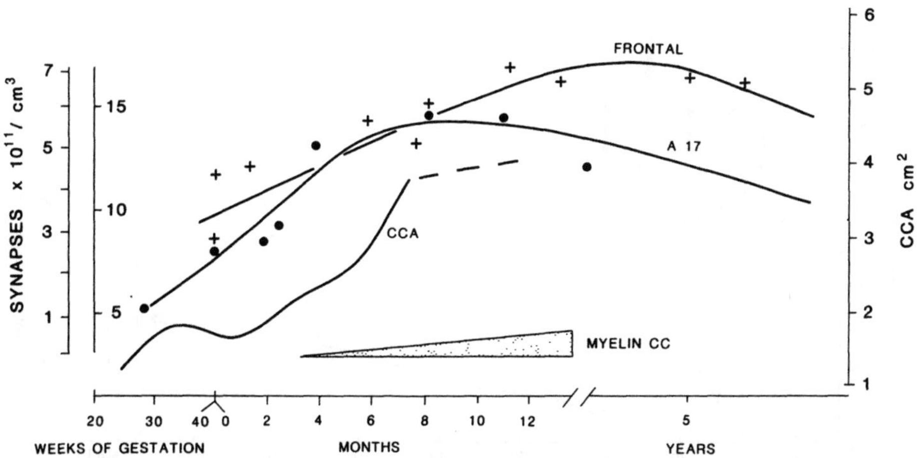

Fig. 10. Temporal relations of different aspects of cortical and callosal development in man, i.e. midsagittal cross-sectional callosal area (*CCA*, Clarke et al. 1989), synaptogenesis in area 17 (*dots*, curve fitted by eye; Huttenlocher et al. 1982) and in frontal cortex (*crosses*, curve fitted by eye; Huttenlocher 1979), and myelination of callosal axons (light microscopic observation of Yakovlev and Lecours 1969). Notice that CCA appears to decrease between the end of gestation and the first postnatal month. This period may correspond to the massive elimination of transient callosal projections and axons in man: as in the cat, most of the suspected elimination occurs before myelination and during the initial phase of fast synaptogenesis in the cortex

associated proteins Tau also occurs between P19 and P28 (Riederer and Innocenti, unpublished; Fig. 11).

These changes in the cytoskeletal proteins of callosal axons occur simultaneously to ultrastructural changes (Berbel et al. 1989). The number of microtubules and neurofilaments per axon increases, although at different rates. The increase in the number of microtubules precedes that in the number of neurofilaments and also slightly precedes the increase in axon diameter to which the number of neurofilaments and/or microtubules is correlated. Changes in the minimal distance between microtubules or neurofilaments are temporally correlated with the changes in microtuble associated proteins and the appearance of the phosphorylated and partially dephosphorylated variants of the heavy subunit.

Thus, the time of axonal elimination, in the corpus callosum, coincides with profound modifications of the cytoskeleton of the remaining callosal axons. It is tempting to speculate that these modifications may correspond to their transition from a juvenile-labile state into an adult-stable state (for a similar concept in the case of synaptogenesis see Changeux and Danchin 1976). This hypothesis requires evidence that the fate of a juvenile callosal axon is not already decided before the elimination begins and also, that the fate of an axon and its cytoskeletal modifications are regulated by the same factors. This evidence exists (see below) since the period of elimination of callosal axons coincides with a "sensitive period" during which various manipulations can either stabilize axons which would

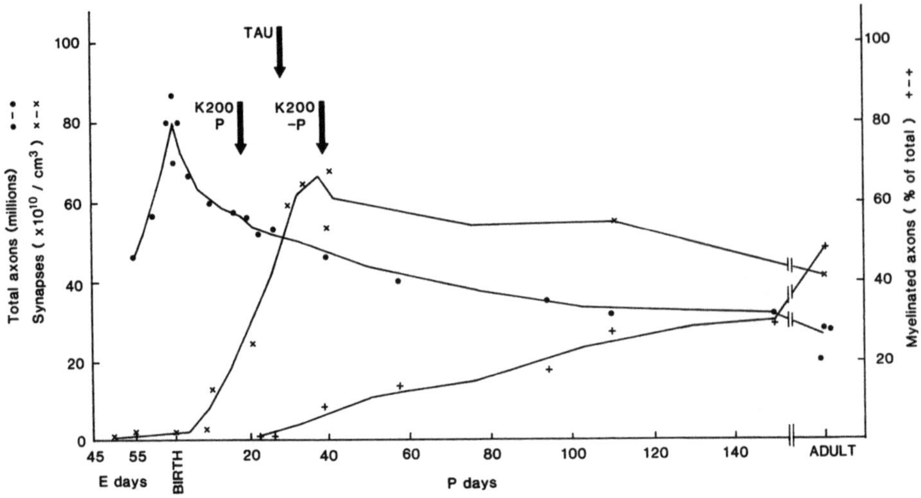

Fig. 11. Temporal relations of morphological and biochemical aspects of cortical and callosal development in the cat. The curves of number and myelination of callosal axons and synaptic density are the same as shown in Fig. 9. *Arrows* point to the first unequivocal appearance of three cytoskeletal proteins in callosal axons: the phosphorylated heavy subunit of neurofilaments (K200 P; Figlewicz et al. 1988), the adult forms of the juvenile microtubule associated proteins Tau (Riederer and Innocenti, unpublished) and a partially dephosphorylated form of the heavy neurofilament subunit (K200-P Guadano-Ferraz et al. 1990) or possibly of the medium neurofilament subunit

normally be eliminated or eliminate axons which would normally be maintained. At least one condition, hypothyroidism, appears to interfere both with the elimination of callosal axons and the maturation of their cytoskeleton, in particular the appearance of the heavy neurofilament subunit (Gravel and Hawkes 1990).

Less is known about cellular events underlying the elimination of transient cortical axons. Although often referred to as "axonal retraction", the phenomenon most probably involves axonal degeneration followed by phagocytosis of the degenerating elements. Indeed, at the time of axonal elimination, macrophages appear in the white matter at sites traversed by transient axons of cortical origin. Electron-microscopically, some of these macrophages appear in the process of phagocytosing axons (Innocenti et al. 1983a, b; Berbel and Innocenti 1988). However, no clear instances of electron opaque degeneration of callosal axons were noticed in a study where they were specifically sought (Berbel and Innocenti 1988). The degeneration may take more subtle forms as suggested by the appearance of vacuoles and swollen mitochondria in some callosal axons at the time of their elimination (Berbel and Innocenti 1988). Alternatively, the degenerating debris may be cleared before they acquire distinctive ultrastructural features.

6 The Regulation of Axonal Survival in the Developing Cortex

Probably, much of the interest evoked by the discovery of the existence of transient projections in the developing cortex derived from the fact that they appeared to reveal an enormous potential for developmental plasticity in this structure. This view, if justified, could provide new tools for understanding brain development as well as the relations between structure and function in the cerebral cortex.

Alternatively, however, the transient projections could be inexorably doomed to elimination, either because they subserve a transient function or because they represent a phylogenetic relict.

This alternative has implications for the difficult question of the teleology of developmental exuberance (Changeux and Danchin 1976; Katz and Lasek 1978; Innocenti 1981b, 1988; Katz 1983; Ebbesson 1984; Cowan et al. 1984 inter alios). If transient projections are inexorably eliminated, then the possibility that transient structures "may allow developmental decisions to be made when the necessary information is not available or not usable in the system" (Innocenti and Clarke 1984b) or, differently phrased, may play the role of an "ontogenetic buffer mechanism" (Katz 1983) could be rejected. Indeed, developmental exuberance may have no function. Its widespread occurrence across species could just indicate that evolution preserved this mode of development as an extravagant but not detrimental way of putting a brain together.

The available evidence is clear-cut. Neither the fate of transient projections, nor that of the projections which are normally stabilized is rigidly predetermined in ontogenesis. Nevertheless, in spite of nearly 10 years of efforts we seem far from understanding what determines this fate and how.

A posteriori, the theoretical basis for most of the work may have been naïve in the sense that we searched for "the" factor responsible for the elimination/stabilization of juvenile projections. The more or less explicit assumptions were that, by appropriate perturbation experiments, as in the case of the formation of ocular dominance columns, one or few factors would be identified and that they would be the same across systems and species. The studies reviewed below weaken both assumptions.

First, the same perturbation experiment has occasionally had different outcomes in different species and systems. Second, we appear to be dealing with several, possibly interacting factors. A more realistic view may be that which juvenile axons are maintained and which are eliminated "emerges" as a property of a network of causal interactions (for the roots of this concept see Weiss 1955), to a large extent determined by the neural network the axons are embedded in, but also by more general signals, for example hormones. Key properties of this network, are that (a) the effects of local perturbations may spread through the network and thus affect distant brain sites and (b) activity may regulate the development of its structure.

This hypothesis does not exclude the possibility that one crucial step, for example, a biochemical modification of cytoskeletal components, may control the transition from the juvenile to the adult connectivity. But this hypothetical step appears to be regulated by multiple, possibly interdependent factors, some of which at least have been identified. These are:

1. Sensory experience
2. Integrity of the sensory periphery and of the pathways originating from it
3. Factors dependent on the target and/or the afferents it receives.

6.1 Sensory Experience

The first experiments aimed at testing the role of visual experience on the development of corticocortical connectivity were inspired by the fact that callosal connections between the area 17/18 borders in the two hemispheres appear to be involved in establishing anatomical and functional continuity across the vertical meridian of the visual field (Choudhury et al. 1965; Berlucchi et al. 1967; Berlucchi and Rizzolatti 1968; Hubel and Wiesel 1967). In normal vision, the line of decussation of the ipsi and contralateral retinofugal projection, i.e. the line which divides the visual world of the two hemispheres and the geometrical midline of the binocular visual field (the vertical meridian) coincide and are superposed in visual space. The effects of horizontal strabismus were studied, since it shifts the geometrical midline of the binocular visual field onto the nasal or temporal hemiretina, depending on whether the eyes diverge or converge. This new vertical meridian is separated from the decussation line proportionally to the degree of squint. The underlying hypothesis was that synchronous activation of cortical loci in the two hemispheres may be necessary for the stabilization of the juvenile callosal connections and that strabismus allowed synchronous stimulation, although through different eyes, of cortical sites whose callosal connections would normally be eliminated because they "viewed" different parts of the visual field.
These experiments produced moderately enlarged callosally projecting zones in area 17 of the cat (Innocenti and Frost 1978, 1979), and the finding was confirmed (Bermann and Payne 1983; Elberger et al. 1983). Expansion of the callosal terminal territory in the same area was also observed (Lund et al. 1978). However, no systematic relation was found between the degree of expansion of the callosally projecting zone in area 17 and the angle of strabismus induced in the animal. The enlargement of the callosal zone was modest when compared with the wealth of callosal projections at birth, and most of the projection still originated at the 17/18 border. Finally, similar expansion of the callosal zone near the 17/18 border was produced by other rearing conditions such as monocular deprivation or enucleation, binocular enucleation (Innocenti and Frost 1979, 1980) and short periods of normal vision followed by binocular deprivation (Innocenti et al. 1985). Rearing conditions which induce expansion of the callosally projecting zone seem to have in common that they interfere with the normal binocular vision during the 2nd postnatal month (see Innocenti et al. 1985 for discussion). One tentative explanation for these results is that the stabilization of callosal connections requires synchronous activation of the callosal axon and of its target neuron and that this is achieved by proximity or overlap, as well as similar orientation and direction specificity of the

receptive fields of the callosally projecting neurons and of the neurons which receive from them. This mechanism is similar to that proposed for the maturation of binocular properties of cortical neurons (Wiesel 1982). As discussed previously (Innocenti et al. 1985), during the 1st and 2nd postnatal months this mechanism may refine callosal connections by stabilizing certain connections and eliminating others. Abnormal binocular vision interferes with this refinement either because normal binocular vision is required or because it results, at the cortical level, in broader and less sharply orientation tuned receptive fields (Chino et al. 1983). Modifications of callosal connections, broadly related to those described in kittens, may occur in humans. John and Timney (1986) described increased interhemispheric transmission times for targets within 5° of the fixation point, i.e. within a part of the visual field whose cortical representation is callosally connected in most animals thus probably in man.

A decrease in the number of retrogradely labeled callosally projecting neurons in area 17 was obtained in the cat by binocular eyelid suture (Innocenti and Frost 1980; Innocenti et al. 1985; Fig. 12) and by dark rearing (Frost and Moy 1989). This

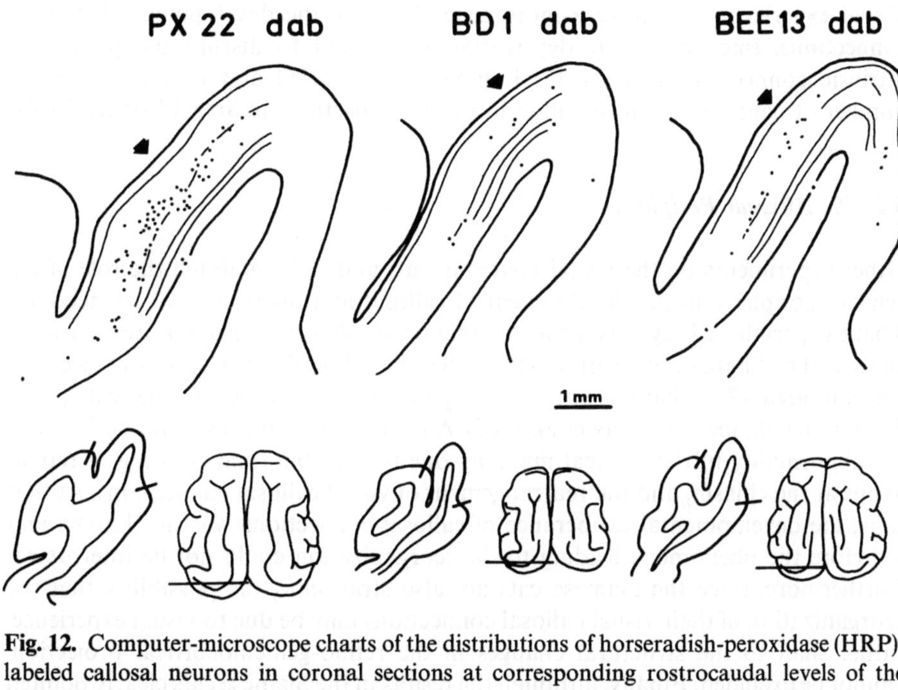

Fig. 12. Computer-microscope charts of the distributions of horseradish-peroxidase (HRP)-labeled callosal neurons in coronal sections at corresponding rostrocaudal levels of the postlateral gyri of a normal adult cat (*PX 22*), a cat (*BD* 1) raised with bilaterally sutured eyelids and a cat (*BEE* 13) which had been bilaterally enucleated on postnatal day 2. All sections were reacted with *DAB*. Each HRP-labeled neuron is represented by one dot. The drawings at the top are enlargements of the regions denoted in the corresponding low-power drawings. The rostrocaudal levels are indicated on drawings made from dorsal view photographs of the corresponding brains. In each section, *thinner lines* mark the lower boundaries of layers I, III, IV, V; a *filled arrow* points to the cytoarchitectonic boundary between areas 17 and 18. (From Innocenti and Frost 1980)

confirmed the role of vision in regulating maintenance or elimination of juvenile projections. However, there were limits to the influence of visual experience: (a) a non-negligible complement of callosally projecting neurons was stabilized in primary and secondary visual areas even under the severest conditions of visual deprivation; (b) these neurons remained centered around the 17/18 border; (c) in some of the experiments, different effects were obtained in supragranular and infragranular layers, the latter being, for example, more resistant to bilateral eyelid suture (Innocenti et al. 1985).

This unequivocally points to the existence of control mechanisms independent of vision, and this conclusion agrees with the recent finding that maturation of callosal connections in primary visual areas may be a largely prenatal event in monkey (Dehay et al. 1988b) and probably man (Clarke et al. 1989).

These control mechanisms are not necessarily independent of activity. Spontaneous activity of retinal or central origin may provide the necessary synchronization in the activity of cortical sites which become callosally connected. Spontaneous retinal activity appears early enough in development for this to be possible (Galli and Maffei 1988).

Visual experience may have a more general role in the development of cortical connections, since binocular deprivation was found to disrupt the pattern of intrinsic connections in area 17 (Luhmann et al. 1986) and may decrease the number of neurons which maintain intrinsic connections in area 18 (Price 1986).

6.2 The Afferent Periphery

Other experiments on the visual system in cats and rodents tested the role of the sensory periphery in the development of callosal and other cortical connections. That the periphery may play a role in organizing callosal connections was strongly suggested by Shatz's observations (1977) that the callosally projecting and receiving zones in area 17 of Siamese cats were expanded compared to normal cats (for a different result, see Tremblay et al. 1987). A systematic relation was found between the abnormality in the cortical mapping due to the abnormal crossing of retinal axons at the chiasm, and the rearrangement of visual callosal connections. At that time, the developmental exuberance of callosal connections was not known and therefore the phenomena leading to the reorganization could not be interpreted. Furthermore, since the Siamese cats are also strabismic, the possibility that the reorganization of their visual callosal connections may be due to visual experience, rather than to the structural changes in the retino-geniculocortical projection, cannot be excluded. Finally, although the results in the Siamese cats clearly pointed to a role of either the periphery or of vision in the organization of callosal connections, the possibility that the same genetic defect which is responsible for abnormal crossing at the chiasm may, independent of the latter, induce abnormalities in either the visual cortex or the corpus callosum could not be ruled out.

The role of afferents from the sensory periphery in the regulation of the selection of juvenile callosal axons could be deduced from experiments on binocularly or monocularly enucleated kittens (Innocenti and Frost 1979; 1980), mouse, rat and hamster pups (Rhoades and Dellacroce 1980; Rothblat and Hayes 1982; Rhoades and Fish 1983; Olavarria and Van Sluyters 1984) and monkey embryos (Dehay et al. 1989) as well as on anophthalmic mice and rats (Olavarria and Van Sluyters 1984; Olavarria et al. 1988a; Rhoades et al. 1984). All these conditions produced enlargement of the callosally projecting zone (Fig. 13), usually interpreted as stabilization of callosal projections which would otherwise be eliminated, but also more subtle changes in the distribution of callosally projecting neurons (Olavarria et al. 1987). The results of the various experiments differed in certain details. Binocular enucleation in the cat also caused a decrease in the number of retrogradely labeled neurons projecting into the corpus callosum (Innocenti and Frost 1980). This is apparently not the case in rodents and monkey, where, however, no quantification was attempted. In rodents, but probably not in cats (Innocenti and Frost 1979 and unpublished), the callosally projecting zone and the callosal terminal territory increase on the side of the remaining eye but remain normal on the other side (Rhoades and Dellacroce 1980; Rothblat and Hayes 1982; Cusick and Lund 1982; Olavarria et al. 1987). In the monkey, the increase seems to be due exclusively to stabilization of projections coming from area 18, while in the cat projections coming from area 17 were stabilized; they were, however, the only ones studied.

To some extent, these differences may reflect the particular emphasis of the different studies. However, they may also be related to differences in the organization of the visual system of monkeys, cats and rodents and in particular to the different degree of crossing of the optic pathways in these three species. In addition, in rodents but not in cats, the callosal connections contribute significantly to the binocularity of cortical neurons (see Innocenti 1986 for references). Differences in the function of callosal connections and in their mode of development, in particular for the monkey (see Sect. 3.1), could also account for differences in the results of peripheral lesions.

To what extent the concept of a role for the periphery and/or sensory experience in callosal maturation can be generalized is unclear. Auditory callosal connections appear not to be modified by neonatal bilateral destruction of the cochlea (Brugge et al. 1983). Similarly, the development of the transient auditory to visual connections is not significantly modified by bilateral enucleation of the eyes (Innocenti et al. 1988).

In rodents, callosal connections develop before eye opening suggesting that the changes after peripheral lesion may not be due to altered visual experience (see for discussion Rhoades and Fish 1983; Olavarria et al. 1987). In the cat, however, binocular visual deprivation by eyelid suture, a condition which prevents form vision but not the detection of changes in diffuse illumination (Innocenti and Frost 1980; Innocenti et al. 1985), dark rearing (Frost and Moy 1989) and binocular eye enucleation (Innocenti and Frost 1980) led to qualitatively and quantitatively distinguishable abnormalities in the callosal connections, suggesting that integrity of the periphery and vision may play distinct roles. As discussed above these results do not exclude activity of retinal origin as a mechanism.

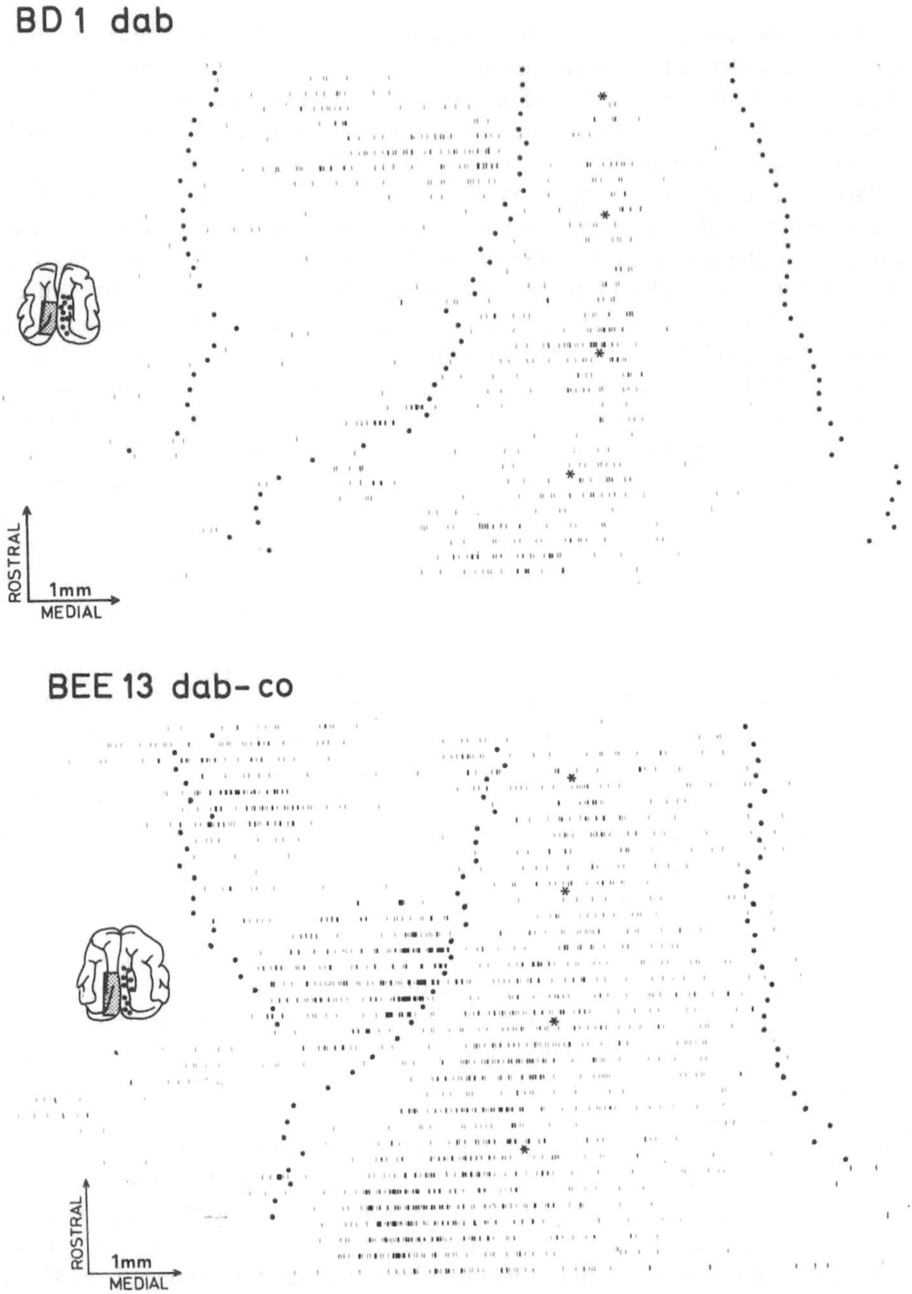

Fig. 13. Computer reconstructions showing labeled callosal neurons in layers II–IV in a cat (*BD* 1) raised with bilaterally sutured eyelids, and in another (*BEE* 13), which had been binocularly enucleated on P2. Brains processed with *DAB*. Flattened representations of postlateral and lateral gyri (corresponding to *stippled area* in the inset). *Dotted lines* indicate, from left to right, fundi of the lateral, postlateral and suprasplenial sulci. The *asterisks* mark the boundary between areas 17 (medially) and 18 (laterally).

As in the visual deprivation experiments, peripheral lesions resulted in minor changes of callosal connections, compared to their wealth at birth. Since most of the enucleations were performed at birth, one possibility is that callosal connections are too mature to be modified further, i.e. that by birth the fate of most juvenile projections is irreversibly determined. This possibility seems to be ruled out by the finding that the stabilization is no greater in congenitally anophthalmic than in neonatally enucleated mice and rats (Olavarria and Van Sluyters 1984; Olavarria et al. 1988a).

Thus, even higher controls of the developmental selection of juvenile corticocortical axons probably exist. Activity-based controls cannot "a priori" be ruled out. In particular, the synchronization of EEG activity between the hemispheres in the adult appears to be, at least to some extent, independent of callosal connections (Berlucchi 1966; Singer and Creutzfeldt 1969; Susic and Kovacevic 1974). If this is the case in development, a "diffuse" cortical activator located below the cortex could provide sufficient synchronous input to the hemispheres to stabilize a somewhat normal complement of callosal connections. A condition which presumably reduces or eliminates this hypothetical activator as well as other possible controls of thalamic origin is the transection of the thalamic radiation.

This experiment produced different results in different species and systems. In the somatosensory system of the rat the callosal termination to SI was studied on the side of a large thalamic lesion including the ventrobasal complex and was found to be unaffected (Wise and Jones 1978). In contrast, a comparable study in the visual system reported tangential expansion of the callosal terminal territory in 17/18 (Cusick and Lund 1982). In the hamster, transection of the optic radiation also resulted, in the deafferented hemisphere, in expansion of the territory of callosal termination, loss of the callosal projection from supragranular layers and maintenance of the projection from the infragranular layers (Rhoades et al. 1987). This finding seems to rule out the hypothesis that the expansion of the callosal terminal territory found on the side of the remaining eye in the case of monocular enucleation may be due to "abnormal instructions emanating from the thalamus ipsilateral to the remaining eye" (Olavarria et al. 1987). Alternatively, the expansion of the callosal terminal territory in both types of lesion may indicate that the intact thalamus normally sharpens the tangential distribution of callosal axons, either through a competitive interaction or through synchronous firing as proposed above. On the other hand, the suggestion that the "initial development of callosal axon" by supragranular callosal neurons, "may depend on thalamic input" (Rhoades et al. 1987) seems based on a somewhat implausible developmental mechanism whose demonstration requires analysis of the callosal projections at short survival after the lesion in order to rule out the alternative mechanism, i.e. that the callosal projections from the 17/18 border are formed, but not maintained, in the absence of thalamic

Fig. 13 (*continued*). The neurons were projected onto a line running parallel to the pial surface and 400 μm deep; the line was divided into bins of 50 μm and the number of neurons in each bin represented by a corresponding number of vertical line segments. Each *horizontal row* of line segments represents one section. Sites of injection are represented by *dots* on the inset. (From Innocenti and Frost 1980)

input. Contrary to what was found in the hamster, similar experiments in the cat led to maintenance of the projection from the deafferented to the intact hemisphere, but striking reduction of the projection from the intact to the deafferented hemisphere (Melzer et al. 1987).

Taken at face value, these differences may indicate that the stabilization of a juvenile callosal axon depends in cats more on the thalamic input at the target and in rodents more on the thalamic input at the origin of callosal axons. Unfortunately, the results of neither experiment are easy to interpret since the transection of the optic radiation interrupts not only the geniculocortical projection but also other subcortical afferents to the visual cortex, and it also transects the axons of corticosubcortically projecting neurons. The consequences of these lesions are unknown but may differentially affect the development of callosal connections in the two species.

6.3 The Target

Although the role of the afferent periphery and/or visual experience on the maturation of visual callosal connections appears well established, the mechanism of action is far from being clear. One uncertainty is whether the afferents affect callosal neurons and axons, or their target, or both. The recent work of Stanfield and O'Leary (1985a) showed that occipital cortical neurons, presumably the same which normally establish transient projections to the spinal cord, maintain this projection if grafted in the somatomotor region from where the normal projection to the spinal cord originates. If it could be convincingly shown that the neurons which establish maintained projections to the spinal cord are the same that would have established the transient projection had they remained in the occipital cortex, these experiments would strongly support the hypothesis that the tangential position rather than some other "innate" quality of a cortical neuron may be critical in determining whether its axon will be maintained or eliminated.

A number of experiments tested the possibility that the fate of a juvenile axon may depend on events at its target and in particular on competition with other axons. The task is difficult, since the fact that one projection is maintained when another is experimentally eliminated is probably not sufficiently rigorous evidence that competition occurs in normal development between the two projections (Guillery 1988). A modest stabilization of normally transient callosal axons from the forepaw representation in SI to contralateral SII was obtained when the SI on the same side was neonatally lesioned. Lesion of the remaining SII did not provoke a similar effect nor did it increase the effects of the SI lesion (Caminiti and Innocenti 1981). These experiments suggested that competition between callosal and ipsilateral axons from SI to SII may affect survival of the former. Unfortunately, the possibility that the SI lesion may also have modified the thalamic input to SII and that this might have caused the abnormal stabilization of the callosal projection could not be ruled out.

Stabilization of the transient cortical projection to the deep cerebelar nuclei was obtained by neonatal lesions of the cerebellar cortex (Panneton 1986), suggesting that competition with the axons of the Purkinje cells may normally be involved in the elimination. Unfortunately, as in the case of SI lesion, the possibility of more indirect effects secondary to modifications of other inputs to the cerebellar nuclei or even unspecific trophic effects of the lesion (see below) could not be ruled out. The fact that crossed corticothalamic and corticorubral projections are maintained into adulthood in cats with neonatal unilateral cortical lesion (Leonard and Goldberg 1987) may also be interpreted as evidence that competition plays a role in the normal elimination of this transient projection.

Intriguing evidence of activity-driven competition between visual callosal projections and thalamocortical projections or the neurons on which the latter impinge was reported by Cynader et al. (1981), who raised monocularly deprived, split chiasm kittens and reported loss of callosal inputs and probably of projections originating in the hemisphere receiving from the deprived eye.

The studies above illustrate the characteristic difficulty of interpreting perturbation experiments in complex systems with a network structure. Because it cannot be excluded a priori that the local disturbance may have remote and cascading effects elsewhere in the network, the network must be fully characterized both in its final structure and in its development. Furthermore, since the function of the network can condition its structure information about the latter is also necessary.

An attempt in this direction was made in a recent series of experiments (Innocenti et al. 1987; Innocenti and Berbel 1989a, b; Assal et al. 1989) whose initial motivation was to study whether trophic dependence of transient axons on their transient target existed, by means of the early destruction of the latter. Areas 17 and 18 of newborn kittens were injected with ibotenic acid, an "axon-sparing" excitotoxin which binds to aspartate receptors. Unexpectedly, these neonatal injections did not provoke complete destruction of the cortex but rather its reorganization into a microcortex consisting of neurons normally destined for layer II and III, in the absence of granular and infragranular layers. These neurons have not finished migration at the time of injection and appear to be spared by the injection. The structure of the microcortex is similar to that of microgyria, a congenital malformation of the human cortex whose most probable cause appears to be an ischemic insult around the end of the period of neuronal migration. The ibotenic acid-induced microcortex possess several features of normal visual cortex, including connections with the lateral geniculate and with other cortical areas and orientation- and direction-specific responses to visual stimuli. Surprisingly, the microcortex maintains the normally transient projections from ipsi- and contralateral auditory cortex (Fig. 14) but not the transient projections from contralateral area 17. The specificity and apparent long duration of the stabilization suggests that it may not be due to generalized trophic effects of the lesion (for this concept see Nieto Sampedro et al. 1983), although this possibility cannot be fully ruled out in view of the increase in the number of astrocytes in the white matter of the microcortex. It appears more probable that the absence of the granular and infragranular layers is directly responsible for the stabilization. Layer VI or perhaps subplate neurons may be more specifically involved, since parts of cortex where only these structures are injured maintain axons from the auditory cortex.

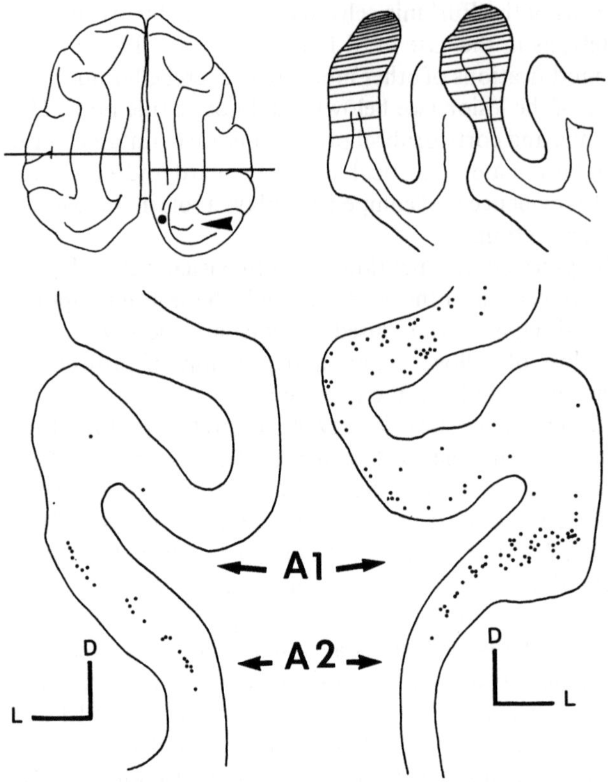

Fig. 14. Stabilization of projections from auditory cortex to frontal cortex in an adult cat in which microcortical regions were created in areas 17 and 18 by neonatal injections of ibotenic acid. The microcortex was injected with wheat germ agglutinin–horseradish peroxidase (*dot* in brain drawing; note narrower gyri in the right occipital cortex). The injection site is shown by the *hatching* on the low-power frontal sections. Labeled neurons are shown in the auditory areas (*A1* and *A2*) ipsi- and contralateral to the injection, which would normally have been eliminated. (See Innocenti and Berbel 1990b for details)

6.4 Others

The factors discussed above are almost certainly not an exhaustive list of those which may influence the fate of juvenile axons. For example, lesioning the superior colliculus was also found to maintain callosal projections which would otherwise be eliminated (Mooney et al. 1984). If in the hamster, as in the cat, visual callosal projections transfer information to the superior colliculus (Antonini et al. 1979), probably by contacting corticocollicular projection neurons in layer V, this finding may indicate that, in development, the fate of an axon is not only affected by its target neurons but also, indirectly, by events affecting the neuronal population to which the latter project.

Factors intrinsic to the neuron undoubtedly can affect both formation and maintenance of axonal arbors (Schneider et al. 1987). These factors may include intrinsic differences among axonal branches in sustaining critical neuronal functions such as survival and target recognition (Sharkey et al. 1986; Tolbert 1987).

General signals, in particular hormones, may play a role. Both the elimination of exuberant callosal axons and the maturation of their cytoskeleton, i.e. the expression of the heavy neurofilament subunit, were prevented by hypothyroidism (Gravel and Hawkes 1990). Gender may affect callosal size and shape in animals and man (Berrebi et al. 1988; Clarke et al. 1989), suggesting a possible role of steroids in callosal maturation.

Understanding the respective roles of the factors discussed above and their interactions will be not be an easy task. But it may be a worthwhile one. Failures in the elimination of normally transient projections may occur in circumstances other than the well-documented situations described above, in for example an experimental model of the fetal alcohol syndrome (Miller 1987) or even in schizophrenia (Feinberg 1982). The opposite type of pathology, i.e. an exaggerated elimination of axons in pathways of cortical origin such as the corpus callosum and the pyramidal tract, can also occur (Fig. 15) and might have a genetic origin (Lyon et al. 1990).

7 Conclusions

One may wonder if, besides their contribution to the understanding of the development of cerebral cortex, and therefore ultimately of higher brain functions, the studies above may also be relevant for the old and more general problem of the formation of neural connections.

Certainly, the formation of exuberant projections linking in a transient way neural structures which will be disconnected in the adult is not restricted to the cerebral cortex. For example, major transient retinofugal (Land and Lund 1979; McLoon and Lund 1982; Frost 1984; Bagnoli et al. 1987 inter alios) and retinopetal (Clarke and Cowan 1976; O'Leary and Cowan 1982; Catsicas et al. 1987) projections have been described, together with others at higher levels in the visual pathway (Stein et al. 1985). Although some of these projections may not involve the formation of long transient axons but only of locally widespread terminal arbors (Rakic 1976; Sretavan and Shatz 1987), the regulation of the survival of some of these projections appears to obey some of the same factors, for example competition and activity, which are crucial in the development of cortical connectivity (Rakic 1981; Frost 1986; Sretavan et al. 1988). One of the fascinating perspectives emerging from these studies is that of redirecting major sensory pathways within the brain with at least partially preserved function (Frost and Metin 1985).

The artificial control of developmental exuberance by selective lesions, constrained sensory experience, chemical manipulations and grafting, in addition to the more obvious potential of genetic manipulations, make a new type of "neuronal engineering" possible, and through the latter the repair of the developing and adult brain.

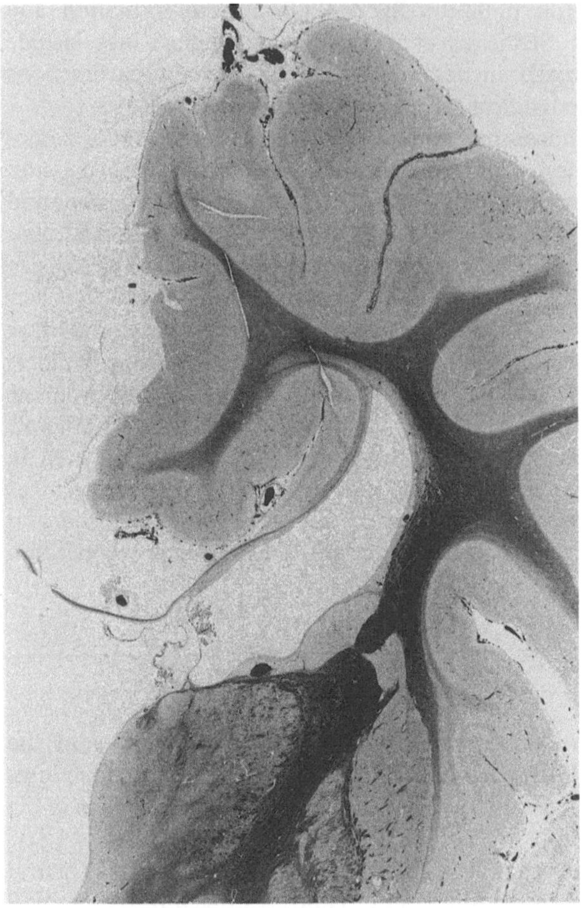

Fig. 15. A case of human congenital atrophy of the corpus callosum accompanied by atrophy of the white matter and absence of the pyramidal tract, considered to be a primary disorder of axonal development, occurring during the late fetal and early postnatal period, i.e. roughly during the time when normal elimination of cortical projections is suspected to occur in man (see Fig. 10). Girl deceased at 10 months. Coronal sections (dorsal is up, medial to the left) of the hemispheres. Myelin stain. Notice the white matter atrophy in the gyrus cinguli and the extremely thin, but myelinated corpus callosum. No Probst bundle exists, and the pathology is therefore different from the classical agenesis of the corpus callosum. This anomaly may be a pathological exaggeration of the normal developmental elimination of cortical axons, possibly a cytoskeletal defect. (From Lyon et al. 1990)

Still, one seems to be only nagging at the main theoretical question: to what extent is brain connectivity determined by prespecified axon-target recognition mechanisms?

At first glance, the development by exuberance of many neural connections seems to exclude prespecification as a plausible mechanism for their formation. Indeed, this mode of development seems at odds with the parsimonious and infallible growth to the target that prespecification, in particular chemoaffinity, seems to imply (see also Innocenti 1988, for discussion). Furthermore, the selection of juvenile axons for maintenance or elimination depend on multiple factors, including functional criteria whose logic seems hierarchically higher than that of the molecular interactions at the site of neuron-target interaction.

Appealing as it may be, this view of neural development must nevertheless be qualified by several other considerations. First, in some systems, projections topographically similar to those of the adult are established from the beginning. This is the case with afferent projections to the cerebellum (Sotelo et al. 1984; Arsenio Nunesa and Sotelo 1985; Mariani et al. 1987), although recently transient projections to this structure were found as well (Bower and Payne 1987). In the case of neuromuscular connections, the elimination of multiple innervation results only in small topographical rearrangements (Bennett 1987). Thus, in these systems, some kind of "prespecification" cannot be excluded and its role may be preponderant in the formation of the connections. Second, even in the case of corticofugal projections, where exuberance is impressive, there are early signs of specificity in the projection, including topographic specificity. Third, the possibility that some "preformed" specific cell recognition mechanism, presumably chemical in nature, may be involved in the selection of the juvenile connections cannot be excluded, although contrary to what chemospecificity theories suggest, it probably does not play the primary role.

Possibly the most important lesson of the developmental studies of cortical connectivity is the demonstration that no unique mechanism is responsible for the final shape of a neural circuit. Possibly, like the brain itself, the mode of its development is the piecemeal product of pragmatic evolution. In the course of phylogenesis, the complexity of regulatory developmental mechanisms may have increased because new mechanisms were introduced and were maintained when and where they had adaptive value.

The isolation and analysis of fragments of the whole developmental process is certainly a legitimate procedure, as far as these pieces can then be modeled in the complex network of causal interactions they belong to. This consideration may warn against the continually resurrected reductionistic vs holistic sectarianism in the field of science.

Acknowledgments. This work was supported by the Swiss National Foundation Grant No. 3.359-0.86. I am grateful to Mrs. C. Vaclavik for typing the manuscript and to Mr. S. Daldoss for his help with the illustrations.

8 References

Adams CE, Mihailoff GA, Woodward DJ (1983) A transient component of the developing corticospinal tract arises in visual cortex. Neurosci Lett 36:243–248

Alberch P (1984) A return to the Bauplan. Behav Brain Sci 7:332

Antonini A, Berlucchi G, Marzi CA, Sprague J (1979) Importance of corpus callosum for visual receptive fields of single neurons in cat superior colliculus. J Neurophysiol 42: 137–152

Arsenio Nunes ML, Sotelo C (1985) Development of the spinocerebellar system in the postnatal rat. J Comp Neurol 237:291–306
Assal F, Melzer P, Innocenti GM (1989) Functional analysis of a visual cortical circuit resembling human microgyria. Eur J Neurosci [Suppl] 2:256
Bagnoli P, Porciatti V, Fontanesi G, Sebastiani L (1987) Morphological and functional changes in the retinotectal system of the pigeon during the early posthatching period. J Comp Neurol 256:400–411
Bates CA, Killackey HP (1984) The emergence of a discretely distributed pattern of corticospinal projection neurons. Dev Brain Res 13:265–273
Bennett MR (1987) Trophic interactions at synapses and the survival of central neurons. Pont Acad Sci Scr Varia 59:141–216
Berbel P, Innocenti GM (1988) The development of the corpus callosum in cats: a light- and electron-microscopic study. J Comp Neurol 276:132–156
Berbel P, Innocenti GM, Prieto JJ, Kraftsik R (1989) A quantitative study on the development of the cytoskeleton of callosal axons in cats. Eur J Neurosci [Suppl] 2:47
Berlucchi G (1986) Electroencephalographic studies in "split brain" cats. Electroencephalog Clin Neurophysiol 20:348–356
Berlucchi G (1972) Anatomical and physiological aspects of visual functions of corpus callosum. Brain Res 37:371–392
Berlucchi G, Rizzolatti G (1968) Binocularly driven neurons in visual cortex of split-chiasm cats. Science 159:308–310
Berlucchi G, Gazzaniga MS, Rizzolatti G (1967) Microelectrode analysis of transfer of visual information by the corpus callosum. Arch Ital Biol 105:583–596
Berman NE, Payne BR (1983) Alterations in connections of the corpus callosum following convergent and divergent strabismus. Brain Res 274:201–212
Berrebi AS, Fitch RH, Ralphe DL, Denenberg JO, Friedrich VL Jr, Denenberg VH (1988) Corpus callosum: region-specific effects of sex, early experience and age. Brain Res 438:216–224
Billings-Gagliardi S, Chan-Palay V, Palay SL (1974) A review of lamination in area 17 of the visual cortex of Macaca mulatta. J Neurocytol 3:619–629
Blakemore C (1978) Maturation and modification in the developing visual system. In: Held R, Lebowitz HW, Teuner HL (eds) Handbook of sensory physiology, vol 8. Springer Berlin Heidelberg New York, pp 377–436
Bower AJ, Payne JN (1987) An ipsilateral olivocerebellar pathway in the normal neonatal rat demonstrated by the retrograde transport of True blue. Neurosci Lett 78:138–144
Bruce LL, Stein BE (1988) Transient projections from the lateral geniculate to the posteromedial lateral suprasylvian visual cortex in kittens. J Comp Neurol 278:287–302
Brugge JF, Feng JZ, Reale RA, Chan JCK (1983) Topographic distribution of primary auditory cortical field (AI) callosal neurons in cats reared with unilateral or bilateral cochlear destruction. Neurosci Abs 9:377
Cajal SR y (1909, 1911) Histologie du système nerveux de l'homme et des vertébrés. Maloine, Paris, pp 655–664
Caminiti R, Innocenti GM (1981) The postnatal development of somatosensory callosal connections after partial lesions of somatosensory areas. Exp Brain Res 42:53–62
Caminiti R, Zeger S, Johnson P, Urbano A, Georgopoulos AP (1985) Corticocortical efferent systems in the monkey: a quantitative spatial analysis of the tangential distribution of cells of origin. J Comp Neurol 241:405–419
Catsicas S, Thanos S, Clarke PGH (1987) Major role for neuronal death during brain development: refinement of topographical connections. Proc Natl Acad Sci USA 84:8165–8168

Catsman-Berrevoets CE, Kuypers HGJM (1978) Differential laminar distribution of corticothalamic neurons projecting to the VL and the center median. An HRP study in the cynomolgus monkey. Brain Res 154:359–365
Caviness VS (1977) Reeler mutant mouse: a genetic experiment in developing mammalian cortex. In: Cowan WM, Ferrendell JA (eds) Approaches to the cell biology of neurons. Society for Neuroscience, Bethesda MD, pp 27–46
Caviness VS, Frost DO (1980) Tangential organization of thalamic projections to the neocortex in the mouse. J Comp Neurol 194:335–367
Chalupa LM, Killackey HP (1989) Process elimination underlies ontogenetic change in the distribution of callosal projection neurons in the postcentral gyrus of the fetal rhesus monkey. Proc Natl Acad Sci USA 86:1076–1079
Chalupa LM, Killackey HP, Snider CJ, Lia B (1989) Callosal projection neurons in area 17 of the fetal rhesus monkey. Dev Brain Res 46:303–308
Changeux JP, Danchin A (1976) Selective stabilisation of developing synapses as a mechanism for the specification of neuronal networks. Nature 264:705–712
Chino YM, Shansky MS, Jankowski WL, Banser FA (1983) Effects of rearing kittens with convergent strabismus on development of receptive-field properties in striate cortex neurons. J Neurophysiol 50:265–285
Choudhury BP, Whitteridge D, Wilson ME (1965) The function of the callosal connections of the visual cortex. Q J Exp Physiol L:214–219
Chow KL, Baumbach HD, Lawson R (1981) Callosal projections of the striate cortex in the neonatal rabbit. Exp Brain Res 42:122–126
Chun JJM, Nakamura MJ, Shatz CJ (1987) Transient cells of the developing mammalian telencephalon are peptide-immunoreactive neurons. Nature 325:617–620
Clarke PGH (1981) Chance, repetition, and error in the development of normal nervous systems. Perspect Biol Med 25:2–19
Clarke PGH, Cowan WM (1976) The development of the isthmo-optic tract in the chick, with special reference to the occurrence and correction of developmental errors in the location and connections of isthmo-optic neurons. J Comp Neurol 167:143–164
Clarke S, Innocenti GM (1986) Organization of immature intrahemispheric connections. J Comp Neurol 251:1–22
Clarke S, Innocenti GMI (1990) Auditory neurons with transitory axons to visual areas form short permanent projections. Eur J Neurosci 2:227–242
Clarke S, Kraftsik R, Van der Loos H, Innocenti GM (1989) Forms and measures of adult and developing human corpus callosum: is there sexual dimorphism? J Comp Neurol 280:213–230
Cook JE, Horder TJ (1977) The multiple factors determining retinotopic order in the growth of optic fibres into the optic tectum. Philos Trans R Soc Lond [Biol] 278:261–276
Cowan WM (1973) Neuronal death as a regulative mechanism in the control of cell number in the nervous system. In: Rockstein M (ed) Development and aging in the nervous system. Academic, New York, pp 19–41
Cowan WM, Fawcett JW, O'Leary DDM, Stanfield BB (1984) Regressive events in neurogenesis. Science 225:1258–1265
Cragg BG (1975) The development of synapses in the visual system of the cat. J Comp Neurol 160:147–166
Cusick CG, Lund RD (1982) Modification of visual callosal projections in rats. J Comp Neurol 212:385–398
Cynader M, Leporé F, Guillemot JP (1981) Inter-hemispheric competition during postnatal development. Nature 290:139–140
D'Amato CJ, Hicks SP (1978) Normal development and post-traumatic plasticity of corticospinal neurons in rats. Exp Neurol 60:557–569

Dehay C, Bullier J, Kennedy H (1984) Transient projections from the frontoparietal and temporal cortex to areas 17, 18 and 19 in the kitten. Exp Brain Res 57:208–212
Dehay C, Kennedy H, Bullier J (1988a) Characterization of transient cortical projections from auditory, somatosensory, and motor cortices to visual areas 17, 18 and 19 in the kitten. J Comp Neurol 272:68–89
Dehay C, Kennedy H, Bullier J, Berland M (1988b) Absence of interhemispheric connections of area 17 during development in the monkey. Nature 331:348–350
Dehay C, Horsburgh G, Berland M, Killackey H, Kennedy H (1989) Maturation and connectivity of the visual cortex in monkey is altered by prenatal removal of retinal input. Nature 337:265–267
Distel H, Holländer H (1980) Autoradiographic tracing of developing subcortical projections of the occipital region in fetal rabbits. J Comp Neurol 192:505–518
Ebbesson SOE (1984) Evolution and ontogeny of neural circuits. Behav Brain Sci 7:321–331
Edelman GM (1987) Neural Darwinism. The theory of neuronal group selection, Basic, New York, pp 371
Elberger AJ, Smith EL, White JM (1983) Spatial dissociation of visual inputs alters the origin of the corpus callosum. Neurosci Lett 35:19–24
Feinberg I (1982/83) Schizophrenia: caused by a fault in programmed synaptic elimination during adolescence? J Psychiatr Res 17:319–334
Feng JZ, Brugge JF (1983) Postnatal development of auditory callosal connections in the kitten. J Comp Neurol 214:416–426
Ferrer I, Hernandez-Martí M, Bernet E, Calopa M (1989) Formation and growth of the cerebral convolutions. II. Cell death in the gyrus suprasylvius and adjoining sulci in the cat. Dev Brain Res 45:303–308
Figlewicz DA, Gremo F, Innocenti GM (1988) Differential expression of neurofilament subunits in the developing corpus callosum. Brain Res 42:181–189
Finlay BL, Slattery M (1983) Local differences in the amount of early cell death in neocortex predict adult local specializations. Science 219:1349–1351
Fleischhauer K, Schlüter G (1970) Ueber das postnatale Wachstum des Corpus callosum der Katze (Felis domestica). Z Anat Entwicklungsgesch 132:228–239
Floeter MK, Jones EG (1985) The morphology and phased outgrowth of callosal axons in the fetal rat. Dev Brain Res 22:7–18
Frégnac Y, Imbert M (1984) Development of neuronal selectivity in primary visual cortex of cat. Physiol Rev 64:325–434
Frost DO (1984) Axonal growth and target selection during development: retinal projections to the ventrobasal complex and other "nonvisual" structures in neonatal Syrian hamsters. J Comp Neurol 230:576–592
Frost DO (1986) Development of anomalous retinal projections to nonvisual thalamic nuclei in Syrian hamsters: a quantitative study. J Comp Neurol 252:95–105
Frost DO, Metin C (1985) Induction of functional retinal projections to the somatosensory system. Nature 317:162–164
Frost DO, Moy YP (1989) Effects of dark rearing on the development of visual callosal connections. Exp Brain Res 78:203–213
Gadisseux JF, Evrard P, Misson JP, Caviness VS (1987) Fascicular organization of the radial glial fibers in developing murine neocortex; an immunohistochemical study combined with [^{3}H]-thymidine autoradiography. Soc Neurosci Abstr 13:1117
Galli L, Maffei L (1988) Spontaneous impulse activity of rat retinal ganglion cells in prenatal life. Science 242:90–91
Gilbert CD, Kelly JP (1975) The projections of cells in different layers of the cat's visual cortex. J Comp Neurol 163:81–106

Gilbert CD, Wiesel TN (1981) Laminar specialization and intracortical connections in cat primary visual cortex. In: Schmitt FO et al. (eds) The organization of the cerebral cortex, MIT Press, Cambridge MA, pp 163–191

Glassman RB (1987) An hypothesis about redundancy and reliability in the brains of higher species: analogies with genes, internal organs, and engineering systems. Neurosci Biobehav Rev 11:275–285

Glickstein M, Rizzolatti G (1984) Francesco Gennari and the structure of the cerebral cortex. Trends Neurosci 7:464–467

Goldman-Rakic PS (1982) Neuronal development and plasticity of association cortex in primates. Neurosci Res Prog Bull 20:520–532

Goldman-Rakic PS, Schwartz ML (1982) Interdigitation of contralateral and ipsilateral columnar projections to frontal association cortex in primates. Science 216:755–757

Gravel C, Hawkes R (1990) Maturation of the corpus callosum of the rat: I. Influence of thyroid hormones on the topography of callosal projections. J Comp Neurol 291:128–146

Graybiel AM, Berson DM (1981) On the relation between transthalamic and transcortical pathways in the visual system. In: Schmitt FO et al. (eds) The organization of the cerebral cortex, MIT Press, Cambridge, pp 285–319

Guadano-Ferraz A, Riederer BM, Innocenti GM (1990) Developmental changes in the heavy subunit of neurofilaments in the corpus callosum of the cat. In: Developmental Brain Research (in press)

Guillery RW (1988) Competition in the development of the visual pathways. In: Parnavelas JG et al. (eds) The making of the nervous system, Oxford University Press. Oxford, pp 356–379

Hamburger V, Levi-Montalcini R (1949) Proliferation, differentiation and degeneration in the spinal ganglia of the chick embryo under normal and experimental conditions. J Exp Zool 3:457–501

Heffner CD, Lumsden AGS, O'Leary DDM (1990) Target control of collateral extension and directional axon growth in the mammalian brain. Science 247:217–220

Heimer L, Ebner FF, Nauta WJH (1967) A note on the termination of commissural fibers in the neocortex. Brain Res 5:171–177

Heumann D, Leuba G (1983) Neuronal death in the development and aging of the cerebral cortex of the mouse. Neuropathol Appl Neurobiol 9:297–311

Hubel DH (1982) Exploration of the primary visual cortex, 1955–78. Nature 299:515–524

Hubel DH, Livingstone MS (1987) Segregation of form, color, and stereopsis in primate area 18. J Neurosci 7:3378–3415

Hubel DH, Wiesel TN (1967) Cortical and callosal connections concerned with the vertical meridian of visual fields in the cat. J Neurophysiol 30:1561–1573

Hubel DH, Wiesel TN (1969) Anatomical demonstration of columns in the monkey striate cortex. Nature 221:747–750

Hubel DH, Wiesel TN, LeVay S (1977) Plasticity of ocular dominance columns in monkey striate cortex. Philos Trans R Soc Lond [Biol] 78:377–409

Huttenlocher PR (1979) Synaptic density in human frontal cortex-developmental changes and effects of aging. Brain Res 163:195–205

Huttenlocher PR, Courten de C, Garey LJ, Van der Loos H (1982) Synaptogenesis in human visual cortex-evidence for synapse elimination during normal development. Neurosci Lett 33:247–252

Imig TJ, Brugge JF (1978) Sources and terminations of callosal axons related to binaural and frequency maps in primary auditory cortex of the cat. J Comp Neurol 182:637–660

Innocenti GM (1980) The primary visual pathway through the corpus callosum: morphological and functional aspects in the cat. Arch Ital Biol 118:124–188

Innocenti GM (1981a) Growth and reshaping of axons in the establishment of visual callosal connections. Science 212:824–827
Innocenti GM (1981b) Transitory structures as substrate for developmental plasticity of the brain. Dev Neurosci 13:305–333
Innocenti GM (1986) General organization of callosal connections in the cerebral cortex. In: Jones EG, Peters A (eds) Cerebral cortex, vol 5. Plenum, New York, pp 291–353
Innocenti GM (1987) Structure, specificity and discontinuities in the development of corticocortical connections. Pont Acad Sci Scr Varia 59:91–108
Innocenti GM (1988) Loss of axonal projections in the development of the mammalian brain. In: Parnavelas JG et al. (eds) The making of the nervous system. Oxford University Press, Oxford, pp 319–339
Innocenti GM, Berbel P (1990a) Analysis of an experimental cortical network: i) architectonics of areas 17 and 18 after neonatal injections of ibotenic acid; similarities with human microgyria. J Neur Transplant (in press)
Innocenti GM, Berbel P (1990b) Analysis of an experimental cortical network. ii) connections of areas 17 and 18 after neonatal injections of ibotenic acid. J Neur Transplant (in press)
Innocenti GM, Caminiti R (1980) Postnatal shaping of callosal connections from sensory areas. Exp Brain Res 38:381–394
Innocenti GM, Clarke S (1984a) Bilateral transitory projection to visual areas from auditory cortex in kittens. Dev Brain Res 14:143–148
Innocenti GM, Clarke S (1984b) The organization of immature callosal connections. J Comp Neurol 230:387–390
Innocenti GM, Frost DO (1978) Visual experience and the development of the efferent system to the corpus callosum. Neurosci Abst 4:475
Innocenti GM, Frost DO (1979) Effects of visual experience on the maturation of the efferent system to the corpus callosum. Nature 280:231–234
Innocenti GM, Frost DO (1980) The postnatal development of visual callosal connections in the absence of visual experience or of the eyes. Exp Brain Res 39:365–375
Innocenti GM, Fiore L, Caminiti R (1977) Exuberant projection into the corpus callosum from the visual cortex of newborn cats. Neurosci Lett 4:237–242
Innocenti GM, Koppel H, Clarke S (1983a) Transitory macrophages in the white matter of the developing visual cortex. I. Light and electron microscopic characteristics and distribution. Dev Brain Res 11:39–53
Innocenti GM, Clarke S, Koppel H (1983b) Transitory macrophages in the white matter of the developing visual cortex. II. Development and relations with axonal pathways. Dev Brain Res 11:55–66
Innocenti GM, Frost DO, Illes J (1985) Maturation of visual callosal connections in visually deprived kittens: a challenging critical period. J Neurosci 5:255–267
Innocenti GM, Clarke S, Kraftsik R (1986) Interchange of callosal and association projections in the developing visual cortex. J Neurosci 6:1384–1409
Innocenti GM, Berbel P, Melzer P (1987) Stabilization of transitory corticocortical projections following lesions provoked by neonatal ibotenic injections. Neurosci Abstr [Suppl] 22:S227
Innocenti GM, Berbel P, Clarke S (1988) Development of projections from auditory to visual areas in the cat. J Comp Neurol 272:242–259
Iriki A, Nozaki S, Nakamura Y (1988) Feeding behavior in mammals: corticobulbar projection is reorganized during conversion from sucking to chewing. Dev Brain Res 44:189–196
Ivy GO, Killackey HP (1981) The ontogeny of the distribution of callosal projection neurons in the rat parietal cortex. J Comp Neurol 195:367–389

Sretavan DW, Shatz CJ (1987) Axon trajectories and pattern of terminal arborization during the prenatal development of the cat's retinogeniculate pathway. J Comp Neurol 255:386:400

Sretavan DW, Shatz CJ, Stryker MP (1988) Modification of retinal ganglion cell axon morphology by prenatal infusion of tetrodotoxin. Nature 336:468–471

Stanfield BB, O'Leary DDM (1985a) Fetal occipital cortical neurones transplanted to the rostral cortex can extend and maintain a pyramidal tract axon. Nature 313:135–137

Stanfield BB, O'Leary DDM (1985b) The transient corticospinal projection from the occipital cortex during the postnatal development of the rat. J Comp Neurol 238:236–248

Stanfield BB, O'Leary DDM, Fricks C (1982) Selective collateral elimination in early postnatal development restricts cortical distribution of rat pyramidal tract neurones. Nature 198:371–373

Stein BE, McHaffie JG, Harting JK, Huerta MF, Hashikawa T (1985) Transient tectogeniculate projections in neonatal kittens: an autoradiographic study. J Comp Neurol 239: 402–412

Sulston JE, Horvitz HR (1977) Post-embryonic cell lineage of the nematode Caenorhabditis elegans. Dev Biol 56:110–156

Susic V, Kovacevic R (1974) Sleep patterns in chronic split-brain cats. Brain Res 65: 427–441

Swadlow HA (1983) Efferent systems of primary visual cortex: a review of structure and function. Brain Res Rev 6:1–24

Symonds LL, Rosenquist AC (1984) Corticocortical connections among visual areas in the cat. J Comp Neurol 229:1–38

Thanos S, Bonhoeffer F (1986) Course corrections of deflected retinal axons on the tectum of the chick embryo. Neurosci Lett 72:31–36

Thanos S, Bonhoeffer F, Rutishauser U (1984) Fiber-fiber interaction and tectal cues influence the development of the chicken retinotectal projection. Proc Natl Acad Sci USA 81:1906–1910

Thong IG, Dreher B (1986) The development of the corticotectal pathway in the albino rat. Dev Brain Res 25:227–238

Tolbert DL (1987) Intrinsically directed pruning as a mechanism regulating the elimination of transient collateral pathways. Dev Brain Res 33:11–21

Tolbert DL (1989) Somatotopically organized transient projections from the primary somatosensory cortex to the cerebellar cortex. Dev Brain Res 45:113–127

Tolbert DL, Panneton WM (1983) Transient cerebrocerebellar projections in kittens: postnatal development and topography. J Comp Neurol 221:216–228

Tolbert DL, Panneton WM (1984) The transience of cerebrocerebellar projections is due to selective elimination of axon collaterals and not neuronal death. Dev Brain Res 16: 301–306

Tolbert DL, Dunn RC, Vogler GA (1984) The postnatal development of corticotrigeminal projections in the cat. J Comp Neurol 228:478–490

Tremblay F, Ptito M, Lepore F, Miceli D, Guillemot JP (1987) Distribution of visual callosal projection neurons in the siamese cat: an HRP study. J Hirnforsch 28:491–503

Tsumoto T, Suda K, Sato H (1983) Postnatal development of corticotectal neurons in the kitten striate cortex: a quantitative study with the horseradish peroxidase technique. J Comp Neurol 219:88–99

Van Essen DC (1985) Functional organization of primate visual cortex. In: Peters A, Jones EG (eds) Cerebral cortex, vol 3, Plenum, New York, pp 259–320

Van Essen DC, Maunsell JHR (1983) Hierarchical organization and functional streams in the visual cortex. Trends Neurosci 6:370–375

Walsh C, Cepko CL (1988) Clonally related cortical cells show several migration patterns. Science 241:1342–1345
Weiss P (1955) Nervous system (neurogenesis). In: Analysis of development. Section VII, Special vertebrate organogenesis. Harper, New York, pp 346–401
Wiesel TN (1982) Postnatal development of the visual cortex and the influence of environment. Nature 299:583–591
Williams RW, Rakic P (1985) Dispersion of growing axons within the optic nerve of the embryonic monkey. Proc Natal Acad Sci USA 82:3906–3910
Winfield DA (1981) The postnatal development of synapses in the visual cortex of the cat and the effects of eyelid closure. Brain Res 206:166–171
Wise SP, Jones EG (1976) The organization and postnatal development of the commissural projection of the rat somatic sensory cortex. J Comp Neurol 168:313–344
Wise SP, Jones EG (1978) Developmental studies of thalamocortical and commissural connections in the rat somatic sensory cortex. J Comp Neurol 175:187–208
Wise SP, Hendry SHC, Jones EG (1977) Prenatal development of sensorimotor cortical projections in cats. Brain Res 138:538–544
Wolpert L (1969) Positional information and the spatial pattern of cellular differentiation. J Theor Biol 25:1–47
Woolsey TA, Van der Loos H (1970) The structural organization of layer IV in the somatosensory region (SI) of mouse cerebral cortex. The description of a cortical field composed of discrete cytoarchitectonic units. Brain Res 17:205–242
Yakovlev PI, Lecours A-R (1967) The myelogenetic cycles of regional maturation of the brain. In: Minkowski A (ed) Regional development of the brain in early life, Blackwell, Oxford, pp 3–70

Neurocognitive Models of Information Processing and Knowledge Acquisition*

O.R. Anderson

Science and Mathematics, Columbia University Teachers College, and Biological Laboratories Lamont-Doherty Observatory of Columbia University, 525 W. 120th Street, New York, NY 10027, USA

* Lamont-Doherty Observatory Contribution No. 4690.

1 Introduction

The quest for a comprehensive and systematic theory of human cognition is a long-established goal of psychology and the neurosciences. However, we are only beginning to acquire sufficient knowledge of brain functional neuroanatomy to comprehend the magnitude of the task, and begin to place some boundary conditions on theory building. The complexity of this quest becomes ever more apparent as additional insights are gained into the neuroanatomy and physiology of the brain. Recent advances in neurophysiology and current interdisciplinary research on neurocognitive processes may improve our understanding of some basic aspects of cognition. Cognition is defined here broadly as perception of sensory input, its transformation, integration into memory, and application during recall. The acquisition of knowledge in a form that enhances its stability in memory, facilitates application to new cognitive tasks, and provides a sound base for further learning is a significant aspect of human information processing. Our understanding of the cognitive processes mediating knowledge acquisition has improved substantially through the application of principles from cognitive psychology and information processing sciences. It is becoming increasingly clear that information acquisition is a dynamic process involving the active participation of the recipient whereby new information is linked to and accommodated within prior existing knowledge structures or schemata[1] (Ausubel 1963; Kintsch 1974; Anderson 1976; Novak 1977).

The connectedness and stability of information in memory, therefore, is substantially determined by the facility of the learner to mobilize relevant prior knowledge and relate it in an unambiguous and logically meaningful way to newly acquired information. This is thought to be a complex and dynamic process of transforming both existing knowledge and incoming information to make them mutually compatible and interconnected for stable storage in memory.

---

[1] The term schema is used in a variety of ways in the psychological and neuroscientific literature. It is used here as a generalized internal representation of experience, akin to a cognitive map, including knowledge structures, maps for motor activities, and expectations resulting from prior experiences.

Although a substantial amount of knowledge gain is attributable to this dynamism, associated with the accommodation of old and new information, it is equally clear that the organization of incoming information may interact strongly with this process to either enhance or diminish the effectiveness of information encoding and storage. The role of organization in sensory input during information processing is controversial and varies among psychological theorists. Behaviorists, who follow a strict empirical tradition, explain behavior largely as a set of sensory-input, contingent responses. Thus in a classical operant conditioning paradigm, only observed responses and concomitant reinforcers (sensory inputs) are major variables. In less reductionistic paradigms, internal processing of information is assumed to occur, but is driven largely by the organization of sensory input. This is sometimes categorized as a "bottom-up" model of information processing. On the other extreme, strict constructivist theorists assume that cognitive organizing activities are the major determinants of information processing and storage. Less emphasis is placed on the organization of sensory input. These models are classified as "top-down" implying that higher order internal organizing functions largely drive information processing. In middle-ground theories, the dynamic interaction between sensory input and internal organizing activity is the main focus of inquiry. This review emphasizes a middle-ground perspective and is intended for a wide audience. Some biological determinants of cognitive processing of sensory input, based on neurophysiological and cognitive psychological principles, are discussed and related to recent developments in mathematical modeling of information acquisition.

2 An Overview of Neuroscientific Concepts

Several summaries of basic neurobiological and neuropsychological concepts relevant to information processing have appeared in recent years (Szentágothai and Arbib 1975; Eccles 1973; Teyler 1977; Wittrock et al. 1977; Thompson et al. 1980a; Trevarthen 1980; Wittrock 1980; Lambert et al. 1982; Ottoson 1983; Glees 1988; Arbib 1989). Therefore, only some basic and essential information is summarized here.

2.1 Fundamental Brain Structures and Their Functions

The human brain (Fig. 1) is conventionally divided into four major regions: (1) the forebrain, dominated in mass by the convoluted cortex, (2) the midbrain, lying immediately beneath the cortical mass and approximately in the middle of the brain, composed of the anterior projections of the reticular formation and the hypothalamus/pituitary, (3) the brain stem, projecting from the dorsal surface of the brain, including major tracts of nerves exiting from the brain (efferent nerves) and incoming nerves from the periphery (afferent nerves); it also has nerve centers that regulate involuntary responses of the viscera, blood vessels, and ducted glands, and (4) the spinal cord, containing a complex network of ascending and descending nerve

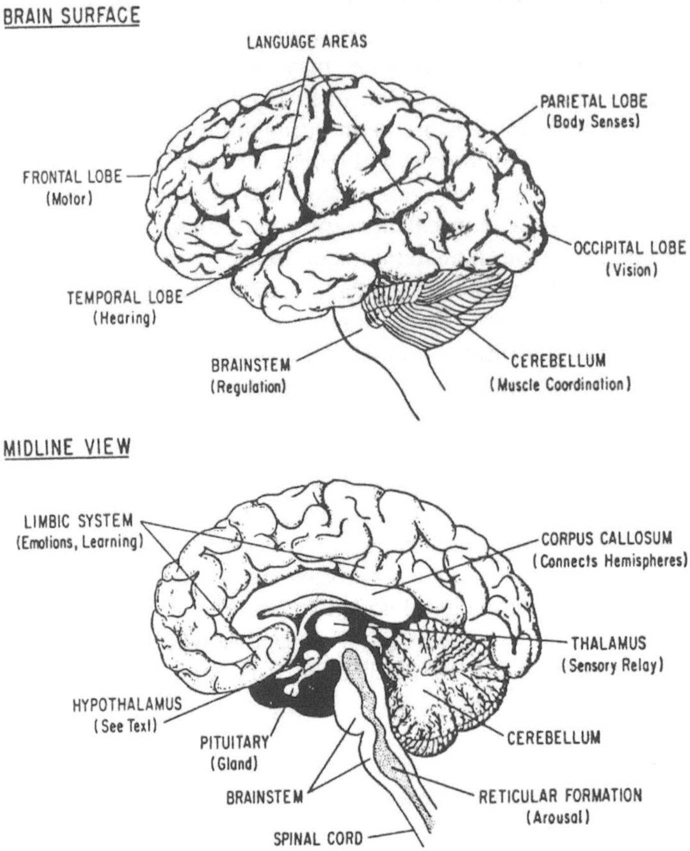

Fig. 1. Major brain structures and some of their functions. *Top*, surface view of the left hemisphere of the human brain showing the large convoluted cortex, smaller cerebellum and brainstem. *Bottom*, a section through the midline of the right hemisphere with the major areas/structures and their functions labeled. (From Teyler 1977, with permission of Prentice-Hall, Inc.)

tracts including large neurons (motoneurons) mediating muscular contraction, and smaller neurons which in part contribute to motor control of bodily movements and facilitate reflexive behavior. The cerebellum (a large convoluted bulge projecting from the brain stem) is involved in regulation of muscle activity including coordination of complex movements.

2.2 Neurons and Neuronal Networks

The fundamental unit of structure and function is the neuron, a cell mediating sensation through electrical or electrochemical responses to stimulation. The remarkable plasticity and variety of response patterns of the central nervous system is undoubtedly mediated in part by the numerous interconnecting processes (axons and dendrites) that link the neurons into complex networks. Recent research with model systems, such as simple neural circuits in invertebrates and higher organisms, promises to yield additional insights into the complicated signal processing that must occur within these networks (Kandel and Spencer 1968; Kandel 1976; Kandel and Schwartz 1982; Davis 1986; Montarolo et al. 1986; Mattheis 1989). Signal transmission within a neural network is accompanied by changes in the electrical properties of the surface membrane of the activated neurons, and by chemical molecules produced in the cell and transported within vesicles to the points of contact known as synapses (Eccles 1973; Weisenberg 1987). Internal cytoskeletal structures including microtubules and microfilaments form the scaffolding that establishes the organization of the radiating cell processes, and hence must account at a molecular level for the pattern of arborizations that determines the spatial relations among interconnected cells (Lasek et al. 1981; Changeux and Dennis 1982; Bereiter-Hahn et al. 1987). Biochemical studies using cytoskeletal disrupting compounds injected into the brain of mice indicate that intact microtubules may be essential to conditioned learning and permanence of memory (Murakami 1980). Some evidence points to the role of contractile (actin) microfilaments in changes of shape and area of synaptic contacts during synaptic strengthening associated with neural plasticity (Fifkova 1987). Shape and size of neurons vary considerably according to location in the nervous system and in relation to function. In general, as one progresses rostrally from the spinal cord toward the cortex, the network of neurons becomes more complexly organized and exhibits greater plasticity in responding. Hence, the massive pair of cerebral hemispheres dominating the top of the human brain represents an evolutionarily advanced development correlated with the remarkably diversified and complex adaptive behavior characteristic of humans and higher mammals.

2.3 Cerebral Cortex

In broad perspective, the cerebral cortex is divided into several topographical regions, each associated with a major function: (1) the frontal lobe (control of motor functions; i.e., locomotion and fine muscular activity), (2) temporal lobe (hearing and nearby language areas) and parietal lobe (mediating body senses), and (3) occipital lobe (vision). Each of these regions, however, is far more complex in organization and function than implied by this generalized categorization. The convoluted surface of the cortex increases the total area available for neurons and their interconnections thus enhancing the complex information processing of this part of the brain. The limbic system is a collection of topographically widespread, yet closely interconnected, group of structures (amygdala, hippocampus, and septum) whose functions are not fully understood, but appear to be intimately involved in learning and emotional responses. Lesions of the hippocampus, due to trauma or disease, decrease ability to store information permanently in memory, although the individual may recall events from the distant past rather lucidly. The reticular formation of the midbrain and some forebrain structures, including the amygdala, appear to control arousal of the central nervous system and mediate attention and wakefulness (Sokolov 1960, 1963; Lindsley 1961; Hobson and Brazier 1980; McGinnis and Pribram 1980). Stimulation of the reticular formation results in a diffuse excitation of the cerebral cortex. Nerve fibers project from the reticular formation in a fan-like array toward the cortical surface. The corpus callosum, immediately beneath the region of the limbic system, contains a massive tract of nerve fibers interconnecting the two cerebral hemispheres and among other functions coordinates perceptual and representational processes between the two hemispheres. Current evidence indicates that each of the hemispheres may be specialized for different functions, but not exclusively. Among other differences, the right represents visual experience and the left verbal experience, a condition known as lateralization. Differentiation of functions between the two cerebral hemispheres, and other evidence of specialized information processing in visual and auditory centers, suggests that the brain is organized in functional modules (Gazzaniga 1989). Modularity refers to the collective action of several subprocesses that produce a covert or overt predictable set of actions or behaviors. Functional modules may be represented structurally by groups of interconnected neuronal modules (Fig. 2). Integration of modular activity occurs through reciprocal innervation. The corpus callosum is a notable example; signals generated in one hemisphere can be transmitted to the other where differential analysis occurs. For example, emotional and visual states aroused in he right hemisphere can be transmitted to the left hemisphere where a more analytical and formal logical analysis is made, although there is also autonomy of cognitive control within each hemisphere (Tucker 1987). Much attention has been given to hemispheric specialization, but increasing evidence points to the importance of interhemispheric interaction in performing cognitive tasks, including variations across individuals and in relation to task requirements (Helige 1987). Within the callosal tract, there is also evidence of differentiation; among other distinctions, the posterior tract carries visual information, and the more anterior tract auditory/tactile information (Zaidel 1987; Gazzaniga 1989). Surgical transection of the posterior tract can cause disrupted verbal analysis of visual images presented to the right hemisphere, apparently due to interruption of the transfer of visual information between the two hemispheres. In general, coordinated cerebral function is achieved through a complex set of interconnecting fiber tracts that relay information among the subprocessing centers. For example, the thalamus, a part of the forebrain, serves as a relay center processing information at a lower level before it is projected toward the cerebral cortex. Interestingly, while it was once thought that each cytoarchitectonic area of the cortex received input from only one major "specific" thalamic nucleus, recent advances in tracer techniques indicate that localized regions of the cortex receive input from several thalamic nuclei including the ventral anterior mediodorsal, medial pulvinar, limitans, and suprageniculate nuclei (Kievit and Kuypers 1977; Goldman-Rakic 1988). This evidence of additional complexity in precortical projections suggests, among other sources of evidence, that the interaction of brain centers in representing experience may be less dependent on a rigid hierarchical pattern than previously supposed, and involves multiple processing pathways in parallel distributed cortical networks.

In our subsequent discussions, we will be particularly concerned with two major functions of the brain (1) the role of the forebrain in representing experience, or, as we

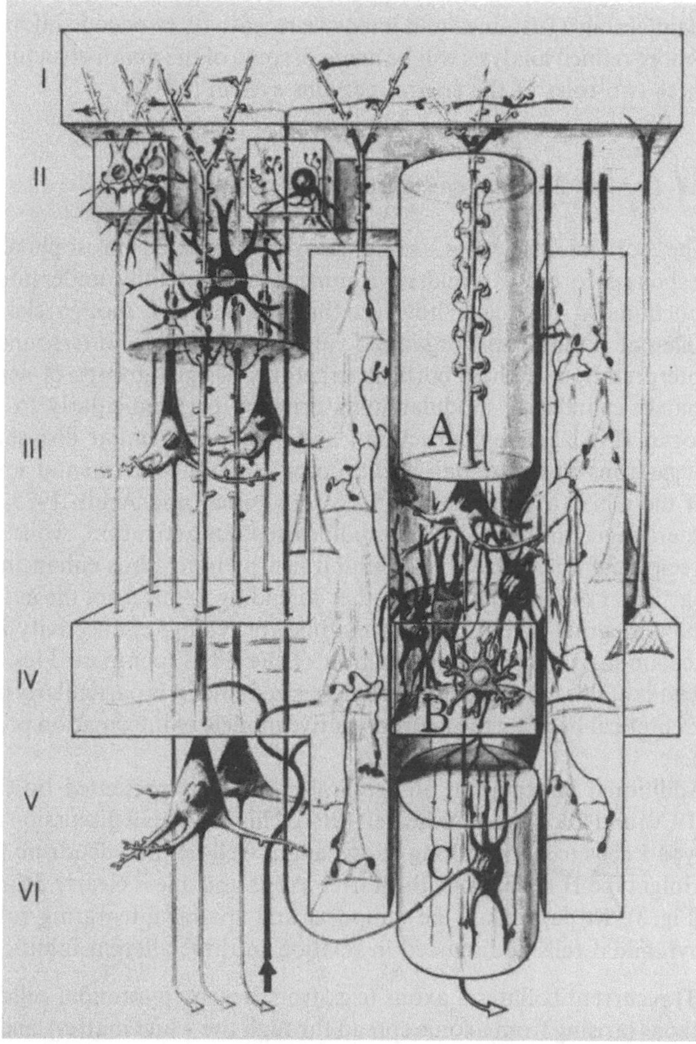

Fig. 2. Some basic neuron circuits of the neocortex delineated as modules (*cylindrical vertical units*) extending from lamina II to VI. Each module consists fundamentally of (1) a pyramidal cell (*A*) with long ascending apical dendrites that branch in lamina I, (2) Golgi type II cells (*B*, with *stippling*) that synapse extensively with dendrites of the pyramidal cell within the cylindric space and perhaps others nearby, thus increasing local excitation, and (3) basket cells (*C*, *solid black*) that make inhibitory synapses in lamina III, IV, and V with some surrounding pyramidal cells thus shaping (or partially cutting down) the primary and secondary foci of excitation. Smaller "octopus-shaped" basket cells (of small-range action) in lamina II could create a functional mosaic of smaller module size by inhibiting the small pyramidal cells of this layer. Suitable convergence of impulses (*solid arrow*) in lamina IV induces an excitatory focus in the cyclinder space (at *right center*) of this layer; this could be translated into a vertically ascending and/or descending volley through excitatory interneurons of the Golgi type II interneurons. Efferent impulses (*open arrows*), shaped by the interaction of the modular units, exit at the base of lamina VI through descending axons of the pyramidal cells. (From Szentágothai and Arbib 1975 with permission of the Massachusetts

will call it, modeling sensory phenomena, and (2) the role of the arousal centers in maintaining a sustained and intensive receptivity to encode information. Therefore, a more refined analysis will be made of some of the major structures associated with these two roles of the central nervous system (CNS).

2.4 Cortical Modules and Localization of Function

The cortical structures, while clearly implicated in sophisticated associative responses to the external environment, are not fully understood structurally or functionally. It is clear however that the neurons, though classified into several different kinds, are organized into recognizable interconnected units. One interpretation of these cortical structures assigns groups of neurons to approximately cylindrical, modular units oriented perpendicularly to the surface of the cortex (Fig. 2). These structures, and somewhat similar disc-shaped assemblages of neurons in the spinal cord, appear to be a fundamental textural component of the CNS (Mountcastle 1957; Szentágothai and Arbib 1975; Jones 1981). The interconnections within the modules sustain activation, while cross-connections (reciprocal innervation) may inhibit nearby units, thus enhancing the activity of a particular module. However, other spreading axons from the modular components may integrate activity among modules providing connectivity in responding and accounting for coordinated action of the CNS (compare Figs. 2,3, and 4c). The details of this model have been emphasized since it is particularly useful as part of the theoretical base for the neurocognitive models of information processing presented here.

Additional information on cerebral modules is presented by Chow and Leiman (1970) and Globus and Scheibel (1967). This includes a discussion of the role of Golgi type I cells (cells with long axons and a well-defined dendritic field) compared to Golgi type II neurons (with shorter axons and a less clearly defined dendritic field) (Fig. 3). In particular, the temporal and spatial integrating role of Golgi type I pyramidal cells is discussed in relation to three afferent inputs:

(1) recurrent collateral axons (e.g., from nearby pyramidal cells), (2) associational axons (arising from axonic spread through the white matter), and (3) collosal axons (e.g., mediating interhemispheric integration). Within a module, each pyramid combines its own unique sample of *extracortical* synaptic drive (a function of the length of the apical dendrite and concomitant density of afferent synapses) with a fixed level of *intracortical* synaptic drive, a function of the width of the dendritic skirt and apical arch (Globus and Scheibel 1967). The afferent innervation is diffused among several neurons resulting in overlapping fields of activation and greater complexity in neuronal functional organization (Fig. 4). Depending on the source of input, pyramidal cells may be excited by more than one afferent pathway, but owing to the different pattern of afferent synapses among sets of cells, only those neurons sharing a common afferent source of innervation will be activated at a given time (Fig. 4C). This is consistent with the view of Abeles (1982; 1988) presented later and of other theorists who propose models of cortical integration based on activation of groups of neurons, sometimes organized in overlapping

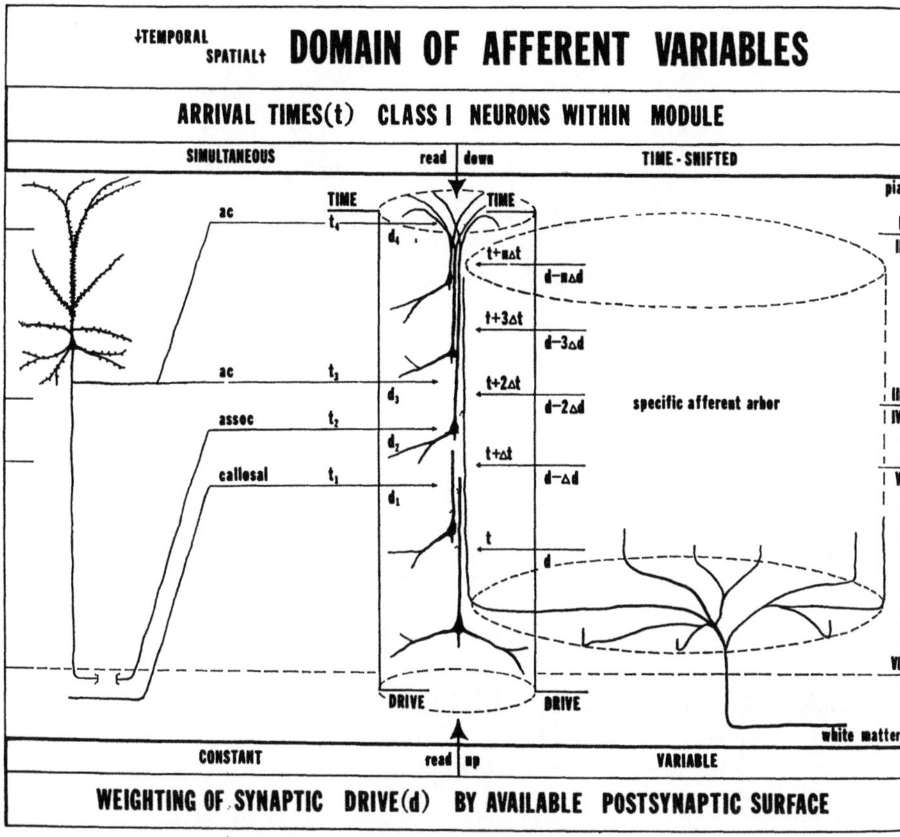

Fig. 3. Temporal and spatial domains of afferent variables. As derived solely from the geometry of the pre- and post-synaptic elements, this diagram illustrates some of the possible limits of the afferent patterns imposed by Class I neuronal alignment within the cortical module. For the temporal domains of afferent variables one reads *down*; for spatial, *up*. For temporal factors, the *left-hand panel* indicates that recurrent collateral, associational, and callosal axons, all arising from pyramids, cross the module in such a manner that they synapse on all the dendritic elements at that level of the class I modules. The *right-hand panel* shows a twig of the specific afferent arbor arising from the lateral geniculate body coursing along apical dendrites, presenting to them identical action potential trains, but at successively delayed times $t + n\Delta d$ (time-shifted) as it nears layer I. Reading up for the spatial factors, the weighting of synaptic drive or relative effectiveness of afferent systems is suggested by the available post-synaptic surface. On the *left*, collateral associational and callosal axons cross a constant diameter of the module, and therefore have a constant amount of post-synaptic surface to affect, indicated by the linear drive series d_1, d_2, d_3, d_4. On the *right*, the twig from the specific afferent arbor has the greatest area of post-synaptic surface available to it on deep pyramids with successively less as it nears layer I as indicated by "$d - n\Delta d$." (From Globus and Scheibel 1967, with permission of the Wistar Institute)

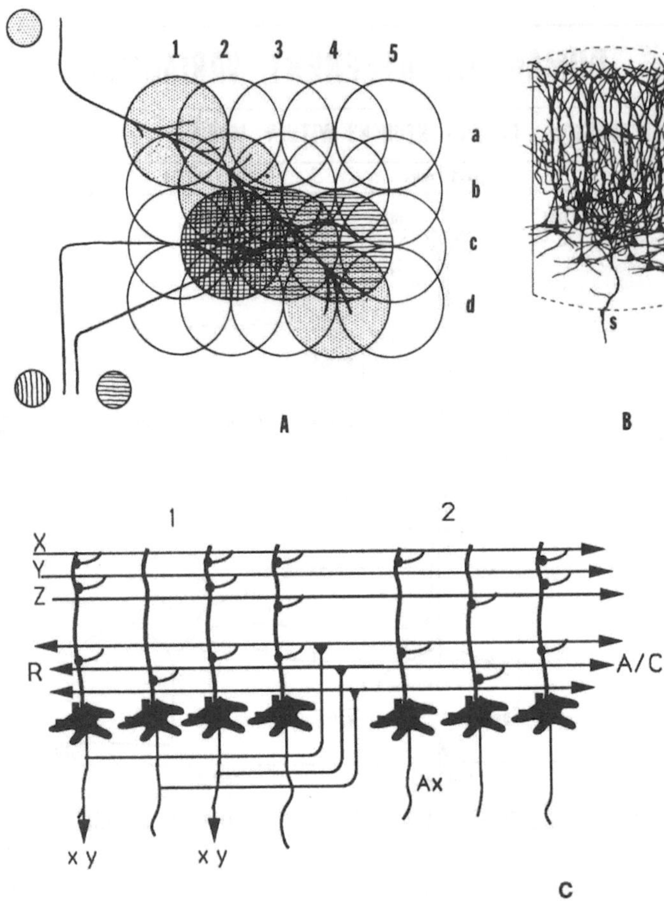

Fig. 4. A Schematic model of overlapping cortical modules forming a matrix with three afferent pathways (*filled circles*). Each module is conceived as delimited by the terminal domain generated by a single specific sensory afferent, *s*, as shown in **B**. The three afferents of different sources and different terminal extent are thought to produce a complex pattern of interactions as suggested by the interweaving and overlapping of the three patterns which represent their terminal domains. In this portion of the cortex, it is clear that the cells of module *1a* are exposed to the simplest input (a single afferent) while most of those of module *2c* are exposed to all three inputs. **C** A composite diagram synthesized from several sources (Chow and Leiman 1970; Scheibel and Scheibel from Chow and Leiman 1970; Kohonen et al. 1981; Lynch 1986) shows the vertical profile of groups of neurons (e.g., modules as in Fig. 2) with long apical dendrites and a single basal axon (*Ax*), receiving multiple parallel innervation along the apical dendritic axis. Some of the synaptic connections originate from sensory afferent neurons (*X, Y, Z*); others are from localized, recurrent axons (*R*) from the innervated group of neurons, and from more distant sources via commisural/association neurons (*A/C*). Through feedback effects, and the activation of groups of neurons with similar synaptic patterns of multiple input, only certain combinations of output occur. For example, in set 1, neurons with output marked *xy* receive synaptic input by afferents *X* and *Y*, and also have recurrent feedback synapses strengthening the input. Thus, the system, forms a combinatorial array to filter or select input to produce a specific output signal (e.g., *x*, *y*). See Fig. 5 for a schematic overview of broader cortical connections. (Models **A, B** by Scheibel A.B. and Scheibel M.E. from Chow and Leiman 1970, with permission of the massachusetts Institute of Technology Press)

sets, thus increasing the diversity and volume of information that can be processed by the cerebral cortex.

2.5 Cortical Integration

As a complement to the role of cortical modules, Braitenberg (1977) and others (Rakic and Singer 1988; Arbib 1989) focus on the widespread integrative, or internal organizing capacity of the cerebral cortex. Braitenberg notes that the cerebral cortex (the major integrative center of the brain mediating higher-level information processing) is symmetrically organized within any plane, but markedly asymmetrical in a vertical section (Fig. 5). While fiber connections are distributed uniformly in all directions within a given plane of the cerebral cortex, the "up-down" direction is distinctively asymmetric. Very different types of nerve fibers extend in the two opposite directions (i.e., upward toward the surface of the cortex versus downward toward the underlying tissue). Moreover, most neurons of the cortex exhibit a definite functional and morphological polarization in the vertical direction. This is reflected in the asymmetry of the pyramidal cells (Figs. 2–5). Long apical, dendrites extend toward, and branch within the surface layers of the cortex, whereas basal dendrites are arranged fundamentally as a sphere around the body of the cell. A main axon emerges from the base of the neuron and sends extensions laterally and at some distance vertically toward the upper surface of the cortex (Fig. 5). This is only a skeletal view of cortical organization to highlight some major organizational

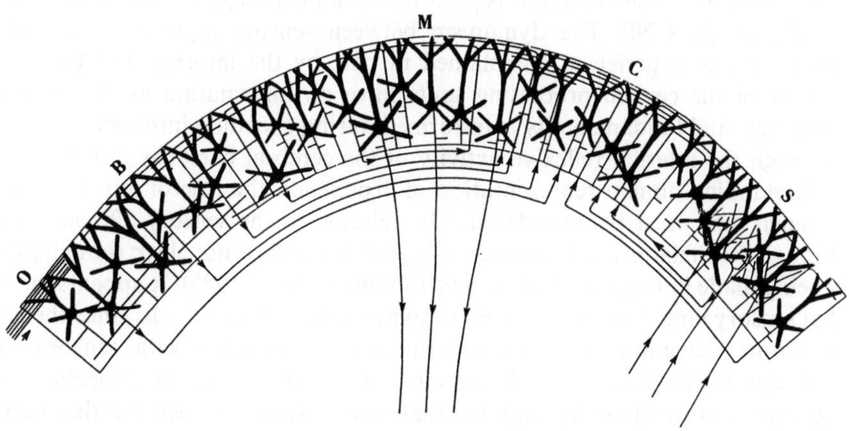

Fig. 5. An outline of some major cells and their pattern of innervation in the "skeleton cortex." There are some exceptional regions: *O*, the olfactory cortex, where the afferent fibers enter the uppermost layer; *M*, the motor cortex, which sends output fibers to the motor periphery of the nervous system; *S*, the primary sensory areas that receive a special set of afferent fibers. The general case is the cortex with long-range, ametric subcortical connections and with short-range, metrically dependent, intracortical connections (A and B system in Braitenberg's terminology). *B*, axon collaterals forming the B-system; C, convergence of fibers from the whole cortex onto a small region (A-system). (From Braitenberg 1977)

patterns. Other significant cells, such as glial, stellate, and Martinotti cells are not included in this generalized analysis. Glial cells, which surround neurons and serve a supportive and trophic role, among other cellular constituents of the cortex contribute to the total integrated activity of the system. The lack of further discussion at this point is due only to the emphasis on major structural relations among neurons.

The general pattern of symmetry within a plane and asymmetry vertically suggests that the fundamental cortical operations on sensory input involve an interaction between layers of the cortex (Braitenberg 1977). Some layers serve as input and others as output stations. This is in contrast to the popular view of the cortex as being composed of sensory, motor and associational regions. In this traditional view, incoming information reaches the cortex within the sensory regions, traversing the associational regions horizontally and finally leaving the cortex by descending fiber tracts in the motor cortex. The major integrative function (reflexive mode) of the cerebral cortex is further reflected in the ratio of sensory input fibers to those within the cortex. All of the c. 10^{10} neurons in the human cortex have axons that emerge from the gray matter, but most of these reenter the cortex, so that the number of input fibers is again of the same order of magnitude. By comparison, the number of fibers in the sensory channels is c. 10^6. Hence, the ratio of intracortical fibers to sensory input fibers spans four orders of magnitude. The numerically large network of cross-linking fibers within the cerebral cortex (Fig. 5) implies that a major component of cortical activity is processing of information generated within the cortex. This concept of the cerebral cortex as a fundamentally self-regulating, internally reflexive system is consistent with modern constructivist concepts, i.e., that representation of sensory input is profoundly influenced by the internal organizing activity of the CNS. The dynamism between sensory input and internal representation of experience is explained in part by the impress that the reflexive activity of the central processing units have on information as it is selectively registered, and fundamentally transformed during encoding into memory. Clearly, although the internal reflexive activity of the cerebral cortex is a major issue in understanding human representation of experience, the role of input and output channels also must be considered. The relation of the sensory afferents, relayed through the thalamus, to the system of pyramidal cells is not clear; though progress is being made in mapping fiber tracts (Goldman-Rakic 1988). It appears, however, that sensory input has direct access to some of the pyramidal cells since the activity of feature detectors has been experimentally correlated with sensory events. Although it was once believed that most, if not all, of afferent projections to the association cortex pass through the thalamus, it is now recognized that there is a wealth of nonthalamic subcortical projections to the cerebral cortex (reviewed in Avendaño and Llamas 1984). These include projections from telencephalic structures (e.g., claustrum and amygdala), and more distant regions such as the pontine tegmentum, and the substantia nigra and ventral tegmental area. Investigation of "direct versus indirect" sensory innervation of cortical neurons raises the important question of the extent of "peripheral" versus "central" processing of sensory input during information analysis and storage. Indeed, within the cortex proper, there may be less specific processing centers that feed into, and are reciprocally controlled by higher, more task-specific assemblages of neurons. This is partially indicated by the successive levels of visual information processing during sensory input to the parietal cortex.

Assuming that sensory input activates assemblages of cortical neurons, several key questions are posited by Braitenberg (1977, p. 119); "If cell-assemblies are the terms of the cortical logic, there must be the possibility of extinguishing cell assemblies. There must be a timer that regulates the succession of activated cell assemblies. There must also be a mechanism of attention that inhibits entire regions of the cortex once something interesting has ignited somewhere." The role of attention and internal timing mechanisms is addressed more specifically at a later point in Sect. 5 "A Model of Communications Structure and Knowledge Acquisition" and Fig. 15.

A variety of models have been published to explain the adaptability and finely controlled representational processes of the CNS. Some of these models assume that a reflex of the classical Pavlovian type is fundamental to the process (Asratyan and Simonov 1982) while other models use complex assemblages of neurons forming fields or network to explain the specialized and coordinated actions of the components of the cerebral cortex (Grossberg 1982; Aertsen et al. 1985; Johannesma et al. 1985; Shaw et al. 1985). It is unlikely that complex experiences are represented by single neurons or their immediate connection with other neurons. The remarkably diffuse activity of the cerebral cortex, and its well-documented plasticity in recovering from some forms of local trauma, indicate, among other sources of evidence, that larger assemblages of neurons must be involved in the functioning of the higher brain centers. Analyses of the overall texture of the brain (Braitenberg 1985) suggest that the approximately 10^{10} neurons, interconnected by numerous cell processes (axons and dendrites), provide astronomical numbers of possible combinations of neurons in forming responding and controlling entities during information processing. Theoretical discussions of the potentiating and limiting factors for cortical integration are presented by Wickelgren (1987) and Glassman (1988), including the following example. If memory is viewed as an input-output associational mechanism, and we assume as a simple boundary condition that the c. 2×10^6 sensory neurons make all possible connections with the 3.5×10^5 motor neurons, this would require 7×10^{11} connections, exceeding the cortical number of neurons by nearly one order of magnitude. This is clearly too simplistic a view since it is known that each neuron can send and receive numerous synaptic connections. Glassman (1988) proposes, among other solutions to the problem of cortical integration and memory capacity, that each neural functional unit (e.g., group of neurons, single neuron, or physico-chemically distinct membrane area of a neuron) can respond either individually or in combination with other units. By positing combinatorial associations of variable size, allowing for a range in state values, and permitting each neural unit to participate in several larger units, it is possible to generate sufficient capacity to explain memory storage encompassing a human lifespan. How the spatio-temporal patterning of activity by these hypothetical entities correlates with cognitive processes at a detailed level remains unsolved.

However, it is becoming increasingly clear that the manifold integrating capacities of the CNS must be considered at several levels of organization (Bullock 1984). This includes: (1) molecular and subcellular organization correlated with membrane excitation, transmembrane ion conduction, labile coupling with cytoplasmic metabolic machinery, and protein stereo-behavior, (2) Single cell variables such as intrinsic rates of activity, reception of neural input, release of transmitters, pacemaker oscillation, etc., and (3) mechanisms at the intercellular level whereby interacting neurons become organized into assemblages mediating specific functions.

The significance of levels of organization may vary substantially among different brain centers and across different neurocognitive functions (Churchland and Sejnowski 1988). Fundamental sensations such as localized pain may be registered as a low-level perceptual state, whereas attention to complex sensory input may occur at the level of larger neural systems located in many different parts of the brain. In a very fundamental and overly simplified view, the greater the distance from sensory receptors (i.e., increasing synaptic distance), the higher the level of information processing. Hence, with respect to visual information, processes localized in the visual area of the neocortex are at a higher level than those occurring in the geniculate nucleus (LGN); and in turn these are at a higher level than those occurring in retinal ganglion cells (Churchland and Sejnowski 1988). However, once the information reaches the cerebral cortex, it spreads diffusely through cortico-cortical projections into numerous parallel paths of processing. While it is possible to clearly segregate earlier stages of information processing into hierarchical levels of operations, the neural organization of the association areas and prefrontal cortex appears to support simultaneous and interactive modes of processing, perhaps mediated by webs of interconnected neuronal networks. Thus, control is more diffusely distributed and not so easily localized spatially. It may be productive to analyze the problem from the perspective of temporal patterns of activity in interconnected networks of neurons.

Abeles (1982, 1988) has proposed a model of neuronal integration that explains coordinated CNS activity as a process of temporally excited groups of interconnected neurons producing a vectorial pattern of activity. Groups of neurons are activated in a pattern that yields a serial input-output relationship. This model, termed a "synfire chain," assumes that spread of activation occurs through sequential activation of synaptically cross-linked groups of neurons, not by a linear sequence of neurons in a one-to-one chain (Fig. 6). Thus, although sensory input arriving from the periphery may be conducted along single neuronal chains, the spread of activation in the cortex may be much more diffuse involving the sequential activation of sets of neurons in sequence, each set identified not necessarily by spatial proximity, but by the geometry of the axonic cross-linkages (Fig. 6C, D). For example, in Fig. 6C, a set of three neurons is organized such that the axons from each of the exciting neurons (far left) innervates each of the other activated neurons (far right) in a sequence. Sensory input sequentially activates *sets* of neurons throughout the cross-linked network. Hence, this model assumes a kind of *temporal modularity* complementing the *spatial modularity* of some of the preceding models (Szentágothai and Arbib 1975; Glassman 1988). By extension of this series of

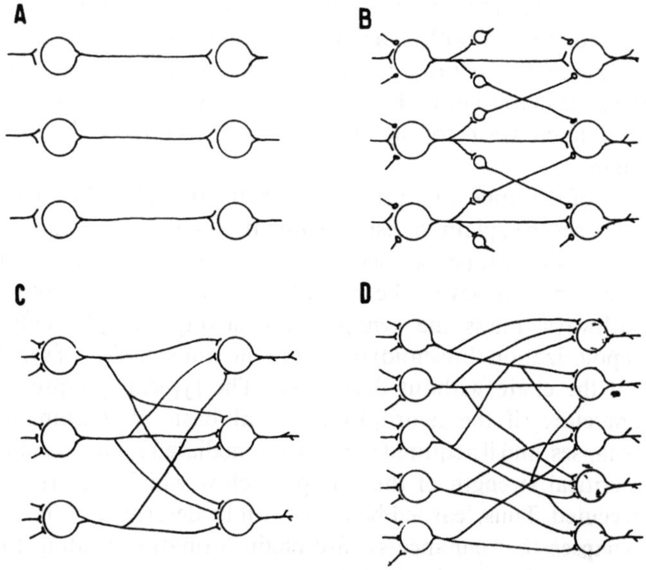

Fig. 6 A–D. Schematic representation of neural chains. **A** One-to-one chain representing dedicated afferent pathways. **B** One-to-one chain with lateral inhibition, **C** synfire chain. Each neuron in one stage excites all neurons in the next stage, **D** modified synfire chain. Each of the five neurons in one stage excites only three neurons in the next stage, hence providing greater differentiation of activation. (From Abeles 1982)

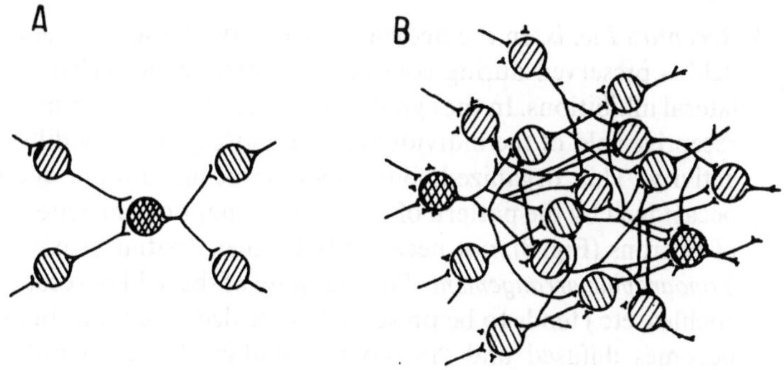

Fig. 7. A, B. Cross-over the neural chains **A** one-to-one, and **B** in synfire chains. **A** Cross-over of two one-to-one chains produces confusion of output and hence destroys the specificity of the conduction pathway. **B** Cross-over of two synfire chains. Such a cross-over can function as long as the two chains are not active simultaneously, which will lead to confusion in the signal. (From Abeles 1982)

reticulated connections, it is possible to explain a particular path of excitation throughout the neuronal network (Fig. 7). The synfire model permits a given neuron to function in more than one excitatory network depending on the pattern of its synaptic linkages with other neurons (Fig. 7B). Afferent input arriving at the upper left of the figure excites a synfire chain that spreads through the network of neurons and emerges at the lower right. Likewise, afferent input from the lower left excites a synfire chain that emerges at the upper right. Certain neurons (cross hatched) can be common to both synchains without distortion of the signal as long as *only one synchain is activated at a particular moment in time.* According to Abeles (1982) the code in this network is the "synchrony of firing among the members of the same set" in contrast to the linear "trace" produced by neurons in a one-to-one chain.

One of the appealing properties of this model is that it requires strengthening of synapses to explain the establishment of synchains. The cortical network of neurons contains the proper connections to support the spread of activity along networks of synchains. However, the neural activity will not become organized along these pathways unless the synapses are strong enough (sufficiently strong to reach depolarization threshold) to ensure the transmission of synchronous firing throughout the chain without decrement. The typical synapse is initially far too weak. Therefore, if the neurophysiological basis of learning is the strengthening of synapses, and if acquired behavior is mediated by synfire chains, then there would be a critical strength of the synapse below which a learned behavior could not be executed. Thus, learned behavior can be described as the selective strengthening of synapses that enhance synfire chain formation leading to vectorial input-output patterns of cortical acitivity.

The following comparative analysis of one-to-one (dedicated) sensory input chains and cortical synfire chains is summarized from Abeles (1982):

(a) *Code.* The simplest code in the dedicated line arrangement is the firing rate of the cell. In the synfire chain, the code is the firing among the members of the same set.
(b) *Receptive Fields.* In the dedicated pathway, the size of the peripheral receptive field is preserved during conduction centrally, or perhaps reduced if there are lateral inhibitions. In the synfire chain, the network of connectivity blurs out the receptive field of the individual neuron giving rise to a diffuse field of neuronal activity. This organized, but topographically diffuse, activity of neurons is occasioned by the pattern of excitatory synaptic connections among the groups of neurons (Fig. 8), not necessarily by their spatial proximity.
(c) *Topographic arrangement.* The mapping of the field of receptors (retinal surface, cochlea, etc.) tends to be preserved in the dedicated line, but in the synfire chain becomes diffused and "mixed up" within the temporally organized sets of topographically dispersed neurons.
(d) *Multifunction.* In the dedicated line, neurons can be used to transmit only one kind of information since crossing-over between lines (Fig. 7A) could produce a confused signal; it is not possible to say where the signal will end up. In the synfire chain, an individual neuron may participate in many different chains depending on the pattern of interconnections (Fig. 7B).
(e) *Immunity to Damage.* In a dedicated one-to-one pathway, the loss of a given linking neuron will disable the sequence. In a synfire chain, there is sufficient "redundancy" so that occasional loss of a network neuron will not necessarily disable the sequence of activation.

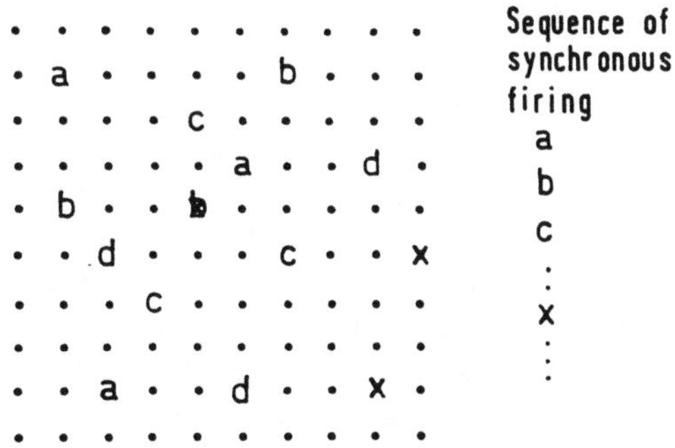

Fig. 8. Distribution of synfire activity within a population of neurons. The order of the synfire chains of Figs. 7 and 8 is given to stress the functional relations. Although the neurons fire in a predictable pattern, anatomically, the links of the synfire chain may be distributed quasirandomly. If one could see the activity in the cortex while a synfire chain is active, a spatiotemporal organization of activity would appear as drawn here. While some neuronal units may fire only once (those shown *without an overprint*), others may fire more than once (*fourth row b overprinted with x*). The net result however is to produce a sequential pattern a, b, c, --- x --- etc. (From Abeles 1982)

(f) *Parallel Processing.* Since the dedicated lines do not cross-over, several different lines can be activated simultaneously without interference. The synfire chains can be activated simultaneously only if they utilize separate sets of neurons. [*Comment.* This suggests that if parallel processing is a major characteristic of cognition and if synchain networks are nominally correct models of cerebral cortical organization, then there must be some coordinating mechanism to ensure synchain autonomy. In a hierarchically ordered system of these networks, this could be performed by an executive operator as described hereafter (Fig. 15)].

(g) *Localization.* Transmission of sensory information in the periphery is completed in stages, each of which is localized in clearly identifiable, different anatomical structures. By contrast, sets of neurons in a synfire chain are likely to be distributed spatially in a quasi-random pattern, but are organized temporally into a predictable pattern of activity. Thus, the structure of stages in these networks represents the time of activation of the successive sets of neurons, not their anatomical position.

Network models of this kind are attractive since they can explain the diffuse, cortical, electrophysiological activity recorded by surface electrodes and the apparent plasticity and relatively delocalized quality of cerebral cognitive functions. However, they also raise intriguing issues of how coordination of activity, differentiation of classes of responding, and temporal patterning of complex behaviors are achieved. Hierarchical systems theory (Laszlo 1972; Pattee 1973; Szentágothai and Arbib 1975, pp. 10–14) may provide some insights into these questions.

2.6 *A Hierarchical-Systems Perspective*

In general, the anatomy of the brain is hierarchically organized, and although there are deviations from hierarchical control theory in brain functioning (Szentágothai and Arbib 1975; Goldman-Rakic 1988), incoming stimuli are partially processed and refined at each level of brain organization. However, as observed previously, greater complexity of organization may be produced by parallel pathways projecting toward higher brain centers. Thus, there is not a direct (one-to-one) projection of sensory experience onto the cerebral cortex. The representation of experience involves complex interactions of neuronal assemblages among the several component parts of the brain (e.g., brain stem, midbrain and forebrain). We may designate this function as the formation of a model of sensory experience, i.e., to represent sensations, and to coordinate the activity of the body in adapting to variations in the internal and external environment.

A theory of CNS activity that intends to represent the dynamics of this modeling should include the following principles (Szentágothai and Arbib 1975): (1) The theory must be action-oriented. The modeling process is not any one of internally representing the environment, but also includes the capacity to interact with the environment in some reasonably structured fashion. Perception is a process of responding to, as well as representing, sensation; (2) Perception is not only what is there, but where it is and how we interact with it; an operational component or active responsive role must be posited; (3) An adaptive system must be able to correlate sensory input and action in a way that permits it to transform or update an internal model of the world. In other words, there must be memory. Complex systems must be sufficiently plastic to modify this internal model to accommodate aspects of the environment that were not initially provided for in the model; (4) The system must be hierarchically organized with adequate "feedback loops" to coordinate the subsystems. The feedback loops need not be directed entirely from lower to higher centers or vice versa, but probably involve a complex reciprocal feedback flow of information among several levels within the system. This can account for the remarkable plasticity and fine controlled responses of the modeling system.

Further distinctions can be made, based on the direction of information flow and control (Johannesma et al. 1985). Systems exhibiting only forward-directed influences (as from stimulus input to CNS encoding) are classified as "reactive." They have a capacity only to react to input, but not to transform it and synthesize it into novel representations. Systems with feedback loops and reciprocal networks, permitting refinement of signals through transformation of input and the internal generation of diverse outputs, are classified as "creative." Qualities of the input signal that improve the signal to noise ratio, such as predictable patterns or other recurrent markers that increase signal saliency for the encoding system, tend to enhance registration and storage of information.

Increasing evidence points to the facilitatory role of patterning in stimuli as an aid to construction of internal representations (models) of experience. Although there is good evidence that groups of cells peripherally and centrally may represent elemental components of complex stimulus arrays (e.g., vertical line detectors, angle detectors, etc. in the visual system; Hubel and Wiesel 1963, 1965, 1968), efficiency of encoding of the total stimulus set is enhanced if a predictable pattern is present (Boddy 1978). Novel stimuli or unpredicted elements within an otherwise ordered set of phenomena produce arousal (orientation reflex). Thus, patterned and predictable input enhance registration and construction of internal models of experience, whereas novelty sustains arousal (primary attention to the stimulus input). The balance between predictability of input and novelty is a significant factor in maintaining an alert and efficient encoding state. This viewpoint has been incorporated in an arousal model (Sokolov 1960; Lynn 1967) with two interactive components (Fig. 9). The system for formation of model (upper rectangle) is the encoding system of the cerebral cortex. The system for arousal (lower rectangle) is the complex network of the reticular formation, and, as subsequently documented by McGuinness and Pribram (1980), the amygdala and frontal cortex. According to this model, afferent conduction [1] from peripheral receptors (e.g., auditory or visual receptors) bifurcates. One pathway [1a] leads to the cerebral cortex and the other [1b] to the arousal centers. It is not necessary to assume that these are direct pathways, and indeed they probably are not. Ascending neurons from the arousal centers to the cortex [2] stimulate the cortical cells and maintain readiness to encode. Other pathways [3] descending from the cortex to the arousal centers provide feedback connections. A particular inhibitory feedback pathway [4], shown as a stippled line, inhibits activation of the arousal centers. This pathway is activated continuously during patterned afferent input, hence blocking stimulation of the arousal centers and suppressing cortical activity. The decline in cortical arousal accounts for habituation (attenuation of responses to repeated or familiar stimuli).

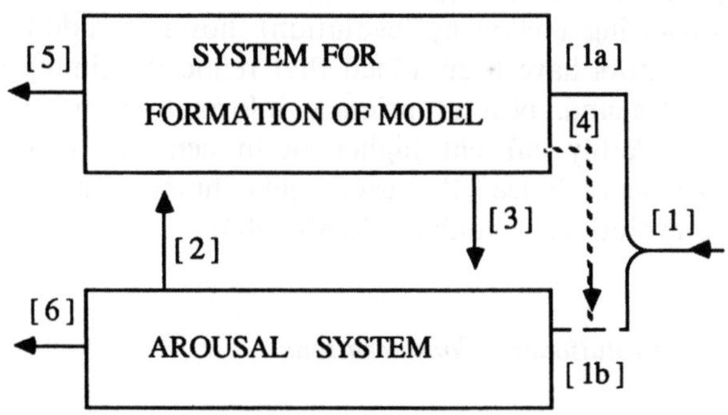

Fig. 9. Sokolov's model of cerebral arousal and habituation. The afferent innervation (*1*) bifurcates. A pathway ascends to the cerebral cortex (*1a*) where appropriately patterned input is registered as a cognitive model of experience. Another pathway (*1b*) innervates the reticular formation and associated arousal centers, but under conditions of patterned input, an inhibitory descending pathway from the cerebral cortex (*4*) blocks stimulation, thus the arousal centers are not stimulated and cortical excitation is reduced producing habituation. Ascending pathways from the reticular formation (*2*) activate the cortex during arousal, while feedback pathways (*3*) carry impulses which inform the arousal centers when continued patterned impulses are being encoded by the cortex. Efferent pathways carry outgoing impulses from the cortex (*5*) and arousal centers (*6*). (From Sokolov 1960)

At the onset of a novel or unexpected stimulus, however, the encoding process is momentarily interrupted, and information input is not registered in the cerebral model of experience. Therfore, the inhibitory pathway is not activated and the arousal centers are stimulated leading to alertness and orientation of the organism. In general, optimum encoding of a signal requires sufficient pattern to enhance encoding, and appropriate novelty to sustain arousal. Although this model of CNS activity is appealing in its simplicity, some additional modifications are required to represent adequately human behavior. For example, the internal regulatory capacity of the human CNS provides an additional complicating factor. Arousal and attentional states are mediated by prior experience registered in cognitive schemata that determine the saliency of stimuli and selectively direct attention to different aspects of the stimulus field. Some improvements have been made by McGuinness and Pribram (1980) to take these factors into account and make the model more consistent with current physiological data. They propose that attention (selective information acquisition) involves three components: arousal, activation, and effort. The model focuses exclusively on the process, not the content of awareness. Arousal is a phasic, short-lived and reflex response to input. Activation is a tonic long-lasting and involuntary readiness to respond. The third component, effort, coordinates arousal and activation. It sustains selective acquisition even under stress, and involves energy expenditure. In general, *arousal* is mediated by the amygdala (though not exclusively), *activation* by basal ganglia (surrounding the thalamus), and *effort* by hippocampal circuits. As in the Sokolov model, encoding of information is momentarily interrupted during onset of arousal, but may be followed by enhanced encoding of subsequent patterned input. In overview, arousal and encoding processes in humans can be viewed as extensions of primitive neural processes whose origins can be traced to CNS functioning in lower animals. With advancing phylogeny (evolution), however, additional regulatory and selective operators have been added that refine the timing and endurance of attentional factors and, beyond primary information processing, have given rise to the remarkably efficient higher brain centers for representing and transforming experience. Indeed, the very origin of human reason or cognition has been subjected to evolutionary analysis (Riedl 1984).

2.7 Evolutionary Considerations

It is generally accepted that the human nervous system has evolved from lower forms in much the same pattern as other somatic structures (Noback and Shriver 1969; Caspari 1979; Hahn et al. 1979; Edelman 1981). This process has included, among other features, increasing complexity and plasticity of neuronal connections, enlargement of brain mass (especially the cortical tissue), and concomitant increase in efficiency and sophistication of adaptive response to the environment. Paleontological investigations using brain casts from the internal cavities of fossil skulls have provided ample evidence to support the conclusion that phylogenetic advancement is accompanied by increasing cerebral volume and surface folding, a strong correlate of predictable and "intelligent" behavior across species (Jerison

1976, 1977, 1979). However, it should be emphasized that there is no evidence that this applies within a species, such as across human groups. Comparative experimental studies of conditioned learning across diverse groups of animals (primitive to more advanced) indicate clear evidence of increasing capacity to encode experience flexibly and efficiently and develop environmentally contingent behavior patterns with evolutionary advancement (Bitterman 1965; Weinberger et al. 1985). For example, major groups of animals corresponding to different phylogenetic levels of development require substantially different amounts of food reinforcement during operant conditioning. Angermeier (1984) concludes that it would be fallacious to equate the number of rewards required to reach criterion with intelligence. He suggests differences in performance across diverse groups of animals may be explained by "(1) the complexity of neural networks; (2) the time that information remains within such networks; (3) the ratio of instinctual to variable behavior; (4) the wealth of sensory inputs; (5) the relationship of brain weight to body weight; and (6) ecological factors, including the avoidance of poisonous substances, the speed with which information has to be available, etc."

Substantial experimental evidence from studies of invertebrates to humans indicates that lateralization of the brain can also be traced through phylogenetic lineages (Dimond and Blizard 1977). Comparative neuroanatomy further reinforces the conceptualization that the brain of higher mammals, and indeed some of their characteristic behaviors, can be traced to origins in lower mammals (Noback 1959, 1975; Olton 1977). In general, increasing development of cortical mass has been accompanied by transfer of some functions from lower brain centers to the cerebrum. However Noback and Shriver (1969) note that increasing encephalization does not necessarily result in complete suppression of lower brain center functions by higher brain centers. They conclude, "Encephalization is a phylogenetic expression of those anatomical and physiological changes during evolution specifically related to the hierarchical organization of centers in the pathways of the nervous system." The interaction of the major centers is viewed as a synergistic interaction of primitive and more advanced brain regions through feedback circuits (loops), integrated with intercalated neurons into both the ascending and descending pathways, thus forming a complex dual hierarchical organization.

Based on paleontological studies, and other evidence of the evolutionary lineages leading to higher mammals and humans, theories have been proposed to explain higher brain functions, including language. For example, there is increasing evidence that the brain of mammals is composed of several stratified anatomical regions that can be traced phylogenetically to very primitive origins. Thus encephalization (the development of increasingly massive and complex forebrain tissue) and hierarchical organization of brain centers can be traced at least to origins in the reptiles (MacLean 1972). Comparative studies of reptiles, birds and mammals using histochemical methods for localizing acetylcholine and dopamine have been particularly useful for identifying corresponding structures of the protoreptilian forebrain. These include components of the basal ganglia, i.e., olfactostriatum, corpus striatum, globus pallidus and satellite collections of gray matter (MacLean 1986). Overlying this is the paleomammalian (limbic) system that exhibits much more sophisticated mechanisms for correlating sensory input and behavior patterns. Finally, encompassing these lower centers, and intricately interlinked with them by ascending and descending pathways, is the cerebral cortex which represents the highest level of development during encephalization. This elaboration of cortical tissue accounts in large measure for the enhanced capacity of mammals to respond in predictable yet flexible patterns to sensory input. Indeed, language, often cited as the highest expression of phylogenetic development in mammals, is mediated by cerebral cortical functions; however, this should be tempered by an understanding that other high level representations of experience including iconic (visual) and enactive (motor-based) schemata have also contributed substantially to the increased adaptive capacity of advanced mammals including humans.

Jerison (1977) theorizes that language could have arisen during encephalization in mammals as a way of marking territory and the boundaries of frequently travelled paths by hunting groups. He assumes that the primitive ancestors of mammals evolved during a time when larger reptiles were active during daylight hours. These primitive mammals were probably nocturnal to avoid competition with the larger reptiles. Therefore, selection pressures favored development of vocal signalling in addition to efficient visual systems. This interactive signalling eventually may have given rise to more sophisticated vocalizations of social animals including human communication. Whether or not one accepts the premise of nocturnal patterns of behavior as a major selective pressure favoring vocalization, it is clear that the development of communications systems of even the most primitive kind would have positive survival value by enhancing group coordination. Thus, the fundamental premises of Jerison's theory of language are appealing and seem thoroughly consistent with classical evolutionary theory. The significant role of language in mediating representations of the external environment through construction of verbal schemata (augmenting visual ones) is a significant component of modern neuropsychological research. Some fundamental neuropsychological principles pertinent to information processing, language and memory are presented as a context for the neurocognitive models discussed at a later point.

3 Some Current Neuropsychological Perspectives on Memory and Knowledge Acquisition

3.1 Fundamental Mechanisms of Memory

Substantial literature exists on the normal and abnormal functioning of human CNS related to cognition, and language development and its usage (Rosenzweig and Bennett 1976; Boddy 1978; Pompeiano and Ajmone-Marsan 1981; Squire and Butters 1984; Will et al. 1985). Processes of memory in normal and dysfunctional subjects have been extensively analyzed to determine what structures of the brain account for the phenomena (Brazier 1979; Ajmone-Marsan and Matthies 1982). Some pertinent publications will be summarized here as a background for a more detailed discussion of neurocognitive models of knowledge acquisition presented subsequently.

With the possible exception of some simple conditioned motor responses, all attempts to localize memory in a particular structure of the CNS have been disappointing (Thompson and Donegan 1986). Indeed, it appears increasingly that human memory is neither unitary (a singular undifferentiated phenomenon) nor localizable (associated with a specific site in the CNS) (Cohen 1985; Crowder 1985; Weinberger et al. 1985, pp. 1–6). The storage of information in the CNS, and various forms of learning correlated with information storage, appear to be mediated by a wide array of advanced and primitive areas of the brain. Simple learned responses, such as conditioned reflexes, may be largely mediated by neural networks in the lower centers of the brain, or even by pairs of synaptically linked neurons (Eccles 1964, 1965), whereas more complex instrumental learning, including language-based learning, appears to require participation of higher brain centers including cortical structures. For example, the intact hippocampus is essential to long-term storage of information in memory, but apparently serves a major role in facilitating transfer of information into stable storage rather than being the site of storage. Thus, while certain centers of the brain can be shown to be essential for memory functions, they are probably not the locus of memory, but rather key participants in the diffuse storage of information in more widely distributed assemblages of neurons in the CNS. As Boddy (1978, p. 338) has noted, diffusivity of memory probably implies that any individual neuron is involved in the representation of many memories. This is evidenced by the psychological observation that evocation of any specific memory lowers the threshold of other memories with which it has some associative link. Extrapolating from this viewpoint, we may conclude that if assemblages of neurons, rather than a single neuron, are the sites of elemental components of information storage, then associative links among these assemblages may represent connectedness in recall of stored information. The individual neurons forming the locus of each elemental assemblage are not unique to a memory trace, but may be involved in combinations with other neurons to form other elemental assemblages representing stored units of information. Such a mosaic of neurons, participating in diverse patterns of combinations, could account for the remarkable storage capacity of the CNS. As a simple analogy, consider a group of tribal drummers. The message in the ensemble of sounds is not unique to any single drum, but rather is produced by the pattern (timing and phasing) of percussions of the group. A particular drum, moreover, can be utilized in a nearly limitless number of combinations with other drums to produce a wide range of sound messages. In this sense, each drum is analogous to a neuron, or elemental unit of neurons. By modulation of activity within each neuronal unit, and by its varied combination with the activity of other units, a remarkably diverse range of information could be stored diffusely within the neuronal system.

3.2 Short-Term and Long-Term Memory

Evidence from cognitive scientific and neurobiological investigations of normal and abnormal memory states suggests that encoding of information in memory occurs in stages, perhaps two, but probably many more. In the simple two-stage view, information can be acquired in an immediately accessible form in short-term memory (STM) lasting for intervals up to seconds or minutes, or in long-term memory (LTM), a more stable and enduring kind, lasting for years (Waugh and Norman 1965; Boddy 1978). Arbib (1989) defines STM and LTM more broadly. Short-term memory is viewed as a working memory with each item categorized as a "schema instance," maintained in immediate memory because of its relevance to the task(s) at hand, not necessarily because it was recently encoded or activated. Items consolidated and committed to long-term storage (LTM) may also remain in STM because of the prevailing demands on the individual, or because the plan or action was not fulfilled. Long-term memory is characterized as an internal model of the world representing a collection of properties that have been stored for subsequent retrieval to guide immediate behavior and to update and improve the LTM store. Hence, in this view, the major difference between STM and LTM is one of schema instantiation for immediate use (STM) versus schema storage (LTM) for future application. The latter is a network of interrelated schemata that provide submodels of objects, tasks, situations, etc. that give coherence and predictability in cognitive representation of experience.

Although the neural mechanisms of information storage in STM and LTM are not known in detail for higher animals, evidence from invertebrates indicates that STM, interpreted as short-interval storage, may be mediated by covalent bond changes (e.g., phosphorylation) within the neuron, whereas LTM (protracted memory) results from long-term change in effectiveness of elementary synaptic connections between neurons (Shapiro et al. 1981). It has also been suggested that a synaptic process known as long-term potentiation (LTP) may be a neuronal mechanism underlying information storage (Teyler 1986; Teyler and DiScenna 1986; Teyler and Perkins 1988). LTP is a long-term change in synaptic efficacy, particularly of monosynaptic junctions in the limbic system of mammals, but also reported in multisynaptic systems (Racine and Kairiss 1987), and spinal cord pathways (Eccles and McIntyre 1951). There is some biochemical evidence to suggest that LTP occurs by activation of a postsynaptic, cytoplasmic protease. This enzyme hydrolyzes cytoskeletal structures and releases additional receptors into the post-synaptic membrane, thus increasing the effectiveness of the synapse and accounting for long-term encoding of the connection (Staubli et al. 1985a, b; Lynch 1986). Some of the limitations of LTP as a basis for LTM, including the relatively short duration of LTP relative to LTM, are discussed by Racine and Kairiss (1987).

Although there is substantial clinical and experimental evidence for differences in the quality and stability of immediately acquired versus long-term memory, some researchers suggest that LTM may develop through several stages, rather than only two. If there are indeed two fundamental forms of memory (STM and LTM), there is some evidence that they are independent and operate in parallel rather than sequentially. Information may enter the long-term store directly, while short-term store is responsible for immediate recall (Boddy 1978; p. 346). Given increasing knowledge of the diverse levels of information processing in the CNS and the diffuse nature of memory storage, these dichotomous distinctions may represent only a convenient categorization of processes that involve many intermediate stages. Thus, it is likely that the highly coordinated and reciprocally linked functions of the CNS provide for a dynamic transformation of information prior to, and during its encoding in long-term storage. Higher order integrative functions may synthesize newly acquired input within previously stored information, hence updating and modifying prior information stores to include the perceptually salient and most readily accommodated new units of information. Thus, what has been called working memory as opposed to long-term storage memory represent a technically convenient, but not entirely correct, separation of two highly linked processes: (1) immediate encoding processes involving perceptual filtering and selection of units of information to be stored (working memory), and (2) consolidation processes of higher brain centers that synthesize this selected input within existing stores of information (LTM). It is assumed that there may be several levels of processing within the CNS for each of these functions (Fig. 15). They are not necessarily localized operations. The implications of this hypothesis for communication processing and storage of information in memory are explored more fully in the following sections.

3.3 Simultaneous and Successive Information Processing

Current interest in the general issue of simultaneous (parallel) and successive (serial) processing of information is growing, partially attributable to applications of computer metaphors to cognitive functions and increasing evidence of the diversity of modes of information analyses used by different individual with different tasks, and by experts relative to novices (Miller et al. 1960; Newell and Simon 1972; Das et al. 1975, 1979; Hunt 1980; Sternberg 1985). Only some pertinent basic concepts will be summarized here and further specific information can be found in a concise review by Äystö (1988).

Luria's (1966a, b, 1973) clinical neuropsychological model included three functional units of information processing (arousal, coding and storing of sensory input, and planning) localized in different regions of the brain. Associated higher order processes are assumed to occur by either simultaneous or successive syntheses depending on the modality and the requirements of the task relative to the individual's experience, cognitive orientation, instructions for the task, etc. (Das et al. 1979). Serial and parallel modes are employed differentially during attention and peripheral processing of sensory input (Näätänen 1982, 1985, 1986). During "routine" or non-task-dependent processing, where selective attention is not required, the information processing occurs in parallel, preconsciously, and automatically. This is a form of basic processing different from the "automatic" responding that develops with increasing experience and mastery of a task. This proficiency-based "automatic" responding is under conscious control and hence is fundamentally different from the innate, non conscious automatic processing mode. When the task requires selective attention, then a serial, controlled, and conscious form of processing is used. Hence, simultaneous and successive modes of processing information are the mechanisms whereby the CNS transforms and encodes information for integration into the system. During simultaneous processing, information is grouped into spatially discrete sets which can be accessed as a whole at any point in time. Successive processing treats information serially and the output is accessible only at an endpoint in time. Although the processes may occur in different parts of the brain, they can be mobilized in various combinations as required by the immediate task and the individual's competency in performing the task.

Das et al. (1979) explain cognition in terms of serial and simultaneous processes in relation to four fundamental components: input, sensory registration, central processors, and output. Information can arrive at the sensory receptors in either a simultaneous or successive pattern and is processed in either mode depending on the demands of the task, other environmental differentiating features, and the individual's state of arousal, expectation, etc. Hence, the two modes are not necessarily hierarchically related nor invariantly ordered relative to one another. A given task can be processed by either simultaneous or successive strategies. Although this model assumes plasticity in employing the two strategies, Das et al. (1979) localize simultaneous (parallel) processing in the posterior (i.e., parieto-occipital) regions of the brain and successive (serial) processing in the anterior (i.e., frontal or fronto temporal) regions.

Äystö (1987) has combined these perspectives with concepts of hemispheric asymmetry (i.e., nonverbal processing in the right hemisphere and verbal in the left) to produce a model that accounts for verbal and nonverbal information and its simultaneous or successive mode of processing. This view assumes information is processed in three stages (1) registered in a sensory buffer, (2) synthesized at the level of a central processor in the cerebral hemispheres, where decision-making and planning outcomes are derived, and (3) passed to an output channel. During cerebral synthesis, four combinations of activity are possible based on cerebral asymmetry (right versus left) and successive versus parallel processes: (1) successive-verbal input is processed in the left anterior region; (2) simultaneous-verbal in the left posterior region; (3) successive-nonverbal in the right anterior region; and (4) simultaneous-nonverbal in the right posterior region. This adds additional texture to the localization of neurocognitive processes while preserving the concept of functional plasticity. It is clear, however, that much additional, experimental, neuropsychological research is required to fully test these models, and to determine the extent of localized versus diffuse activity during serial and parallel information processing. Given the plasticity and diffusivity of neurocognitive functions, some theorists assume that stochastic or other mathematical models may be particularly effective in representing behavior rather than an overreliance on structural-functional models that require increasing precision in functional localization. Some of these mathematical approaches and emergent neurocognitive models are presented in the next sections.

4 Some Historical and Current Perspectives in Neurocognitive Modeling

4.1 Mathematical and Biophysical Models

Some pertinent research on the mathematical modeling of information processing and learning is presented as background for more recent research on neuromathematical models summarized in Sects. 5 and 6.

Bush and Mosteller (1951) modeled simple stimulus-response learning (instrumental conditioning) as a set of probability equations. Their work is within the general framework of statistical learning models as employed by Estes (1950) and Miller and Frick (1949). They conceptualized the process of instrumental conditioning in four stages: (1) sensory input and stimulus perception, (2) arousal of a response or performative act, (3) occurrence of an environmental event (nominally a reinforcing event), and (4) execution of a goal response. Moreover, any sensory input which follows an instrumental response is categorized as an environmental event. Their rationale was to model the probability (p) that an instrumental response will occur in a period of time (h). Assuming that the probability of a response occurring on a given trial is a function of the immediate preceding probability state, then the likelihood that the response will occur is positively dependent on the reinforcement (factor a), and negatively dependent on decremental effects such as fatigue and lack of reinforcement (factor b). By establishing a basic operator (Q of p) for transforming the probability as a function of factors a and b, Bush and Mosteller obtain the following linear equation

$$Qp = p + a(1 - p) - bp \qquad (1)$$

This expression represents the change in probability of a response on each trial as a function of reinforcement i.e., extent to which the maximum possible increment in probability $(1 - p)$ is realized in proportion to the magnitude of a and to the decrement or extinction of the response due to inhibition factors $(-bp)$. Thus as the term $a(1 - p)$ increases, then Qp increases since this term is added to p, the initial probability state before the trial. However, the larger the decremental term $(-bp)$, the less the value of Qp, and, therefore, the lower the probability of the response. All those factors that increase the probability are included in the factor a, whereas all those factors that decrease the probability of a response are included in the factor b. With repetitive trials, the operation Q of p is applied iteratively, yielding the following generalized equation for n trials

$$Q^n p = \frac{a}{a+b} - \left(\frac{a}{a+b} - p\right)(1 - a - b)^n \qquad (2)$$

If the parameters a and b are not both zero, nor both unity, then as n becomes very large, the factor $(1 - a - b)^n$ approaches zero, and the total equation approaches $(a/a + b)$ as a limit. Hence, the equation represents a response acquisition curve with asymptote of $(a/a + b)$. By further elaboration of the fundamental equations, other aspects of conditional learning were modeled, including mean latent time of the response, effects of fixed ratio and random ratio reinforcement, and of aperiodic and periodic reinforcement. No assumptions were made about the neurophysiological correlates of the conditioning, nor were the equations applied to more complex behavior such as human communication and information processing.

Grossberg (1978) applied network theory to produce a mathematical model of behavior described as a "psychophysiological theory of the self-organization and performance of sensory-motor codes, maps, and plans." The model accounts for a wide range of behaviors spanning imprinting of sensory-motor coordinates in infants to complex goal-oriented, serial activity of adults. It includes several types of parallel, nonlinear, self-organizing, hierarchical and feed-back regulated systems that provide a context for modeling more complex aspects of human sequential behavior. While the mathematical arguments become elaborate for more complex networks, a fundamental concept is the strength of a link (z_{ij}) from node i to j within the network. This strength, a time derivative (rate of change of strength with time), is a function of the decay due to memory attrition ($-uz_{ij}$), and the strengthening of association due to co-occurrence of activation of the two nodes i and j within a limited time interval (θ), expressed as $[\beta x_i(t-\theta)x_j(t)]$. For the correlated occurrence to be effective, x_i must occur before j within the interval $(t-\theta)$. Thus, the rate of change in strength of the connection of the two nodes is a function of an exponential decay term for forgetting plus a strengthening term represented by the concurrent activation of the two nodes, thus increasing their linkage. Among the events modeled using these concepts are (1) performance in a maze task, (2) instrumental conditioning, (3) serial learning and the interactions of short-term and long-term memory, and (4) visual versus auditory processing and cerebral dominance. In a similar approach to Grossberg's sensory-motor based paradigms, Lawson (1986) developed a model of problem-solving that combines Piagetian paradigms with neurological network theory. The focus is more descriptive than mathematical, but provides a coherent perspective on possible ontogenetic origins of higher level cognitive functions.

Rashevsky (1945a–c, 1960) assumed an alternative perspective to stochastic theories by mathematically modeling neurobiological phenomena based on the assumption that stimulus conduction in a bundle of neurons, comprising a nerve fiber, is a graded response phenomenon amenable to analysis using differential and integral calculus. This was in contrast to the research of McCulloch and Pitts (1943) who began with the assumption that the all-or-none response of individual neurons could be modeled using boolean algebra. Their method was a more reductionistic approach beginning with numerical analyses of individual neurons, that were combined and synthesized into models of larger neuronal functional units. Some modern interpretations of their work are summarized by Arbib (1989). Landahl (1961) however suggested that even the interaction between two neurons may be approximately described by continuous equations, further reinforcing the concept of the use of differential calculus as applied by Rashevsky and coworkers. Rashevsky proposed that the activity in nerve fibers and nerve networks containing recurrent loops can be modeled mathematically by beginning with the assumption that the intensity of excitation (E) of an afferent nerve fiber varies approximately linearly with the intensity of the stimulus (S), provided the stimulus is sufficiently weak and above threshold. A mathematical expression was derived for the rate of decay of excitation (ε) within nerve networks. If ε is the number of excited cycles within a network containing neural loops and the quantity a is a constant of proportionality, then the rate of change of ε is given by

$$\frac{d\varepsilon}{dt} = -a\varepsilon \tag{3}$$

If we assume that each of the afferent fibers from one recurrent neuronal loop in a network makes excitatory synapses with another loop, thus increasing the spread of excitation and amplifying the number of excited neuronal loops, then the number of excited cycles within a feedback network is given by

$$\frac{d\varepsilon}{dt} = AE - a\varepsilon \tag{4}$$

where A is another constant of proportionality representing the enhancement of excited neuronal cycles due to spread of excitation (E). By rearrangement and integration, the expression becomes

$$\frac{d\varepsilon}{dt} = AE\mathrm{e}^{-at} \tag{5}$$

When $t = 0$, then $\mathrm{e}^{-at} = 1$, and $\varepsilon = 0$. As t increases, e^{-at} decreases to zero and ε tends to AE/a, the maximum value of the number of excited cycles in a neuronal network, (Rashevsky 1961). To simulate differentiation, mediated by inhibition within networks, Rashevsky assumed that some of the afferent fibers from a neuronal network may also make inhibitory synaptic associations with neurons in higher order neuronal loops. The number of excited cycles that yield an inhibitory connection with other neurons is given as follows

$$\frac{dj}{dt} = BE - bj \tag{6}$$

where j is the number of excited cycles yielding inhibiting effects and B and b are constants analogous to A and a in Eq. 4. The mathematical transformations applied to Eq. 4. can be applied to Eq. 6. Combining the equations for excitation and inhibition permits an analysis of the joint effects of spread of excitation and inhibition within a complex nerve network. By varying the four constants A, a, B, and b, Rashevsky and coworkers, Householder A.S. and Landahl H.D., were able to simulate some higher order neurocognitive phenomena such as reaction times, psychophysical discrimination, color vision, some aspects of learning and aesthetic perception (Rashevsky 1960).

To illustrate the elements of their argument, the derivation of an equation representing reaction time (delay between stimulus and response) is presented. A solution to Eq. 6, with the condition that at $t = 0$, we have $\varepsilon = 0$, is given by Eq. 7:

$$\varepsilon = \frac{AE}{a}(1 - \mathrm{e}^{-at}) \tag{7}$$

This expresses the degree of excitation as a function of time. Now, at the onset of a stimulus, the time required to activate an efferent pathway will be determined by the time required to achieve an excitation level ε equal to the excitation threshold (h_2) of the efferent pathway. By setting the right-hand side of Eq. 7 equal to h_2, we obtain an expression (Eq. 8) that can be solved for t, the reaction time of the efferent pathway.

$$\frac{AE}{a}(1 - \mathrm{e}^{-at}) = h_2 \tag{8}$$

Much of this research has been limited to biophysical representation of isolated neural networks rather than a systems level analysis appropriate to building neurocognitive models of information processing.

In a further quest to represent the dynamic interaction between LTM and STM processes, Grossberg (1981) proposed a mathematical model based on a premise that resonant states of activity (apparently somewhat similar to Rashevsky's excited neuronal cycles) develop within neural networks when their feedforward data and feedback "templates" reach consensus through a matching process. Resonant activity in STM systems causes changes in LTM. Reciprocally, these LTM changes alter the resonant STM patterns. These interactive cycles continue until STM and LTM equilibrate. The dynamic interaction between STM and LTM is categorized as "adaptive resonance." Hence, the "code" of a network is the set of stable resonant states that can be generated in response to sensory input. A variety of patterns in visual and auditory information processing can be analyzed from this perspective including how neurocognitive processes achieve code stability while retaining plasticity and adaptability.
Hicklin (1976) took a different approach with the aim of representing human information acquisition as a function of the dynamic interaction between information gain and loss. Thus, the amount of information acquired in a given unit of time is represented by the balance of information stored and that forgotten (Eq. 9).

$$\frac{\mathrm{d}N}{\mathrm{d}t} = p - qN \tag{9}$$

where N is the amount of information acquired in time t, and p and q are constants relative to a given individual and within a specific learning situation. The constant p represents the aptitude of the individual and q is an expression of the rate of information loss during forgetting. By integration, we obtain

$$N = A(1 - \mathrm{e}^{-qt}) \tag{10}$$

where $A = p/q$. This equation yields an information acquisition curve which approaches an asymptote of A with increasing values of t. It is a useful tool for representing the time-course of human learning when only some intrinsic characteristics of the learner (i.e., aptitude and rate of information loss) are considered. It does not provide predictions of learning based on extrinsic factors such as the quality and quantity of the sensory input. These factors are considered in a subsequent model discussed in Sect. 5.

4.2 Neurocognitive Models

A rather comprehensive model of human learning based on activation of neuronal assemblages was offered by Hebb (1949). His seminal papers contain concepts that often seem to be echoed in modern model building, though his idea of the cell assembly containing loops of interconnected neurons is probably too simplistic to represent adequately the complex integration and differentiation functions of the CNS required in modern information processing models. Fundamentally, Hebb proposed that during learning, groups of neurons (that were only randomly activated prior to learning) become linked into an assembly of self-exciting neurons. This hypothetical construct of neuronal assemblies forms the basic unit of association. These neuronal assemblies through fractionation and recruitment, or growth become differentiated and highly organized, thus accounting for the texture of internal representation of experiences. During subsequent information acquisition, sequences ("phase sequences") of neuronal assemblies are generated, representing interlinked self-exciting loops that mirror activities of perception and information acquisition. Consider, for example, three stimuli A, B and C that have been experienced in temporal relation and may have stimulus properties in common. According to Hebb, these become encoded as a series of neuronal assemblies with overlapping neurons, thus accounting for their co-association in memory. The arousal of a memory trace is attributed to activation of neurons in the memory loop (neural assembly). Four factors determine whether a cortical neuron is activated or fires: (1) the number of post-synaptic impulses arriving at a neuron from all sources for the few milliseconds when temporal summation can occur, (2) the strength of the synapses which is positively related to the frequency of activation, (3) whether or not the neuron is refractory, and (4) neuronal fatigue. The concept of the modifiable synapse ("Hebbian synapse") was a significant contribution toward understanding how networks of neurons (assemblies) become stabilized, and this concept frequently recurs in modern terminology. Once stable assemblies have been formed, further learning may occur by "coalescence" of activity among assemblies that previously were not connected. That is, each complex that is reliably aroused by input also has a fringe system of assemblies that are facilitated by (but not reliably aroused with) the complex. When two non-linked complexes have a common pool of facilitated systems, they may become linked through the simultaneous arousal of their fringe components. Symbolically, when two responses A and B are acquired separately but possess some properties in common, A and B may become linked owing to subsequent activation (in memory) of the neuronal elements representing their common properties.

Based on advances in functional neuroanatomy and logic of computer circuitry, the cell assembly concept was improved by including lateral inhibition, producing contrast in local networks (Milner 1957; George 1962). A synthesis of these emergent ideas with those of Hebb's reverberatory circuit is presented as Fig. 10. Stimulation of neurons with long axons (A) leads to propagation of the signal through excitatory synapses at neurons a1, a2, a3, etc. Lateral circuits produce (axo-somatic) inhibition of nearby neurons with short axons (e.g., b2). Inhibitory neurons (I1) stimulated by input (A) have axo-somatic endings on nearby neurons, thus establishing a field of inhibition (contrast) surrounding the stimulated neuron (e.g., a2). The latter, through initial sensitization and repeated activation, becomes specifically associated with the sensory input. Moreover, the activated circuit (A) through extensive cortico-cortical connections can activate other neurons including positive feedback circuit (Σ) that summate with the input signal thus producing a reverberatory assembly associated with a given sensory event. Local inhibitory neurons (opaque circles) damp excitation and terminate reverberatory firing. Learning occurs when cell assemblies activated simultaneously are repeatedly aroused, thus strengthening the linkages. In Milner's view, phase sequences become established when a cell assembly after firing under direct control of a sensory input is under *indirect* control for some minutes, and thus can also become associated with other assemblies of

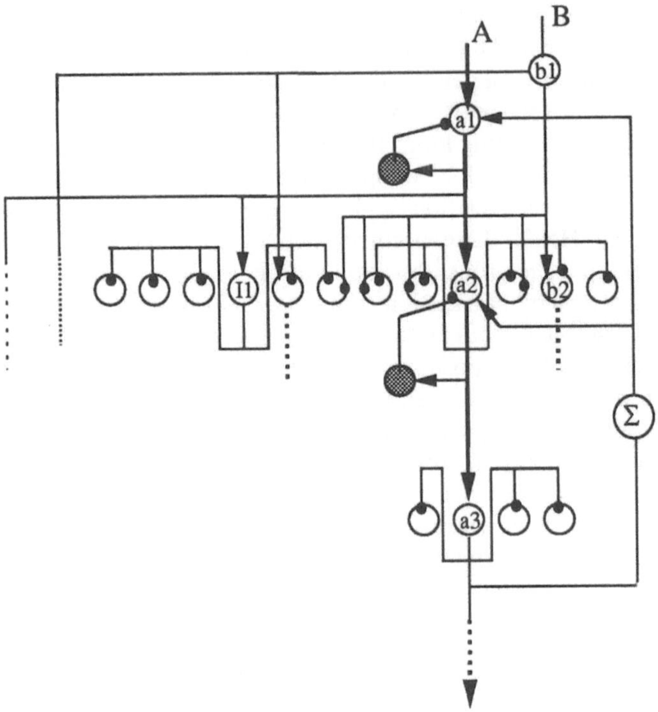

Fig. 10. A diagram of a neural network with feedback circuits and lateral contrast inhibition. Sensory input (*A* and *B*) arrives by afferent innervation. Excitatory synapses from input fiber *A*(*arrows*) arouse focal neurons (*a1*, *a2*, and *a3*). The axons from these neurons have inhibitory synapses (*solid*) on surrounding cells (producing *contrast*), and excitatory synapses on forward innervated cells (e.g., *a2* to *a3*). Input from fiber *B* produces activation of *b1* with a different set of collateral inhibitory connections, thus producing a different pattern of contrast inhibition in surrounding groups of neurons. Recurrent excitatory circuits (Σ) summate with afferent input producing reverberatory activity. This is damped by local inhibitory neurons (*opaque circles*). Consequently, excitation by a given input fiber arouses a limited set of neurons (forming a reverberatory assembly) that by alignment with the input signal and collateral inhibition become predictably and specifically associated with the stimulus. Compare to the multiple input, Lambda module shown in Fig. 12. (Based on data from Milner 1957, George 1962 and Hebb 1949)

neurons that are activated within a sufficiently short time period. Milner's models and others involving binary states of neurons (activated or inhibited) are analogous to states of computer junctions and have been used to relate CNS models with those of computer circuits (George 1962). The analogy is only approximate, however, since hard-wired computer circuits do not mimic one of the more significant aspects of CNS organization, namely, changes in axonal configuration and number of synapses, sprouting of dendrites, and the decrease in total number of neurons especially during early maturation. The latter is thought to be especially important in the gradual differentiation and localization of response patterns within the CNS.

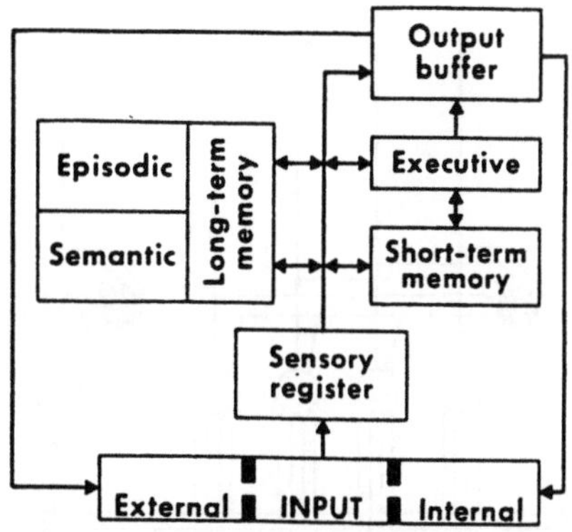

Fig. 11. A model of the cognitive system based on some concepts derived from computer technology. Input information is temporarily stored in a sensory register. This input can be distributed into several subsystems; i.e., long-term memory for permanent storage as semantic (verbal) or episodic (experiential) content, or temporarily placed in short-term memory where it can be mobilized for storage or output along the pathways shown by arrows. The executive function coordinates flow among several of the subsystems as shown by *arrows*. (From Andre and Phye 1986)

Nonetheless, the architecture and design theory of computers have substantially influenced thinking about the spatio-temporal requirements for information reception, analysis and storage by the CNS (Ashby 1952; Minsky 1959, 1968; Minsky and Papert 1988; Arbib 1989). For example, many modern cognitive models of information processing (Fig. 11) include components such as access buffers (STM banks), executive operators, and permanent memory storage (LTM) that reflect the impact of computer technology on the way we think about how the CNS operates. The influence between computer technology and neuroscience is becoming increasingly reciprocal as knowledge of CNS functions are applied to robotics and vice versa (Minsky and Parpert 1988; Arbib 1972, 1989).

Sommerhof (1974), while recognizing the limitations of computer analogies, combined computer logic with neurophysiological theory to produce a simple network model of learning designated a "Lambda module" (Fig. 12). He acknowledges that this is only a hypothetical module and offers it as a "heuristic" device to gain insight into CNS information processing. The model combines recurrent inhibition loops with multiple afferent innervation to yield selective output (responses) based on the strength of afferent (stimulus) input. Sommerhof, in building his model, applies three fundamental properties of afferent input to higher cortical structures: (1) complementary innervation (activation of a center is contrasted by inhibition of non-activated neurons, as also proposed by Milner 1957), (2) afferent convergence (input from widely different sensory channels can be localized within related groups of neurons), and (3) afferent divergence (sensory input from localized peripheral receptors may be distributed among overlapping groups of activated neurons). In this model, multiple afferent pathways (a–d) innervate groups of neurons through convergence (synaptic junctions on open circles, Fig. 12). Each activated neuron

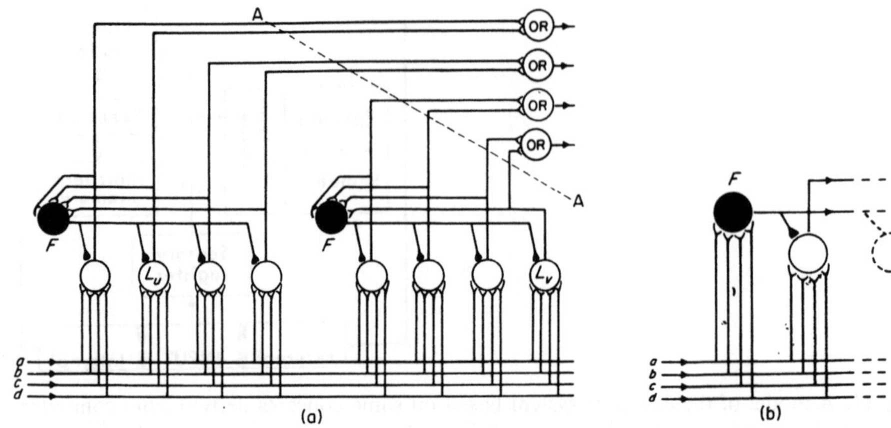

Fig. 12 a, b. Sommerhof's Lambda system is defined as a system of input-sharing neurons which are subject to some form of contrast-enhancing lateral inhibition. **a** Simple Lambda system using recurrent inhibition. Two "modules" are shown, each consisting of four "lambda neurons" governed by a single inhibitory neuron (*F*). The "*Or*" gates shown to the *right* of the *line A-A* are not included in the definition of the system since their function may be performed implicitly by follow-up systems. **b** Inhibitory segments of Lambda system using forward inhibition rather than lateral recurrent inhibition as in **a**. (From Sommerhof 1974)

sends collaterals to inhibitory neurons (F) whereby recurrent inhibition (opaque synaptic endings) is applied to the neurons within the field of its effect. The output from the neurons innervates integrating neurons with the property of an "Or" gate in computer logic. That is, the neurons will fire when activated by any one of the input fibers. There are three interpretations or models presented by Sommerhof to explain simple input-output specificity of the system. Model I is an arrangement that acts as a maximum intensity filter. That is, upon afferent activation of the assembly, the feedback inihibition is sufficiently strong that only the most strongly activated individual neuron reaches threshold and fires. Hence, afferent input becomes associated with a single output signal for any given combination of activated input pathways. Model II provides a more realistic interpretation. Recurrent inhibition is less selective, and it is assumed that a subset of neurons (not a single neuron) within the field of effect of the inhibitory neuron reaches threshold potential. This provides a far greater information potential, since various combinations of neurons in a module may be activated depending on the frequency and strength of input from the multiple innervating pathways. Instead of an output signal being either "on or off" from a single neuron, groups of neurons in various patterns of combination can be activated making the system much richer in responding capacity (compare to a combinatorial array, Fig. 4c). Model III (Fig. 12b) provides for forward inhibition instead of recurrent inhibition. Selective forward inhibition, through input from the afferent pathways, sharpens the contrast within the modules and refines the group of neurons in a module that become associated with a given afferent signal. This produces a system wherein only learned input patterns would be effective while non-learned input (and noise) would be ignored. Sommerhof cites the input-sharing and forward inhibition organization of the Purkinje cells of the cerebellum to support the concept of model III. In summary, "A system functioning according to this model would be effective in a case in which the desired effect of learning is that one source of inputs (the shared inputs) should gradually come to substitute for a second source (the supplementary inputs) as adequate, appropriate and possibly anticipatory releasers of the responses originally triggered by the second source" (Sommerhof 1974, p. 196).

Young (1966) combined computer systems theory with contemporary knowledge of neuroanatomy to propose a cybernetic model of memory. Much of his work is based on experimental studies of the relatively simple nervous system of the octopus (Fig. 13). According to this scheme, sensory input at dendritic fields of "classifying cells" produces categorical transformation of the input. This information is

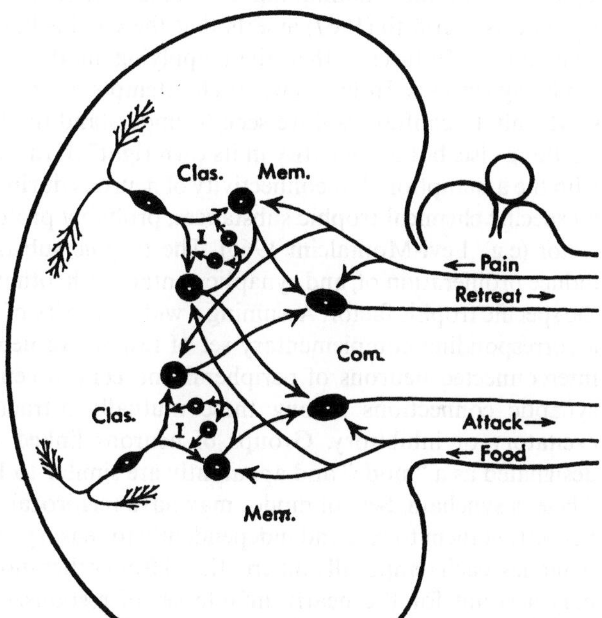

Fig. 13. A model of the memory system in the octopus. Upon initial input, the object seen by the eye is analyzed by dendritic fields of classifying cells (*Clas.*) with large branching dendrites; input fibers not shown. The information is relayed to memory cells (*Mem.*) which in the naïve state, pass on to command or motor cells (*Com.*) the signal to attack. If the attack succeeds in securing food, fibers signalling this result activate the memory cells, increasing the excitability of those cells promoting attack in that context. This differentiation is achieved by activation of small inhibitory cells (*I*), which in this case will prevent firing of memory cells whose output is to the motor cells governing retreat. If the attack is punished, the memory cells receive a pain input. Hence, those cells governing retreat are activated and concurrently activate inhibitory cells that suppress those memory cells that promote attack. (From Young 1964)

transmitted from the classifying cells to memory cells that coordinate input with output and pass along a command to motor cells to effect an "appropriate response." When a classifying cell activates a memory cell to effect an appropriate response, it is assumed that nonrelevant memory cells are inhibited through activation of small inhibitory neurons. Thus, through repeated presentation of sensory input and arousal of effective output by memory cells, a repertoire of learned responses is acquired. By further extension of the model, Young also discusses the requirements for an exploratory computer analog of neural activity, i.e., how information is acquired through actively probing the environment, based on observations of invertebrate exploratory behavior. Although modern research with invertebrate preparations has substantially improved our understanding of the cellular events accompanying learning in neural networks (Kandel 1976; Lynch 1986), Young's model is impressive as an early example of combining experimental knowledge from invertebrate neurophysiology with emergent concepts in computer logic toward understanding higher cognitive functions.

By contrast, Griffith (1967) asserts that the CNS is best understood in terms of its own unique features rather than applying models from computers and other nonliving systems. In his view, "such attempts are not entirely valueless, but they serve only to confuse us if we seek to understand the brain, not as a computer or anything else, but as an entity in its own right" (Griffith 1967, p. 1). Griffith begins with the assumption that connectivity of neurons during ontogeny can be explained by specific chemical trophic substances, probably proteins similar to nerve growth factor (e.g., Levi-Montalcini 1965). The trophic substances, secreted by neurons, induce proliferation of, and synaptic contact with, other neurons having affinity for the specific trophic factor. Assuming a wide diversity of such specific molecules, and a corresponding complementary set of responsive neurons, the regular matrix of interconnected neurons of peripheral and central centers can be generated. The synaptic connections among these mutually attracted neurons can be either excitatory or inhibitory. Groups of neurons linked by excitatory synapses are designated as a "mode" and apparently are similar to Hebb's assembly or perhaps Abeles's synchain. Sets of modes may have reciprocal inhibitory innervation, thus permitting them to respond independently to sensory input. Hence, a large network of modes, each temporally differentiated from other modes by reciprocal inhibition, may account for the nearly infinite set of responses exhibited by an advanced organism. In addition, Griffith hypothesizes that there are two sets of nerve cells – cortical and those of the "center" (perhaps in deeper levels of the brain) that act in concert, thus increasing the complexity of signal processing in the CNS. In Griffith's perspective: "The internal language is the language of the modes – that is, the presence of a particular word or idea in the consciousness corresponds not to some subtle details of the activities of neurons, but to the massive activity of a certain group coupled with the quiescence of all the others in that region" (Griffith 1967, p. 26). To account for coordination of activity in conscious experience, he further postulates the presence of a specialized set of modes known as a moderator, e.g., one of the subcortical gray masses, perhaps part of the hypothalamus that has an independent modal system of its own. The role of the moderators is to control the sequence of activity of modes within the center and can account for generalized states of consciousness and emotion. In brief perspective, memory is explained as strengthening of synapses through repetitive joint firing of interconnected neurons in a mode, and the increase in synaptic area through concomitant uptake of growth substances at the synaptic region. Thus, response patterns become fixed and predictable by the increased synaptic strength produced by elaboration of synaptic junctions occasioned by the uptake of nerve growth molecules concurrent with each activation of the synapse. Applying these ideas more broadly, Griffith offers explanations for more complex cognitive functions such as stream of consciousness, language, and dreaming.

Eccles (1978), building on cortical, structural-functional analyses of Marr (1970), Kornhuber (1973), and Szentágothai (1978), proposed a selection-instruction model for the dynamic interaction of the hippocampus with neocortical associational areas. In his model, ascending fibers from the hippocampus select and potentiate synapse formation by lateral cortical fibers intersecting the dendrites of pyramidal cells (Fig. 2, layer I). Fibers from the hippocampus form cartridge synapses (Fig. 2, sector A) with spines on the lower portion of apical dendrites of a pyramidal cell within a cortical module. Lateral fibers (commissural, associational and Martinotti axons) in lamina I make multiple contacts with apical dendrites of the pyramidal cells. Potentiation of synapses by lateral fibers (representing the memory trace) is produced when input from hippocampal afferents is sufficiently coincident with lateral fiber input to reinforce the connectivity. Thus, the hippocampus acts as a selecting center to determine the pattern of cortical pyramidal connections based on temporal conjunction of cartridge synaptic input with signals arriving at lateral fiber synapses in lamina I. Contrast inhibition within a localized region of the activated pyramidal cells may further differentiate the field of activation. Memory is laid down in a neural network, organized spatiotemporally by the pattern of input from the hippocampus. Hence, the hippocampus serves as a mediator of information encoding in LTM through its selection-potentiation function, but as reported in Sect. 3.1 it is not assumed to be the locus of the memory trace.

An additional model of the dynamic interaction between hippocampus and neocortex has been elaborated by Teyler and DiScenna (1986) who hypothesize that the hippocampus provides an index for the spatiotemporal activation of neocortical neural modules. Thus, the hippocampus serves as a mediator of initial memory encoding and as a memorial index for subsequent rearousal of the neuronal assemblages that form the memory trace. Memory, therefore, is encoded as a set of spatiotemporal patterns of activation of neocortical neurons orchestrated by signals from the hippocampus as posited by Eccles (1978). The hippocampus is not the locus of the memory trace, but the indexing center that determines the orderly arousal of neocortical activity representing prior experience. Moreover, the temporal relations between neocortical patterns of activity and those of the hippocampus are hypothesized to be the basis of historical reference, that is, our ability to assign an experience to a point in time.

A more detailed, and perhaps more complex, interpretation of multiple interactions among brain centers in memory has been proposed by Lynch (1986) based on a structural-functional analysis of olfactory learning. Only the major features of the model are summarized here, and some essential details, required for a full appreciation of the predictive capacity of the model, must be sacrificed for brevity. A major premise of the model is that much of the associative cortex, including memory organizing loci in the hippocampus, contains multiply innervated neurons receiving multiple synapses in a quasi-random pattern from afferent pathways (Fig. 4c). These multiple sensory input pathways, combined with recurrent axonal inputs from the same set of neurons and those from commisural/associational inputs from other distant neurons, form a *combinatorial array*. This is contrasted to a punctate or *topographical* connection where a single synapse is made on each neuron, producing topographical correspondence between a localized peripheral input and its registration in higher brain centers. The combinatorial array, though more diffuse and complex in organization than a topographical arrangement, has certain advantages for complex encoding of sensory input. Sensory input, containing several components, is refined during passage through such an array. At the first level the array serves as a type of filter to encode only those synaptic connections that represent the combined component input, and to eliminate those that contain only partial or individual components of the signal. This is reminiscent of the filtering functions of the Lambda module proposed by Sommerhof (1974). Output from the activated neurons in the first level (the combined component signal, e.g., xy in Fig. 4c) may spread by axonal connections to other combinatorial units. These may further refine the signal to ensure that only the signal containing the stimulus-specific, combined-component properties of the sensory input are registered and propagated to higher centers. By positing a coordinating combinatorial array at a higher level that receives the filtered signal and by feedback can re-excite the intermediate input arrays, Lynch's model provides for recall of encoded information. The hippocampus, richly supplied with combinatorial arrays, may serve as a directing center to combine input from more than one modality and ensure that this combined input is targeted to the appropriate site for encoding in higher centers. Moreover, through a modulating role that suppresses localized inhibitory neurons that damp repeated firing of excited encoding neurons (see Fig. 10, opaque interneurons), the hippocampus may induce sustained bursts of firing by the encoding neurons in a loop. This produces long term potentiation (LTP) at the synapses and can account for LTM formation. Some finer aspects of complex associational learning, and the neuroanatomical basis for differentiation between declarative (data) knowledge and procedural knowledge are also addressed by Lynch (1986).

Further descriptive analyses of CNS function in relation to memory and higher cognitive processes have been presented by Russell (1959), Smythies (1970), Young (1978, 1987), John (1980), Ushakova (1980), contributed papers in Thompson et al. (1980b), and Markowitsch (1988).

The emphasis in some of the foregoing models is on the ultimate role of the CNS in information storage and retrieval and less on perceptual processing of complex sensory input. Royer (1985) presented a model of text perception and comprehension that is based on a biological metaphor. Royer assumes that there are networks of responding elements called collectively "nurogens." These entities have properties similar to neurons, that is, they have thresholds for firing and they can spread activation to other nurogens. Nurogens are organized into a hierarchical

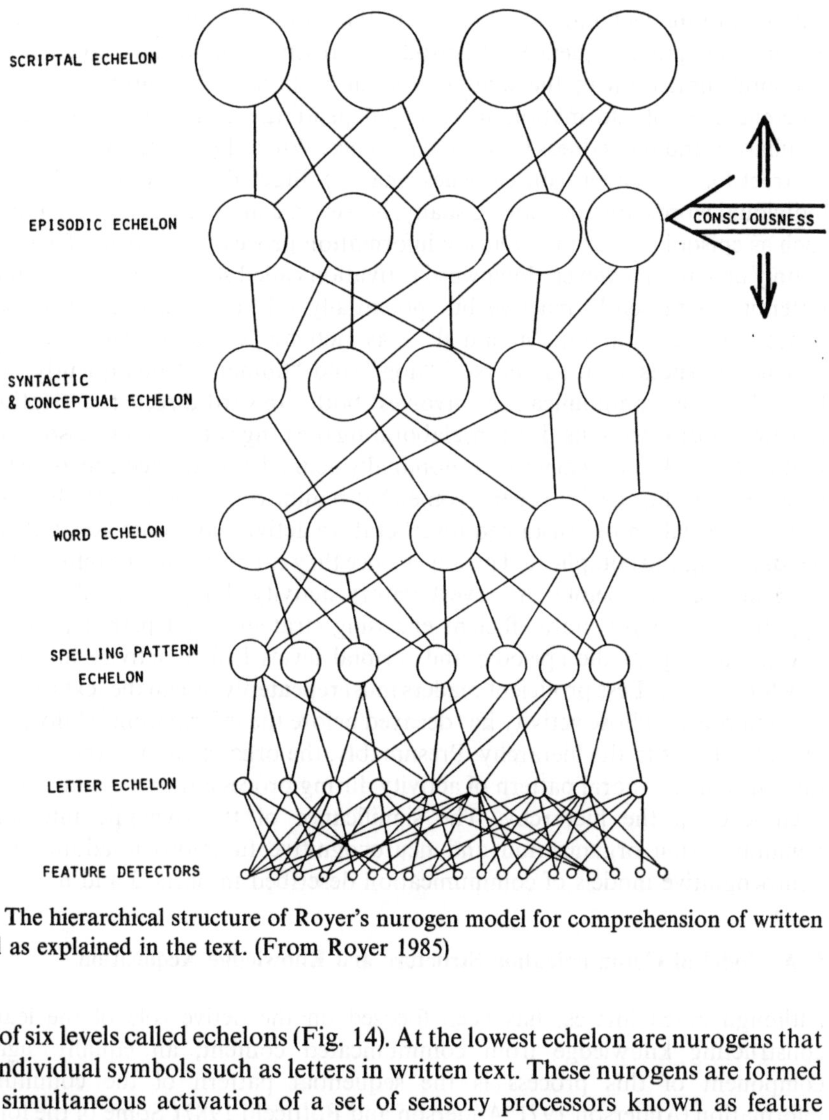

Fig. 14. The hierarchical structure of Royer's nurogen model for comprehension of written material as explained in the text. (From Royer 1985)

system of six levels called echelons (Fig. 14). At the lowest echelon are nurogens that detect individual symbols such as letters in written text. These nurogens are formed by the simultaneous activation of a set of sensory processors known as feature detectors that correspond to a particular symbol or letter. These detectors may correspond to the lowest tier of perceptual processing units in the neurocognitive model diagrammed in Fig. 15. The more refined and differentiated information from the feature detectors in Royer's model is fed into a higher level of processing known as the *spelling pattern echelon* which in turn activates more comprehensive processes mediating word recognition. The next tier of information processing nurogens mediate more conceptual and syntactical categories. At this level, groups of words are parsed into categories such as nouns, verbs, and modifiers, etc. Moreover, the conceptual meanings of each word are established in this echelon. Further processing of the information occurs in the episodic echelon where words are assembled to represent categories of experience, i.e., the unique and individual set of representations each individual relates to the symbolic input. Higher order scriptal echelons integrate sequences of episodic representations to give coherence and a temporal dimension to the written communication. Consciousness can occur at different levels of information processing depending on such factors as complexity, familiarity, and relatedness to prior experience. When the written material is new or abstract for the reader, consciousness may be directed toward lower levels where individual words are analyzed. Usually, however, consciousness is at a higher level such as episodic or scriptal. During information processing, spread of activation is mainly upward. However, downward activation called sensitization raises the firing potential of presently inactive but potentially relevant nurogens. For example, concept recognition for "mammal" may activate word nurogens representing specific instances such as "horse," "dog," and "human." Consequently, comprehension of a communication involves both upward-directed activation and downward-directed sensitization, elaborating the range of meanings associated with written texts. Working memory is nominally defined as sustained encoding through reverberatory firing of nurogens, somewhat reminiscent of Hebb's (1949) assembly. Individuals with more sustained reverberatory activity are more adept at comprehending long and complicated passages since they can integrate the information into working memory before the reverberatory activity dampens. Proficient readers hypothetically have more efficient encoding strategies and permit processing of information up to the episodic and scriptal levels before activity in lower level echelons decays. Less proficient readers must repeatedly rescan the text to reactivate word nurogens whose activity has decayed before the information is fully processed at higher levels in the hierarchy. Presumably, the organization of the text can also influence the temporal pattern of activity during processing of a communication at each level of the hierarchy. The significance of the dynamic interaction of communication organization and neurocognitive functions is addressed in the neurocognitive models of communication described in Sects. 5 and 6.

5 A Model of Communication Structure and Knowledge Acquisition

Although much interest has been focused on the active role of the learner in constructing knowledge from communicated content, an equally significant component of this process is the sequential pattern of the communicated information (Anderson 1971; Anderson and Botticelli 1987). Some of the foregoing theory points toward the facilitatory role of patterned stimulus input in cerebral encoding of experiences. The biopsychological origins of some of these processes are incorporated in this model of human communication and information acquisition.

5.1 Assumptions

The assumptions of the model are:

1. The human nervous system, and the behaviors mediated by it, are the product of evolution. They have been shaped by environmental selection pressures

including effects of patterned stimuli (visual, auditory, and tactile stimuli, periodicity of radiant energy, diel light cycles, and other periodic phenomena) that have become encoded in CNS activity as a way of effectively responding to the environment and hence increasing species survival. A more detailed analysis of the evolutionary aspects can be found in previous publications (Anderson 1969, 1971). More recently, Rapp (1987) has proposed five significant phylogenetically adaptive roles of periodicity in biological systems; (a) temporal organization, (b) spatial organization, (c) prediction, (d) efficiency, and (e) frequency encoding of information (e.g., nerve fiber transmission). Among these a, c, and e are particularly relevant to the model presented here.

2. Patterned stimulus encoding enhances accurate and stable storage of information in memory, while novel stimuli sustain arousal. A proper balance between patterning of input and novelty maximizes information storage.
3. Phylogenetic advancement has been accompanied by increased sophistication of the CNS to regulate itself, mobilize stored information, and selectively attend to information input to be connected with, and assimilated into, prior stores of information.
4. Through hierarchical control mechanisms and multiple feedback loops, sensory input is parsed into encodable units and selectively incorporated into existing knowledge structures by the action of higher order synthesizing operations linked to lower order perceptual processing functions.

5.2 Theory

The organizational scheme for this model is presented as Fig. 15. The higher order executive operator (possibly a complex set of interacting centers) controls the mobilization or prior knowledge by activating encoding centers (E_n) and synthesizing existing knowledge with incoming information. The sensory input is parsed into identifiable units by suboperators (S_n) organized in parallel processing arrays and controlled by executive operator functions. The system is assumed to be hierarchically organized and regulated by forward directed control emanating from the executive operator. Feedback innervation from the suboperators (dashed arrow) provides timing and prompting cues that coordinate the work of the executive operator. This total structure is presented as a refinement of the "System for creation of model" contained in the Sokolov/McGuinness/Pribram model (Fig. 9). Thus, part of the arousal functions presented in the description of Fig. 9 are superimposed on the encoding functions displayed in Fig. 15. The net flow of encoded information, mediated by the executive operator, is from suboperators (S_n) to long-term storage in the encoding centers (E_n). Part of the executive operator functions may be located in the hippocampus and its neocortical connections. The hippocampus has been described as a major center for place-keeping functions, and as an indexing or mapping center of the brain as reviewed in foregoing sections (e.g., O'Keefe and Nadel 1978, 1979; Teyler and DiScenna 1984). As much as 50% or more of the cells in the hippocampus are classified as having a place correlate and are known as place units (O'Keefe 1979). These place units respond to cues from the environment, and

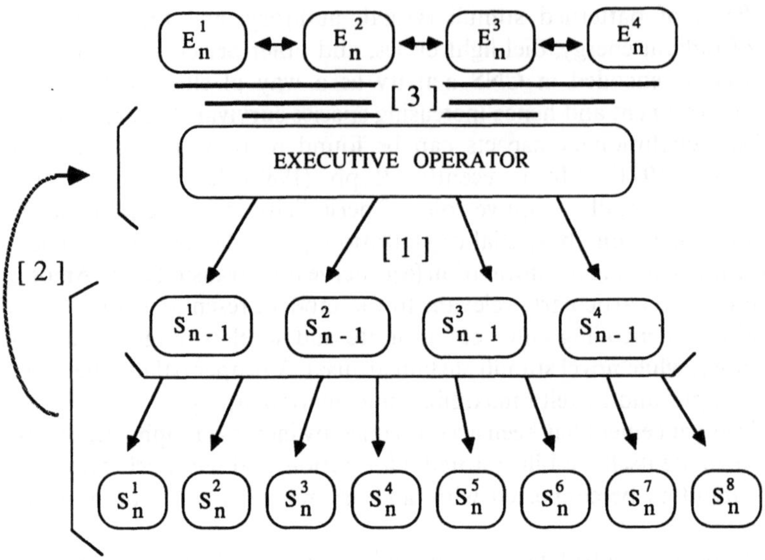

Fig. 15. A hierarchical and parallel-processing model of information analysis and encoding during knowledge acquisition. A coordinating function, designated "executive operator" (probably involving several brain centers, perhaps particularly associated with hippocampal activity), mobilizes long-term memory stores by activating encoding operators (*E*) through controlling neural pathways (*3*). Forward-directed control (*1*) of primary information processing functions mediated by suboperators (*S*), organized in hierarchical (perhaps parallel) arrays, determines the filtering and selection functions during immediate perception of incoming information. Feedback (*2*), through recurrent innervation, signals the executive operator as to the status of the suboperators. Flow of information into long-term memory is from suboperators (*S*) to memory encoders (*E*). Effective encoding requires proper phasing of two processes: (1) activity of the executive operator in mobilizing schemata for encoding new information, and (2) readiness of the perceptual processing units as monitored through the feedback signals. If memory storage functions are not properly synchronized with primary perceptual processing of information, interference or inhibition of information acquisition can occur, as explained in the text

apparently contribute to the place-keeping and mapping capacity of the hippocampus. The possible role of hippocampal function in LTM (Teyler and DiScenna 1986) has also been presented in the foregoing section.

Multiple control mechanisms, labeled (1) and (3) in Fig. 15, integrate the processes of perception and storage of information into LTM. The forward directed control (1), among other functions, determines selective attending, stimulus focusing and other filtering functions associated with early stages of information processing. Presumably, these functions are coordinated by the executive operator with reference to prior information stores in LTM, and thus constitute the active construction phase of perception and the mobilization of schemata for storage of new information. Among other significant spatio-temporal events in the CNS, the timing of, and phase relationship between, the forward directed control signals (1) and the multiple feedback (2) from more peripheral processing centers is significant in the efficiency and smooth forward flow of encoding. For example, if feedback signals, indicating completion of perceptual processing of new input, arrive at the executive operator before the current load of information is transferred into long-term storage, encoding may be disrupted. One characteristic of cognitive ability may be the efficiency and rate of mobilization of encoding strategies by executive operators. The more proficient the learner, the more rapid the mobilization of LTM loci for storage of new information. Hence, the rate of encoding into LTM is sufficiently rapid to keep pace with primary perceptual processes and results in a smooth flow of information into memory. Within this limited interpretation, the more capable the learner, the greater the coordination between executive operator functions and feedback signalling from primary perceptual processors. Learning dysfunctions arise when there is an imbalance in the categorization steps and timing of these operations. During visual processing of written material, for example, the synthesis of semantic elements into schemata during an active construction phase may lag behind the parsing of content into identifiable units, thus leading to phase imbalance between encoding and feedback. This can cause disruption of encoding, requiring rescanning of the written material to bring the encoding functions into proper phase relationship with the feedback cues from suboperators. Under these circumstances, reading behavior is characterized by a seemingly uncoordinated forward movement involving frequent halts and rereading. This phenomenon can occur also when people attempt to read complex material too rapidly and find that they "get ahead of themselves." Efficient strategies of "chunking" or grouping content during perception may contribute to smoother encoding operations by increasing the efficiency of incorporation of the content into schemata, thus maintaining a coordinated action between forward-directed control and feedback. During aural processing of information (oral communicated content) the opportunity to rescan stimulus input is limited, and the likelihood of information loss due to a mismatch between feedback and forward-directed control becomes more significant. Again, the greater the capacity of the individual to group information during input, perhaps by schematal control over categorization processes in the perceptual units (S^n), the more likely that place-keeping will be enhanced and steady state flow of information into LTM will be sustained.

Multiple information processing pathways are indicated in the model (Fig. 15). The spatio-temporal patterning among the parallel processing units may vary among sensory modalities and in relation to the quality of the sensory input. Input with a high signal to noise ratio may be processed by parallel analysis of component parts of the signal to extract the "texture" or meaningful diversity in the information. For example, while listening to musical harmony, the variety of tonal and temporal patterns may be processed through parallel units to extract as much of the spectral diversity in the information as possible. If, however, there is a low signal to noise ratio, e.g., during focal attending to a verbal message in a noisy environment, then the capacity of the multiple processing units may be dedicated to extracting predictable features from a a narrow band of the spectral input in order to detect the useful components of the auditory signals. In this case, the multiple processing units may be directed toward cross-comparative analyses (e.g., by "cross communication") to detect the correspondent or matching signal elements that carry the information content of the signal. This would explain the inner sense of "unified purpose" or "single purpose of intent" that accompanies focal attending. Hence in focal attending, at least some of the mutiple processing pathways become dedicated or unified toward a focused signal analysis aimed at cross comparative signal matching to detect the usable or predictable information elements from the more chaotic components categorized as "noise." In such noisy environments, signal matching is particularly enhanced if the message contains sufficient thematic unity and closely connected ideas among sentences to facilitate perception (i.e., cross-channel detection of the significant signal elements and their transfer to memory encoding centers). Moreover, the presence of a "familiar context" used as a theme can arouse higher order (schematal) organizing functions (via the executive operator) that sensitize lower order perceptual processors (forward directed arrows in Fig. 15) to detect expected features in the communication. This sensitization to anticipated features further reduces the information processing load and indeterminacy of the suboperators (S^n) when encoding a noisy or complex message. Cortical plasticity may be a significant factor in promoting the moment-to-moment dynamism required to switch from broad spectral processing of information to focal processing. Weinberger and Diamond (1987) note that in the auditory cortex, plasticity appears to provide a parallel type of processing whereby features of sensory input are abstracted, and fine details can be ignored *unless they are of importance*. This plasticity may be "learning-induced" (Weinberger and Diamond 1987, p. 42) and is further evidence of the dynamic role of cortical processes in selecting and delimiting input for synthesis into cognitive representations of experience.

Within the context of the neurocognitive models (Sects. 5 and 6), the active contruction phase of encoding information into long-term storage involves minimally two key aspects: (1) the availability of relevant connecting content in schemata to permit stable and reproducible storage, and (2) the proper organization of content to facilitate place-keeping during encoding within updated schemata (models). For long sequences of verbal and figural content, as presented in extended discourse, the role of patterned semantic markers (e.g., connecting ideas of information within contiguous sentences) and the development of clearly identifiable themes to maintain "continuity of thought" become a significant variable in efficiency of learning. Streams of discourse with clear references to recurrent theme elements, and salient linking ideas between utterances, facilitate encoding by providing semantic markers that recur in successive stimulus units ("sentences") thus providing the necessary predictability and place-keeping functions needed for smooth flow of information into long-term storage.

The presence of linking ideas in successive units of discourse (Table 1) is designated "commonality." Commonality contributes to effective communication by providing one form of patterning that enhances construction of schemata. Clearly, if the commonality elements are selected to match existing information units in LTM, their place-keeping functions can be enhanced. To maintain arousal, however, novel or salient content should be introduced at appropriate intervals during exposition.

Table 1. Coded discourse units and coefficients

Discourse units[a]	B_1	D_s	B_2
1. What determines the shape[1] of proteins[2] in solution[3]?	—	—	—
2. Proteins[2] can assume a variety of shapes[1] depending on the solvent[4] of the solution[5] where they occur	0.57	0.86(2)	0.95
3. Some proteins[2] form tightly spiraled[6] helices[7]	0.29	1.42(2)	0.92
4. Other proteins[2] are loosely arranged[8] in a coil[9]	0.33	1.34(2)	0.94
5. The shape[1] of the protein[2] is determined in part by the interaction[10] of the protein side groups[11] with the solvent[4]	0.25	1.50(2)	0.88
6. Now an important distinction[12] we need to make is based on the solubility[13] of side groups[11]	0.25	1.50(2)	0.89
7. Side groups[11] may be polar[14] or nonpolar[15]	0.33	1.34(2)	0.91
8. Nonpolar[15] side groups[11] have an oily property[16]	0.67	0.33(1)	0.96
9. They[11,15] are not soluble[17] in water[18]	0.57	0.86(2)	0.96
10. Polar[14] side groups[11] have a dipole moment[19] and are soluble[17] in water[18]	0.67	0.33(1)	0.95
11. The proportion[20] of polar[14] and nonpolar[15] groups[11] in part determines the shape[1] of a protein[2] in a solvent[4]	0.33	0.67(1)	0.90
Means	0.43	0.92(1.6)	0.93

[a] Codes for each word element are shown as superscripts.

This distributed novel input is labeled "progression". Progression serves the dual purpose of sustaining arousal and increasing the quantity of new information encoded, thus contributing to the efficiency of the communication. We therefore arrive at a central principle of the model:

A proper balance between (1) commonality (continuity of thought and patterning of familiar content) and (2) progression (introduction of salient, orienting, new content) yields an *effective* and *efficient* communication.

Quantitative measures of the patterns of distribution of commonality and progression are significant predictors of knowledge acquisition of relatively unfamiliar (new) information (Anderson and Lee 1975; Ferraro et al. 1977; Johnson and Lockard 1985; Simmons 1980).

5.3 Quantitative Measures

A low-inferential and reliable method of coding communications has been developed to identify the presence of explicit semantic markers contributing to commonality and progression. While it is possible to code implicit meanings, such as underlying principles or concepts (Anderson and Botticelli 1987), we have found that the best indicator of organization and the most reliable predictor of knowledge acquisition is obtained by coding explicit word elements within each sentence. In this procedure, the communication is divided into discourse units (sentences for written discourse or clauses, comparable to sentences, in oral discourse). Each unit is examined for word elements.

These are the major semantic markers (terms, concepts, principles) that are used in the content field being communicated. Rules have been developed (available from the author) to increase the reliability of coding. An example of semantic elements coded in a short communication appears in Table 1. Each semantic element is assigned an identifying number. These numbers are entered in computer programs to compute coefficients of structure for the communication. Every occurrence of each element within the communication is marked by assigning its code number to the discourse unit. The coding results in a list of code numbers of each discourse unit representing the flow of major semantic markers in the communication (Table 1). Although this process may seem relatively trivial or simple, the subsequent, more sophisticated, theory-based, numerical analyses using these codes provide good predictors of knowledge acquisition.

5.4 Commonality and Progression

Commonality is assessed by computing a matching coefficient (B_1) for each *pair* of contiguous discourse units (Table 1). The sentences presented in the table are purposefully simple to provide sufficient examples within a limited space. Commonality is assessed by the following coefficient:

$$B_1 = \frac{n_1}{n_0 + n_1} \tag{11}$$

where n_1 is the number of semantic elements that are matched (repeated) in the pair of discourse units, counting each occurrence in both of the pairs, hence twice for the pair, and n_0 is the number of unmatched (not repeated) semantic elements in the pair. For example, in the first pair of discourse units exhibited in Table 1, elements 1 and 2 occur in both statements, their total occurrence therefore is $n_1 = 4$. Counting the remaining unmatched elements in both statements we obtain $n_0 = 3$. Hence the quotient of matched to total number of elements is 4/7, and $B_1 = 0.57$. The B_1 value for each successive pair of discourse units is also listed in Table 1. A mean value is obtained as one of the fundamental (albeit insufficient alone) quantitative indicators of communication structure. Mean commonality values observed in secondary and college science instructor's communications typically range from 0.20 to 0.40.

A first order measure of progression (P) is obtained by taking the complement of B_1:

$$P = 1 - B_1 \tag{12}$$

A more robust and essential measure of progression and recurrent themes in a communication is obtained by the progression density coefficient (D_s), as in Eq. 13. This is computed for each discourse unit within a segment of discourse comparable to a paragraph in written communication, and a mean value ($\overline{D_s}$) is obtained as reported at the bottom of Table 1.

$$\overline{D_s} = \frac{\sum_{i=1}^{n} [P_i \cdot (N_i)]}{S_n} \tag{13}$$

where P_i is the progression value for each discourse unit (i), N_i is the number of new verbal elements in the discourse unit, and S_n is the total number of discourse units in the segment. The new verbal elements in each discourse unit are those not perviously introduced in the segment of communication being analyzed. In this operation, each progression value is multiplied by the number of new elements introduced at each step, thus giving a density dimension to the coefficient. If there are no new elements introduced in a discourse unit, that is, only previously introduced content is being reorganised and presented, then the value of D_s at that point is zero. In the first pair of discourse units in Table 1, the progression is 0.43 and there are two new elements (4 and 5), hence the progression density is 0.86(2). For purposes of reference, we place the number of new elements in parentheses following the progression density value. The mean of the sequence of values is computed as an overall indicator of progression density in the segment of discourse. Mean progression density values obtained for scientific lectures typically ranges from 0.50 to 1.50. The value of 0.92 obtained for the sequence of discourse in Table 1 is a moderate value.

5.5 *The Kinetogram*

To fully characterize a communication graphically, a kinetogram (Fig. 16) is plotted. Numerically coded human communications consistently yield a predictable pattern similar to that of Fig. 16, whereas nonsense or random input produces a irregular pattern without the marked vertical spikes and intervening elevated traces. The kinetogram is a visual record of the flow of communication structure (encompassing commonality and progression components) produced by a point-to-point line graph of B_2 values (Eq. 14). The B_2 coefficient is a modified B_1 coefficient. The progression value (P) at each discourse unit pair is multiplied by a weighting factor (W) and the product is subtracted from 1 to yield an adjusted B_1 value, hence labeled B_2.

$$B_2 = 1 - [P \cdot (W)^{1/2}] \tag{14}$$

where $W = (f' + f'')/\Sigma f$. It is necessary to take the square root of W to obtain the proper weight for this factor. The denominator, Σf, is the sum of counts of the codes in the communication. The number of coded words in each discourse unit is summed across the discourse units. As an example from Table 1, $\Sigma f = 43$ which is the total count of the code numbers. However, for short communications, it is preferable to normalize the equation by setting the $\Sigma f = 300$. This is in the range where fluctuations of the graph more accurately reflect structural variations in the communication. Hence, it provides greater reliability in cross comparisons among communications of varying length. The Σf quantity is a constant value for a communication. The terms f' and f'' are obtained for each computation of a B_2 and hence are variables.

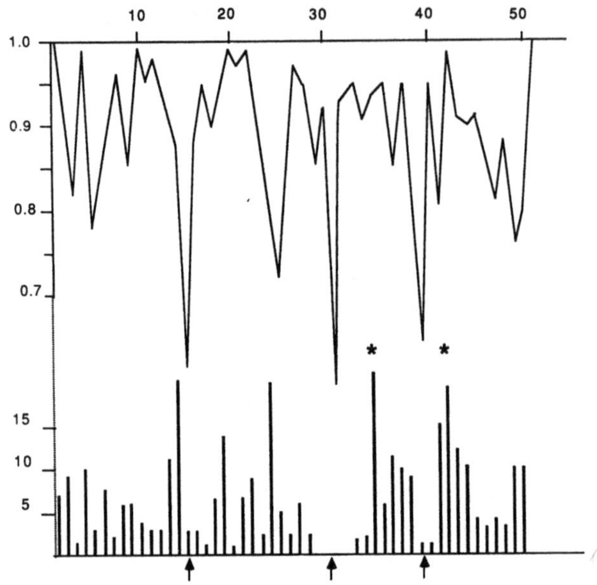

Fig. 16. A kinetogram (*upper* trace) records the fluctuations in communication structure in relation to free recall of communicated content (bar graph *below*). *Arrows* at the base line mark points where there was a sharp transition into new content, indicated by large downward-directed spikes in the graph, and recall was suppressed. Novel scientific information was presented to reduce the effects of prior memory that may interdict the transition-associated depression in recall. Recall increases with greater elevation of the kinetogram trace indicating greater semantic continuity in the communication. This effect is more pronounced for more recently acquired informatin (*asterisks*)

The value substituted for f' is obtained by examining the first discourse unit of the pair (for example discourse unit 1 of pair 1 and 2 in Table 1). We note the elements that are unmatched. In the first discourse unit, code 3 is unmatched, and in the second discourse unit, codes 4 and 5 are unmatched. Next, the frequency of occurrence of these unmatched elements in the *total* communication is noted. The frequency of highest value (most salient by potency) is substituted for f'. Likewise, f'' is obtained by examining the second discourse unit of the pair (discourse unit 2 for pair 1 and 2) and noting the elements that are unmatched. Again, among these noted elements, the frequency of highest occurrence in the total communication is substituted for f''. In the case of the discourse units 1 and 2 in Table 1, element 3 is the sole unmatched element in discourse unit 1, it occurs only once in the total segment, hence $f'=1$. In discourse unit 2, the unmatched elements are 4 and 5 with frequency counts of 3 and 1 respectively. Therefore, we select the larger value and set $f''=3$. The sum $(f'+f'')=4$. When the sum is divided by Σf, this yields a weighting factor representing the potency of the progression step at this point in the communication. Hence, $W=4/300$. Since it is burdensome to do these calculations by hand, a computer program is available from the author to compute the foregoing coefficients. The range of B_2 values is 1.0 at the upper boundary and approaching zero at the lower boundary. B_2 values near 1.0 (especially in the range 0.9 to 1.0) indicate strong continuity of thought within the communication, whereas lower values indicate increasing lack of continuity, or the inception of strongly developed theme ideas without linkage of foregoing thought. In standard practice, the B_2 values are plotted as a function of discourse unit to form the kinetogram. For convenience in interpreting the graph, the abscissa is placed at the top of the graph as a reference line.

5.6 Experimental Analyses of the Model

According to the foregoing theory, large downward-directed spikes in the kinetogram record indicate a sharp transition in the communication, with non-linked introduction of novel material, and therefore should be accompanied by a decrement in encoding due to arousal, lack of predictability, and suppression of cerebral encoding. However, a subsequent development of commonality around a theme, as represented by upward-directed shallow spikes on the kinetogram, should enhance encoding of this integrated information. Therefore, although the content associated with the immediate, initial arousal of a theme may be suppressed, there should be more stable encoding and enhanced recall of semantically connected content following the arousal event. To test these assumptions, a modest experiment was done with adolescent learners who were presented with a tape-recorded communication on life in the open ocean consisting of four short, descriptive paragraphs. Each paragraph focused content on a particular life form, and was introduced with a sharp transition in thought followed by development of continuity around a theme idea. This produced the kinetogram shown in Fig. 16. The arrows at the base line indicate the points of sharp transition between paragraphs as indicated in the kinetogram by the relatively large-amplitude, downward-directed spikes. After hearing the tape-recorded message, each listener was asked to use free recall and write down as many of the statements as they could remember from the listening task. The total recall of the group was recorded as a bar graph at the bottom of the kinetogram (Fig. 16). Thus each bar represents the total recall of the group for the statement corresponding to the discourse unit at that point in the kinetogram. The fluctuation in knowledge recall can be compared to the communication structure at each point by noting the relationship of the fluctuations in bar heights to the changes in elevation of the kinetogram trace. As predicted by theory, the recall is suppressed at points of sharp transition (arrows), but increases substantially for thematically organized communication sequences represented by elevated traces on the kinetogram, especially in more recent segments (asterisks).

In a more formal analysis of theory (Anderson and Lee 1975), two groups of adolescent students were presented tape-recorded communications in science wherein the structure varied alternatively higher to lower from one paragraph to the next (Fig. 17). However, using a type of counterbalanced design, one group received the communication with a pattern commencing with high structure in the initial

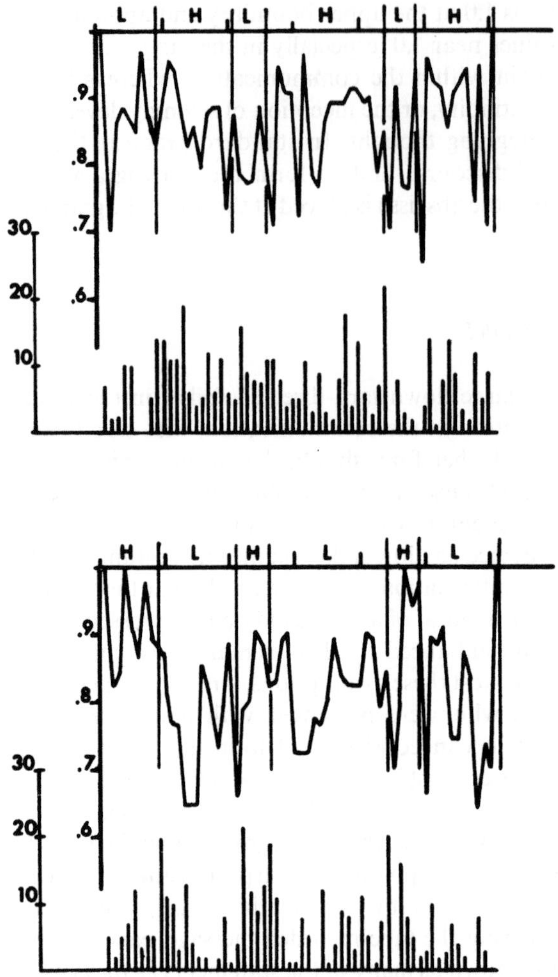

Fig. 17. Free recall (bar graphs) in relation to communication sequential organization (kinetogram traces) for lessons with alternating segments of high and low thematic continuity. *Upper trace*, pattern commencing with high progression density and low-continuity segment (hence low-high pattern). *Lower trace*, pattern commencing with low progression density and high-continuity segment (hence high-low pattern). Comparisons between upper and lower segments, representing the same content with different amounts of continuity, show greater recall for high continuity segments compared to low continuity segments. See Table 2 for quantitative data. (From Anderson and Lee 1975)

paragraph, whereas the second group received an alternative form of the same content commencing with a low-structured paragraph. This can be confirmed by examining the trace in the first paragraph, delimited by vertical lines. The lower trace (high structure version) has a more elevated trace than the corresponding segment in the upper kinetogram trace (lower structure). Thus, it was possible to

Table 2. Mean commonality coefficients and mean recall scores for communications 1 and 2 in Fig. 6

	Communication 1 Low-high treatment (L-H)		Communication 2 High-low treatment (H-L)	
Segment	Mean B_1	Mean recall	Mean B_1	Mean recall
1	0.35 (L)	5.6	0.52 (H)	7.1
2	0.45 (H)	9.2	0.24 (L)	4.9
3	0.23 (L)	10.6	0.37 (H)	14.6
4	0.32 (H)	7.9	0.13 (L)	6.2
5	0.31 (L)	2.6	0.53 (H)	6.2
6	0.48 (H)	7.4	0.19 (L)	3.8
	Total Mean Recall for low segments 5.5		Total Mean Recall for high segments 8.4	

Note: Comparisons are meaningful only within rows between high (H) and low (L) versions of segments with the same content. As is to be expected, learning varies across segments containing different content. However, when content is constant (within a row) there is a consistent direct effect of mean commonality (B_1) on recall. Differences are significant at $p < 0.01$.
From Anderson and Lee (1975)

examine changes in recall of knowledge within a group and to compare recall performance on a given paragraph across the two groups. A comparison of free recall within each group demonstrates once again that on the whole, given the variation in content among paragraphs, the bar heights follow closely to the trace of the kinetogram. Holding content constant, and comparing results between groups for each paragraph, knowledge recall was consistently directly related to structure (Table 2) and the mean differences were statistically significant ($p < 0.01$, by *t*- test).

6 Neuromathematical Model of Information Processing and Knowledge Acquisition

Based on the foregoing theory, a neuromathematical, predictive model of knowledge acquisition was derived (Anderson 1983, 1986). The theory guiding the development of the model is based on the hierarchical model of Fig. 15, and also pertinent concepts of CNS information processing proposed by Sommerhof (1974), Eccles (1978), Szentágothai (1978), and more recently by Lynch (1986). The distillation of significant points is as follows. Nerve networks stimulated by afferent innervation achieve a steady-state level of excitation through the combined effects of lateral inhibition and recurrent excitatory neurons within a network. Spread of excitation from one part of the network to another by branching fibers, making excitatory synapses on other neurons, produces an elaboration of effect resulting in amplification of the afferent input. Modules of neurons, with multiple afferent innervation and contrast enhancement controlled by lateral inhibition, become aligned to afferent patterned input as the contrast effect becomes more differentiated yielding a predictable relationship (connectedness) between the afferent input and the patterned neuronal output. There is initial instability in the predictiveness of the input-output connectivity until sufficient recurrence of the patterned input induces alignment (registration) of the ouput neurons leading to stable encoding of experience.

6.1 Assumptions

The major assumptions of the model based on this theory are:

1. The processes of immediate acquisition of information and its synthesis into LTM within prior-gained stores of information can be modeled as a set of time-dependent equations representing rates of general processes in CNS activity.
2. Theoretically, these rates represent the summative contribution of fundamental neurophysiological processes that cannot at present be analyzed individually, but are amenable to wholistic rate analysis.
3. Initial acquisition of information is characterized by instability of neuronal encoding, perhaps occasioned by insufficient contrast formation among neuronal networks, that decreases during subsequent intervls of responding to patterned input. The decline is proportional to the degree of patterning in the communicated information.
4. Stability of encoding tends to decrease as information load in primary processing centers increases during information reception. The rate of decrease is proportional among other factors to the amount of unpredictability in the communicated information.
5. The mobilization of permanent memory stores for assimilation of newly acquired information during knowledge construction can be represented by an exponential function including variables of intelligence, prior store of information, quality or abstractness of information presented, and the sequential structure of information.

The composite equation contains two parts: a *gain function* representing mobilization of schemata for storage of information into LTM, and a *modulation function* representing the perceptual processing of information and its temporary storage for passage into more permanent memory. The modulation function is composed of a stability function based on assumption 3, and an instability function based on assumption 4. Thus, in contrast to more parsimonious models that utilize only internal states of the learner (Hicklin 1976; Preece 1982, 1984), the neuromathematical model presented here includes the effects of information quality and quantity. Therefore, it holds promise for greater accuracy in predicting knowledge acquisition in complex learning settings.

6.2 Stability Function

It is assumed that the capacity of immediate processing centers is limited. The stability of activity will in part depend on the amount of information held in temporay storage, and hence will decline as the information load increases with time. This assumption must be expanded to include characteristics of the information that may influence its efficient storage and the capacity of the learner to effectively organize and transmit the information to stable, long-term storage. Two properties of stimulus information are considered in a first approximation: (1) information quality, a composite variable representing the abstractness or conceptual complexity of the content symbolized by β, and (2) information quantity (ρ) which is scaled to the progression density (D_s) of the communication. The learner's capacity to organize information is represented by a variable (κ) scaled to standard intelligence quotients.

Let S represent the stability of immediate encoding centers; the rate of decrease in stability with time will be proportional to the amount of activity accumulated. Hence we write

$$-\frac{ds}{dt} = f(S) \tag{15}$$

However, the rate of decrease should be inversely related to intellectual ability (as more intelligent learners should be more facile in immediate organizing of information), and directly related to the load of information acquired including progression density and quality of information. Hence we refine the statement to include a stability coefficient (σ), where α is a constant of proportionality.

$$\sigma = \frac{\alpha\beta\rho}{\kappa} \tag{16}$$

The rate of decay in stability is given by:

$$-\frac{ds}{dt} = \sigma S \tag{17}$$

Rearranging and taking the time integral we obtain

$$\int_{so}^{s} \frac{ds}{S} = -\sigma \int_{to}^{t} dt \tag{18}$$

which yields

$$S = S_0 e^{-\sigma t} \tag{19}$$

where S_0 is the initial value of S at time zero (t_0), and t is the total time from onset of learning. This is a time-dependent function representing the change in CNS encoding capacity. It predicts the amount of residual information processing capacity for coding of information in perparation for its passage to LTM at a given point in time.

6.3 Instability Function

As learning progresses, and behavior becomes more differentiated through greater alignment of processing centers with the patterns of information input, initial instability associated with a new learning task will decrease with time.
Let I represent activity in the CNS associated with instability of the system and λ the coefficient of decay with time. Hence

$$\frac{dI}{dt} = -\lambda I \tag{20}$$

Rearranging the equation and integrating it, as was done with the stability equation of similar form, we obtain

$$I = I_0 e^{-\lambda t} \tag{21}$$

where I_0 is the initial value of the instability factor at $t = 0$. The value of the decay coefficient as a first approximation has been empirically estimated to be $\lambda = 2.1\,\sigma$.

6.4 Modulation Function and Gain

At any point in time (t), the capacity of the CNS to hold immediately perceived and categorized information will be given by the difference between the values of the stability function and the instability function ($S - I$):

$$S - I = S_0 e^{-\sigma t} - I_0 e^{-\lambda t} \tag{22}$$

This expression is the modulation factor to be multiplied by the gain function to yield the composite equation representing total information acquired as a function of time on task. The modulation equation is of a form known as a double exponential decay function. Graphically it is unimodal and skewed (Fig. 18). In general, this function represents the capacity of the learner to process immediately acquired information at each point in time. Hence, the form of the curve is informative as it indicates the time interval for optimum attention to information reception as described in the figure legend.
The gain function, representing the exponential increase in information acquisition as a function of time, is derived as follows. Let N represent the activity of the CNS associated with transfer of information into long-term storage. In this equation, the rate of activity should be directly proportional to the intelligence of the individual and inversely proportional to information load and complexity. We write a coefficient of proportionality (γ) defined as:

$$\gamma = \frac{\kappa}{\alpha' \beta \rho} \tag{23}$$

Now setting the instantaneous rate of increase in N (i.e., dN/dt) as a proportion of N, we write

$$\frac{dN}{dt} = \gamma N \tag{24}$$

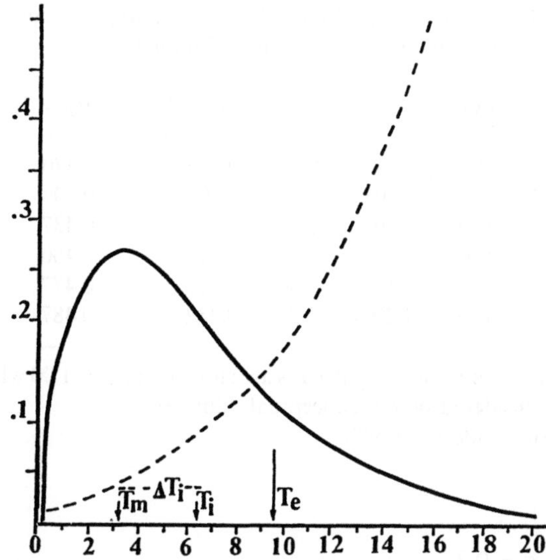

Fig. 18. A plot of the modulation function (*solid line*) and the gain function (*dashed line*). The ordinate values must be multiplied by 40 to represent values of the gain function. For the modulation curve, the time coordinate (T_m) of the maximum point of the curve and the coordinate (T_i) of the inflection point define the interval ΔT_i. The upper limit for the interval for maximum receptivity of encoding of continuous narrative is marked by T_e which is equivalent to $T_m + 2\Delta T_i$. (From Anderson 1983)

By rearranging the equation and integrating it, we obtain

$$N = N_0 e^{\gamma t} \tag{25}$$

where N_0 is the initial gain activity at $t = 0$. Hence, N_0 is a complex variable representing the role of prior information in facilitating transfer of new information into memory and the organizing capacity of the learner at the inception of the learning task.

It is an expression of the antecedent facilitatory factors that potentiate transfer of information into long-term storage. The function generates a positively accelerated curve (Fig. 18) that represents the activity of the learner in mobilizing information in stable memory and reorganizing it to permit connection with new information and its storage in memory. The higher the intelligence of the learner, the more sharply the curve accelerates upward.

6.5 Definition of Coefficients

Based on a broad data set of knowledge acquisition by subjects of varying mental ability and with content of varying organization, the coefficients in the equation have been calibrated to yield the following relationships (Anderson 1983):

Table 3. Some values of variables within a range appropriate for use with the neuromathematical model

κ	I.Q.	ρ	D_s	$\sigma(\beta=1)$	$\lambda(\beta=1)$
1.13	80	0.794	0.50	0.176	0.370
1.17	90	0.908	0.75	0.194	0.407
1.20	100	1.000	1.00	0.208	0.437
1.23	110	1.077	1.25	0.219	0.460
1.26	120	1.145	1.50	0.227	0.477
1.30	130	1.205	1.75	0.232	0.487

Note: The σ and λ coefficients are shown for a $\beta = 1.00$ which is considered of low conceptual difficulty.
From Anderson (1983)

$$\alpha = 0.25$$

$$\alpha' = 6.0$$

$$\kappa = 0.00325\,(\text{I.Q.}) + 0.874$$

$$\rho = (\overline{Ds})^{1/3}$$

where $\overline{Ds}$ is the mean progression density for the communication presented to the learner. The cube root must be used to properly scale the variable for substitution into the coefficients of the equation. A method of estimating β is presented in the original publication describing the model. Some typical values for these variables are presented in Table 3.

6.6 Composite Equation

The product of the gain function (N) and the modulation factor ($S - I$) yields the following composite equation for prediction of knowledge acquisition:

$$N_t = N_0 e^{\gamma t}(S_0 e^{-\sigma t} - I_0 e^{-\lambda t}) \tag{26}$$

where N_t is the net information gain at time t. Some examples of the graphs generated by the equation are shown in Fig. 19.
Under the special circumstances where the gain and stability coefficients are equal (i.e., $\gamma = \sigma$), the composite equation reduces to the following form:

$$N_t = N_0[1 - e^{(\sigma - \lambda)t}] \tag{27}$$

where by definition $\lambda > \sigma$. Conceptually, this represents a case where the gain in knowledge is compensating for the decay in stability of immediate processing functions, thus the equation approaches an asymptote. This represents the net gain in knowlege as the higher order encoding processes compensate for loss of

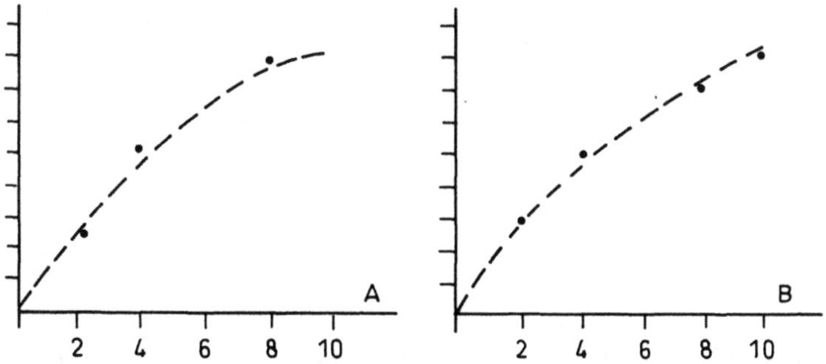

Fig. 19 A, B. Predicted learning curves (*dashed lines*) and experimental means recall scores (*points*) for communications, on **A** biology of *Hydra*, and **B** human health (Group B-1) as presented in Table 4. (From Anderson O.R. 1983)

information during perception and primary processing. Thus, there is a limited amount of information that can be stored. There is a sifting of information resulting in storage of a constant and limited quantity represented by the asymptotal value of the function.

This equation generates a negatively accelerated curve with asymptote N_0. Methods of estimating the value of N_0 are discussed elsewhere (Anderson 1983). The slope of the curve is determined by the magnitude of the exponential coefficient.

6.7 A Test of the Predictive Capacity of the Model

To determine the accuracy of prediction of the model, two listening tasks on biology topics were presented to adolescents for whom intelligence quotients were available. After listenting to the communication for successive intervals of time, they were asked to recall as much of the information as possible at the end of each interval. The results of this test are presented in Table 4 and Fig. 19. There is a strong concordance between predicted and actual recall scores, particularly as shown by the curves in Fig. 19. Note that the progression density is lower for communication B than A, therefore, as predicted, the learning curve for communication B continues to rise while that of A is levelling off. It is particularly important to notice that the period of optimal encoding of information is about 8–10 min. This is interpreted as the period of maximum encoding receptivity for continuous narrative. It is estimated from the plot of the modulation function (Fig. 18), and also by noting the point of maximum acquisition of information from the plot of total information recall (N_t).

6.8 Some Limitations of the Model

The model provides predictions for short-term learning of continuous verbal communications. At present, it is not known how interposition of questioning or

Table 4. Comparison of predicted performances (N_t) and mean recall scores ($\bar{X}$) for students who received communications A and B

	A			B								
				Group 1			Group 2			Group 3		
	$\kappa = 1.28$			$\kappa = 1.24$			$\kappa = 1.28$			$\kappa = 1.32$		
	$N_0 = 17$			$N_0 = 15$			$N_0 = 13$			$N_0 = 19$		
	N_t	$\bar{X}$	S.E.	N_t	$\bar{X}$	S.E.	N_t	$\bar{X}$	S.E.	N_t	$\bar{X}$	S.E.
N_2	6.2	5.6	(0.09)	5.4	6.0	(0.2)	4.7	5.6	(0.4)	6.9	7.0	(0.3)
N_4	10.3	10.6	(0.19)	9.5	9.9	(0.2)	8.5	9.5	(0.6)	12.8	12.4	(0.4)
N_8	15.0	15.7	(0.30)	14.9	14.6	(0.6)	14.2	13.0	(0.9)	22.7	19.5	(0.7)
N_{10}	—	—	—	17.0	16.7	(0.4)	16.6	16.0	(1.1)	27.3	23.0	(0.7)

Note: in group A, $\rho = 1.0$; and in group B, $\rho = 0.8$. Entries labeled $N_2 - N_{10}$ represent recall intervals at 2 to 10 min. S.E. is the standard error of measurement.
From Anderson (1983)

other forms of attentional-directing cues would perturb the accuracy of prediction. The present form of the model accounts for information acquisiton during uninterrupted listening. Presently, we have not attempted to scale κ to other measures of information processing capacity (Sternberg 1985) other than the standard I.Q. Moreover, the accuracy of scaling the exponential coefficients has been tested for a limited I.Q. range, but appears to be appropriate for subjects within the interval of $90 < \text{I.Q.} < 130$. No tests have been done with students outside this interval. Further tests are needed to determine the accuracy of the model under conditions of varying motivation when appropriate coefficients are included to account for this variable using forms of the equation that have been modified for this purpose (Anderson 1986). The presentation of visual material during listening may also alter the form of the information acquisition curve, and additional research is needed to clarify this effect and make appropriate modifications in the model as required. Within the limitations of the model, some interrelationships with other theoretical perspectives and an overall view of current theory are presented in Sect. 7.

7 Conclusions and Implications

7.1 Current Perspectives

It is becoming increasingly clear that higher cognitive functions are less clearly localized than lower order functions. While it is possible to specify the locus of fundamental regulatory mechanisms for somatic homeostasis and to detect electrochemical correlates of some simple conditioned responses, the specific sites mediating language, information processing and memory are at present much less certain. Increasing evidence points to the importance of the hippocampus for immediate encoding of information (STM functions) and the enhancing role of the amygdala and reticular formation in maintaining alertness and readiness to receive information. However, there is increasing evidence that cortical processes associated with language, memory and other highly symbolic representations of experience are diffusely distributed. The plasticity of human cortical neural activity may produce substantial idiosyncrasy across individuals in the pattern of neuronal responses mediating higher cognitive functions. For a given individual, there must be orderly and predictable relations among information input, its cortical representation, and the output at a given moment and in relation to a particular environmental context; however, the set of neurons mediating cortical integration may vary among individuals. Hence, it may be impossible to make a generalizable statement about the locus of a given memory event without detailed knowledge of the individual's history and probability states of the CNS. Various stochastic models have been developed to accommodate this fundamental uncertainty, and a wide range of multineuronal complexes (including neuronal assemblies, modes, synfire chains, and modules) have been posited to account for the organized diffusivity, and the idiosyncrasy of CNS encoding.

Given the plasticity of human information encoding and the apparent diffusivity of cortical activity, an alternative approach to strict structure-functional correlation is presented here in the form of a neuromathematical model (Eq. 26). The objective is to model major functional roles of the CNS as a set of time-dependent equations rather than positing spatial localized functions. Although the model is consonant with several current interpretations of multineuronal mechanisms for encoding and transforming sensory input, the model does not require that these processes be reduced to a set of spatially-dependent states. This is not to imply that studies aimed at localizing CNS functions are fruitless, rather, it is a strategy to develop numerically predictive models that avoid the dilemma of current uncertainty in localization.

While this approach may yield insight into general patterns of information processing, it is clear that a theoretically comprehensive reductionistic interpretation would be desirable, to fully account for the neuronal events correlated with a particular cognitive process. Some criteria (Clark 1980, p. 80) for valid neuroscientific modeling of cognitive processes are summarized as follows:

1. Every variable in the model should be represented by a corresponding collection of neural states which are values of that variable.
2. Every component (box) in a model must correspond to a specific structure in the nervous system.
3. Every input or output link to a component (box) in a model must represent an input or output pathway to the corresponding structure in the nervous system, such that the states of that pathway are the values of the appropriate variable in the model.
4. Every transformation of inputs specified for a component in the model must be carried out by the corresponding structure in the nervous system.

Given the apparent diffusivity of some higher cortical functions, it may not be possible at present to meet all of these criteria in building reductionistic models of higher order cognitive functions. Hence, more generalized models may be required. As with any modeling enterprise, one must carefully evaluate to what extent a generalized model is sufficiently specific to explain adequately and predict the diversity of phenomena we hope to represent. To what extent can we sacrifice specificity without losing the explanatory power and predictive accuracy we hope to achieve? Refined methods of localizing electrical activity in the CNS, and hopefully visualizing "real-time" changes at the microstructural level by noninvasive means, could contribute substantially to the goal of determining the degree of specificity versus variability in CNS events associated with diverse classes of human behavior.

7.2 A Synthesis of Some Current Perspectives on Neurocognitive Functions

Some current theory on localization of neurocognitive functions is assembled in a highly schematized diagram (Fig. 20). This assembly focuses on the modular organization of the CNS and parallel network arrangements of some pathways in reciprocally innervated major centers in the brain. Details of circuitry among brain centers is not shown (Smythies 1970; Eccles 1973). It is only a skeletal view to highlight those functions that are most pertinent to information processing and knowledge acquisition reviewed in this paper. Some of the functions and their relationships are still hypothetical, for example the location of declarative versus performative knowledge, whereas others, such as the role of the hippocampus in place-keeping and encoding of new information into lasting memory, are more clearly established. Differentiation of function within the cerebral hemispheres (upper box) includes possible specialization for visual/iconic representation in the right hemisphere, and predominantly verbal/analytic in the left, although semantic recognition, but not syntactic representation, can also occur in the right hemisphere. Serial processing may be dominant in the frontal areas, and parallel processing in the parieto-occipital areas (Äystö 1988). However, this requires further empirical investigation. The massive fiber tracts of the corpus callosum interrelate left and right hemispheric functions, but there is evidence for parcellation; auditory-tactile transfer occurs in anterior tracts and visual in posterior tracts (Gazzaniga 1989). There are, however, genetic and developmental variations among individuals in the localization of these functions and fiber mass in each tract. In a broader perspective, there is also evidence for modular organization of functions within the cerebral cortex as well as in lower brain centers. Moreover, increasing evidence suggests that the brain utilizes parallel processing pathways and that reciprocal innervation between higher and lower processing centers produces interactive effects during information processing. Hence, in broad perspective, the brain is no longer viewed as rigidly organized in a unidirectional and hierarchical pattern. Rather, current neuroanatomical and physiological evidence (Goldman-Rakic 1988) points to parallel input networks impressed upon a more reflexive hierarchical control system integrated through reciprocal innervation of higher and lower brain centers (e.g., parallel arrows from thalamus to cerebral hemispheres, and recurrent path A, Fig. 20). Localized cortical areas may receive specific, convergent innervation from several nuclei in the thalamus, further suggesting finer modular control in CNS organization. Opposed arrows (Fig. 20) signify multiple pathways with reciprocal

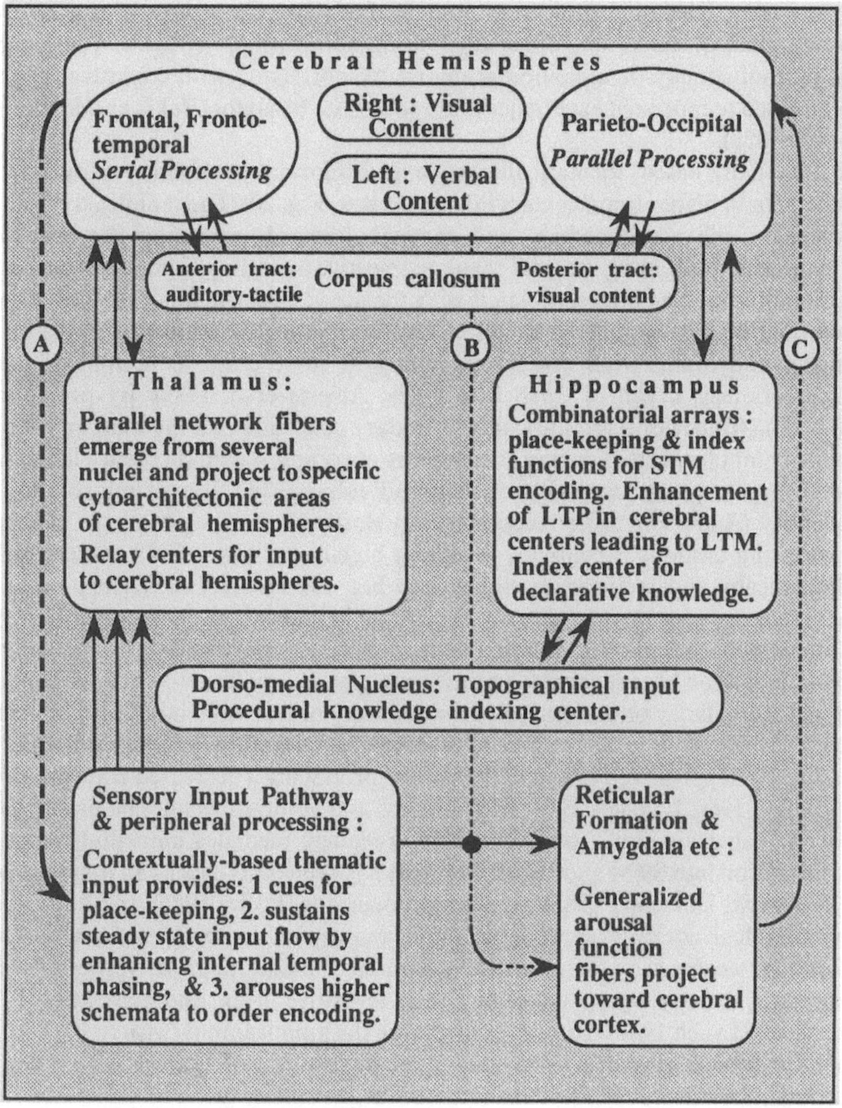

Fig. 20. A synthesis of some current perspectives on modular organization of neurocognitive functions, showing the role of organized sensory input, its projection by parallel network pathways to the cerebral cortex, and the role of higher order schemata in controlling encoding and perceptual processes during construction of knowledge by recurrent innervation (*A*) controlling lower order processing centers. The location of *arrows* indicates targeted centers and is not topologically descriptive. Arousal, generated by input to the cortex from the reticular formation (*C*), is stimulated by novel sensory input, or by mobilization of affective schemata in higher centers (*recurrent arrow*, *B*). However if the sensory input is too patterned, producing habituation, activation of the reticular formation may be blocked (*opaque circle*, *pathway B*) by signals from the cerebral cortex. There is some evidence to suggest that storage of declarative knowlege is particularly dependent on hippocampal functions, and procedural knowledge on dorso-medial nuclear functions (see Lynch 1986), but this requires further investigation. For details, see text (Sect. 7.3)

innervation. Thalamic projections radiate from specific groups of nuclei toward broad areas of the cortex, often innervating nonoverlapping fields. These cortical fields, individually or in combination, my be correlated with cognitive maps and plans used to represent experience and coordinate behavior (Rakic and Singer 1988; Arbib 1989).

Contextually based, thematically sequenced information (sensory input) arouses schemata in higher brain centers (lower left box, Fig. 20). The mobilized organizing schemata enhance encoding, and through forward projection (A) onto lower "perceptual processing centers" regulate selective attending, stimulus focusing, and other filtering functions associated with the active construction phase of encoding information into cognitive structure. This integrating function may spread across higher and lower brain centers extending in some cases to peripheral sensory receptors (e.g., organ of Corti in cochlea, Alberts et al. 1989). By providing an encoding framework and imposing boundary conditions on the kind and range of sensory input admitted during constructive encoding, schemata reduce information load in perceptual processing centers and enhance the flow of information into memory (Anderson 1983; Anderson and Botticelli 1987). Hence, the perceptual processing of input is complex, involving higher and lower CNS centers that are dynamically and reciprocally linked (see Fig. 15). Novelty in sensory input, and aroused emotional states (affective schemata) in higher centers (recurrent arrow B), stimulate the reticular formation, and associated arousal centers, producing broad-scale activation of cortical areas by ascending pathways (C). However, if there is too much redundancy or patterning in the encoded information, inhibitory impulses in pathway (B) block sensory input to the reticular formation (opaque circle) and thus indirectly suppress activation of the cortex, producing a decreased state of alertness (Sokolov 1960; Lindsley 1961; McGuinness and Pribram 1980). The hippocampus, among other functions (including place-keeping), mediates immediate storage of information into STM and LTM. This role has been described as an indexing center (Teyler and DiScenna 1986) or selection operator (Eccles 1978). In addition to its hypothetical indexing function, the hippocampus also may enhance LTM formation through feedback pathways that disinhibit cortical neurons (especially those encoding sensory input), increasing their repetitive firing, and generating LTP at synapses (Lynch 1986). Though conjectural, the hippocampus (with richly stratified, combinational neural arrays) may mediate storage of declarative (verbal.conceptual) knowledge, while the more convergent innervation of the dorso-medial necleus (with one-to-one topographic input) may mediate storage of procedural (operative) knowledge (Lynch 1986). The generative and "deep structural" properties of language, and other generalized representations of experience, may be a product of the coordination of activity in these specialized modules, yielding a coherent and unified substructure to anchor more specific information elements encoded in memory. These properties are discussed more fully in the following section.

7.3 Some Implications for Communication Strategies

We have only begun to understand the complex processes accompanying knowledge acquisition and recall, especially in relation to higher order functions. These include: (1) the role of information processing strategies mediating knowledge gain and recall, (2) the effects of the organization of prior acquired information on acquisition of new information, and (3) the dynamics of the interaction of active phases of knowledge construction by the recipient with the organization and kind of information to be acquired. In the foregoing sections, a theory of the evolutionary and neurobiological origins of the effects of sequentially connected and thematically coherent information on information acquisition has been presented.

With growing interest in parallel and serial processes in computers and human cognitive functions, the question of why human language and streams of consciousness are serial phenomena, at least in their ultimate expression, has become of increasing importance. Calvin (1987) hypothesizes that the brain generates parallel or simultaneous sequences of responses, but at any point only one that has survival value for the organism is actualized. Thus, serial behavior is a product of a selection mechanism that filters stochastically the most favorable sequence from among many simultaneous neuronal serial patterns. Parallel, and sometimes apparently random, processing activity provides a pool of "creative" and potentially adaptive sequences of responses that the organism can refine and individually actualize to best enhance survival in a changing environment. In this view, parallel processing serves in part to provide a source of potentially adaptive responses subjected to natural selection, with only those enhancing survival being emitted.

Parallel processing of visual information is well established (Zeki and Shipp 1988) forming three pathways (mediating color, form, and motion) localizable at several levels in afferent and cerebral cortical structures. The three pathways, however, are not autonomous, but are linked internally by forward and backward connections and externally by cross innervation. Confluence of input and integration across features are achieved by forward connections within each functional pathway and by "cross communication" among the three pathways. Therefore, hierarchical control is enriched by additional input within, and across, levels of the information processing pathways. This pattern may occur throughout the cerebral cortex and explain the remarkably rich and varied capacity of the CNS to impose order on highly diverse sensory input signals, in some cases arriving by parallel afferent pathways. A variety of parallel processing networks has been proposed involving laminar arrangement of intracortical and external-derived synaptic contacts on parallel input processing units (Kohonen et al. 1981; Lynch 1986; Arbib 1989). Some of these resemble a series of cortical cylindrical units (e.g., Figs. 2 and 4c) with parallel input at the distal dendrite surface, and multiple modulating synapses from transverse fibers at varying levels (laminae) along the dendrite. The successful development of computer circuitry based on these principles increases the likelihood that some simultaneous neurocognitive processes and other encoding events can be mediated by parallel organized neural systems.

Integration of parallel processes in the brain raises the conceptual question of how these separate processes are eventually merged to yield higher order representions of experience such as principles and generalizations. Johnson-Laird (1987, p. 13) posits two possible explanations for the emergence of principles from multiple processing modes: (1) "One option is that high-level principles are paramount, and that parallel processes are merely the brain's low-level language into which all mental life must ultimately be translated: high principles are, as it were, 'compiled' into low-level parallel processes", and (2) "At the other extreme, another option is that the mind contains no high-level principles: there is only parallel distributed processing from which all behavior emerges." The latter seems intuitively unsatisfactory since it is patently clear that human behavior is characterized by generalizations derived from predictable but sometimes disparate experiences. Hence, some higher order function seems necessary to generate coherence among lower level parallel processes. One aspect of this integrating function may be to select appropriately and order information generated in parallel processes to yield a connected, serially ordered representation of experience at the conscious level.

The question remains, however, why language and many logical operations are ultimately expressed as a serial sequence of interconnected units. What are the neurocognitive origins of seriation and connectedness in linguistic and symbolic representations of experience? Is serial organization an inherent property of higher processing centers and internal representations of experience, or has this characteristic arisen to facilitate communication and reception of information due to requirements of neural input-output channels? Not all expressions of language are exclusively serial, however, depending on the mode of expression. The overall formational structures of spoken as opposed to signed phonemes (using hand sign language) appears to be largely determined by their modalities: sequential organization for spoken-auditory reception and a "layered pattern" for signed-visual reception (Studdert-Kennedy and Lane 1980; Studdert-Kennedy 1983). Some of the specialized features, moreover, of syntax and semantics may be traced to organizational aspects of the motor functions mediating their expression (Lenneberg 1974). These differences in communication mode (serial or layered) also reflect issues discussed in prior sections under topics of serial versus parallel processing, and their possible hemispheric localization (Ottoson 1987; Äystö 1988). Some of the models presented in the foregoing sections suggest that there are fundamental constraints on signal processing, particularly in "more peripheral" auditory processing centers, that impose limitations on the diversity of information that can be effectively encoded. In general, where hierarchical processing of sensory input occurs, the greater capacity of higher brain centers may be dependent upon and somewhat limited by the constraints imposed by lower level processing units. Some of these lower level processors may also be more phylogenetically primitive, thus accounting for thier more limited capacity (Fair 1988). Through reciprocal innervation, however, higher processing units may exert some control over lower processing units, thus enhancing their processing capacity. It has been suggested, for example, that focal attending (e.g., auditory focal processing in a noisy environment) is a mechanism to reduce the information load to manageable levels during processing (see Sect. 5). These limitations on sensory processing capacity may have favored serially connected forms of communication. While higher brain centers exhibit considerable plasticity and multiple processing capacities, more peripheral centers, mediating perception and transformation of incoming information, may be more limited in their capacity to analyze divergent and low redundant input. This is particularly likely if the information is novel.

Although there may be parallel processing of peripheral sensory signals (as implied by Fig. 15), the properties of the system impose certain limitations on the complexity of the input. Hence, to maintain a steady-state flow of information to higher brain centers, a certain degree of predictability and redundancy in signal input may be required. At the immediate interface with the environment, sensory receptors are limited, among other factors, by their range in detectable energy, their capacity to follow and transduce rapidly changing patterns of energy, and their ability to transmit these signals to neurons in afferent pathway. At higher levels of processing, the many signals generated at the periphery must be integrated and properly encoded for more generalized processing. The limiting step for diversity in input may be at the peripheral and intermediate levels of signal processing where sensory input must be "sorted" and combined in appropriate 'categories" to be made available for higher order processing. Too much diversity of input overloads these centers and reduces the quantity of information passed to higher processing centers. Hence, the apparently strictly serial and connected quality of communication may be an evolutionary adaptive response to maximize reliability and fidelity in transmission of information.

It is clear from a wide range of cognitive studies that information acquisition is an active process of constructing internal representations of experience, often by widely diverse strategies among different learners. However, the research summarized here also indicates that there are some fundamental neurobiological properties of human information processing that are strongly influenced by the organization of communicated information. The amount and stability of acquired knowledge can be substantially enhanced by proper attention to sequential organization of the information. These effects have been observed in experimental studies with diverse subjects spanning a range from educable mentally retarded children to college students (Anderson and Lee 1975; Ferraro et al. 1977; Moché 1976). In general, the more intelligent the individual, the less they are reliant on external organizational features of the communication. Our studies, however, have shown that both low and high I.Q. subjects are benefitted by enhanced thematic structure of the kind discussed here. Likewise, perceptions of clarity of discourse, satisfaction with the communication experience, and the amount of information acquired in experimental and natural learning settings are directly related to the thematic structure of the discourse (Anderson and Lee 1975; Nkpa 1984).

The modulation function of the neuromathematical model, and the point of maximum knowledge acquisition in the predicted learning curve, indicate that with uninterrupted narrative adolescents have an optimal (though not limiting) aural information processing interval of about 8–10 min. This can be extended, however, by the enhancing effect of mobilizing appropriate organizing schemata (including affective arousal) as represented by the gain function (Eq. 25). Our general experience indicates that for complex material, as opposed to story narrative, this interval of optimal encoding is probably not much greater for adults, although mature listeners can use various categorizing or grouping procedures to aid efficient acquisition and storage of information. This requires further experimental evaluation. Moreover, the limited duration of the optimum encoding interval for continuous narrative (reaching a maximum at about 3–4 min) may explain our tendency during human dialogue to interrupt extended discourse with questions or other intervening commentary. The interjection momentarily halts the progression of information, provides some verbal cue to the speaker about the internal coding state of the hearer, and indicates a relevant organizing context to initiate a new segment of communicated information. Thus, information is acquired in sufficiently limited segments to fall within the optimal encoding interval for uninterrupted narrative. In general, as shown by the neuromathematical model, the rate of information gain and the total knowledge acquired during communication can be increased if the communicator uses appropriate themes that provide moderate progression density and embeds the content within a meaningful context.

7.4 Concluding Comment

Current neurocognitive models include a wide range of labels for process and structural variables, including maps, schemata, plans, operators, short-term buffers, channels, successive and simultaneous processing, etc. It is not always clear whether these are intended to represent very fundamental and innate qualities of the system, or derived properties acquired through experience (e.g., learning) that have been built upon fundamental properties and processes. Indeed, a perplexing question from a reductionistic perspective is to what extent can we identify the most fundamental or irreducible set of operations that represent the neurobiological substratum for neurocognitive activity, and hence behavior, as opposed to those acquired by experience and enculturation. One may argue that all verifiable categories of behaviour represent a product of nervous system activity, and therefore must represent to some degree the operational qualities of the system. Granted that this is the case, a heavy burden is borne by the qualification "to some degree," once again implying that there are more fundamental versus derived properties of nervous system activity. At the other extreme, to assert that all behavior is a product of nervous system activity and therefore represents a fundamental property of the system is not particularly enlightening and may become indeterminately circular, namely," we are what we see, which in turn is a product of what we are, etc." Indeed, one of the intriguing challenges of neurobiology is to understand the biological basis for this remarkably reciprocal interaction between neural encoding mechanisms and the diversity of sensory input presented by the environment. These problems have been addressed in various guises in the literature including the mind-body problem, dualism, the enigma of the mind fully understanding itself, and other philosophical issues of neurobiological and neuropsychological epistemology (Eccles 1977; Popper and Eccles 1977; Eccles and Robinson 1984; Churchland 1986; Young 1987; Kornhuber 1988; Moss 1989). From a neuroscientific perspective, these issues are clearly profound, and not only a matter of philosophical interest. The biological research strategy of analyzing systems from a hierarchical, structural-functional perspective often begins with a premise that we can identify fundamental or inherent properties of a living system and explain the occurrence, if not the origin, of secondary or more highly derived qualities based on interactions of these fundamental units within the system or with the external environment. Thus, using a classical biological research paradigm, it is desirable to have a clear delineation of more fundamental versus derived properties of nervous system activity. The diversity of functional categories employed by various researchers, and the absence of a standard vocabulary, may enhance creative modeling but also diminishes our capacity to make clear specifications of the level of analysis. This is often characteristic of pre-paradigmatic phases of scientific theory building. For example, are "plans," and "operators" rudimentary qualities of nervous system activity or derived aspects? To what extent can these higher order operations be explained by more fundamental qualities? Some of these fndamental qualities may include: (1) the multi-associational capacities of higher neuronal centers, (2) temporal and spatial representational capacity based on internal and external cues, and (3) a capacity to extrapolate a series of associations toward a represented future time point. Even the latter function is a very sophisticated operation and may involve contributions by more than one basic process. Given the highly malleable and fluid quality of higher brain processes, we may expect to find very diverse, but ontogenetically explainable, modes of operation that have been impressed upon the system both by enculturation and the biological history of the organism. Parsing the "innate" from the "acquired" aspects of neurocognitive processes promises to be a very significant, but perhaps challenging, aspect of future inquiries in the field. It may be that a totally analytical and categorical-reductionistic approach will not be adequate, and a more wholistic approach based on continuum logic and something akin to field theory will be required. These theories, however, could accommodate reductionistic interpretations of neuronal activity, including modular functional interpretations, in constructing continuum-based rather than quantized, explanations of the physiological basis for cognitive processes.

Current neurophysiological research at the level of individual neurons and simple nerve nets will undoubtedly provide significant insight into fundamental mechanisms toward further theory building (Kandel and Schwartz 1982; Alkon and Farley 1984). Alkon (1987) concludes that the fundamental synaptic mechanisms observed in conditioned learning of invertebrates could account for eleborate interactions among field of neurons in the representation of complex experiences by higher organisms. He provides several models to illustrate how convergence of input on dendrites of cortical neurons, and the patterned arousal of sets of neurons, can account for complex behavior such as recall of imagery. This view is further strengthened by experimental observations of similar conditioning-induced increases in neuronal excitability in invertebrates and the mammalian hippocampus (Coulter et al. 1987). However, as recognized by researchers using invertebrate models, it is not immediately clear whether higher order brain functions such as verbal information processing, generative language processes, and complex logical activity, can be fully represented in terms of complex arrangements of relatively simple nerve networks of the kind investigated in invertebrates. Are the higher brain centers of mammals simply an elaboration of basic neural systems such as those found in invertebrates, or has a much more complex organizational principle arisen during evolution of the cerebral cortex in higher organisms?

Among the higher cognitive functions discussed here, the role of thematic and contextual content in the enhanced processing and storage of communicated information remains an important area for further inquiry. O'Keefe and Nadel (1978, p. 389) summarize two areas of research that indicate the central importance of context in LTM (Bower 1970; Johnson-Laird 1970): (1) Studies on imagery as mnemonic aids indicate that storing information in an interactive, relational form vastly improves retention. The visual imagery is apparently not stored as a separate enabling device for semantic recall, but is part of a generalized representation that can generate both the verbal and visual content during recall; (2) Recall of verbal information (sentences and narrative) takes the form of the more general idea, meaning or sense of the discourse, typically with the loss of specific lexical elements and syntactic information. Analysis of the surface characteristics of language is insufficient to explain the remarkably generative (elaborative) capacity of language and its comprehension. And, based on the foregoing, it appears that we must appeal to some "deep structure" that encodes the meaning of a communication, without necessarily preserving lexical elements. An analysis of word metaphors (MacCormac 1985) also indicates a contextual organizing role in deeper cognitive structures during information processing. Metaphors may be represented in memory hierarchically in three levels: (1) Surface structure (a fundamental analysis of word content), (2) syntactic and semantic (sentence structural analysis), and (3) cognitive (a contextual inferential analysis requiring abstract and generative linguistic representations). All of these functions imply a type of spatio-temporal framework for information encoding, perhaps mediated by hippocampal neuronal networks.

While it is not possible at present to fully explain the enabling role of context and thematic organization of communication based on neurophysiological principles, many of the models summarized here point toward the importance of patterned activity among neuronal modules in representing language and thought. That is, activation of groups of neurons, representing component aspects of sequential phenomena (such as language and visual events), is enhanced when the communication contains some contextual referents that provide place-keeping functions and generate cues to mobilize appropriate subsets of neurons (equivalent to schemata) during information reception. Hypothetically, this enabling function may be due to several contributions including: (1) a more consistent arousal of overlapping populations of neurons thus reinforcing their mutual activation, (2) less complexity in the diversity of cues that an executive center in the CNS must receive in order to arouse and order sets of neurons during encoding of the information, and (3) better temporal integration of sensory input rate with internal timing mechanisms and cycles of neuronal events. In short, a certain degree of predictability in sensory input is required to (1) promote orderly activation of groups of encoding neurons, (2) reduce the complexity of place-keeping functions of CNS executive centers (perhaps especially in neocorticohippocampal interactions), and (3) provide consistent cues that maintain equilibrium among the several timing mechanisms required to sustain a steady-state flow of sensory input into encoding centers.

From an evolutionary theoretical perspective, the information processing models (Sects. 5 and 6; Fig. 15) and other research (Jerison 1976, 1977) suggest that increasing encephalization during mammalian phylogeny was closely linked with communication. Indeed, the increased effectiveness of neurocognitive functions mediating language, and its consequent advantage for group survival, may have been a major driving force favoring neocortical evolutionary advances. However, I hasten to add that it is unlikely that communication-based survival advantages alone account for evolutionary advances in cortical development. Other advantages, including finer control over movement, integration of diverse sensory experiences, plasticity in responding, etc. also would have improved species survival. Natural selection processes usually work on broad suites of characteristics (not unitary traits) in complex environments. Thus, communication advantages may have been significant but not exclusive factors favoring survival of individuals with increasing cortical elaboration. According to theory presented here, at least two advantages for improved communication would result from neocortical elaboration: (1) Information encoding would be enhanced by the improved facility of the executive operators (Fig. 15) to form contextual (schematal) control of perceptual suboperators (S^n), thus increasing their discriminatory capacity and reducing information processing load. The forward-directed arrows (Fig. 15, Fig. 20A) represent in part contextual control of perceptual processing functions. Schemata provide contextual cues for "filtering" information, to admit only the most relevant signals in relation to the readiness of higher processing centers; (2) The effectiveness of language would be enhanced by providing a contextual and thematic framework to order and organize uttered information in a way that improves its reception by the recipient.

From a broader biological perspective, it is clear that many of these properties of information processing have evolved through selection pressures unique to our planet and perhaps solar system. It is very possible that different systems for representing and efficiently responding to sensory input would have evolved with a different history of our geo-planetary system. Our language, formal logical systems of thought, and ways we sequentially and combinatorially represent experiences may be a unique product of the peculiar evolutionary development of our nervous system in response to our planetary experience over geological time spans. At present, we cannot say to what extent the "critical" selection pressures accounting for the evolution of terrestrial human neurocognitive functions, including deep-structural and sequential patterns of communication, are operative elsewhere in the universe.

8 References

Abeles M (1982) Local cortical circuits: an electrophysiological study. Springer Berlin, Heidelberg New York, pp 67–75

Abeles M (1988) Neural codes for higher brain functions. In: Markowitsch HJ (ed) Information processing by the brain: views and hypothesis from a physiological-cognitive perspective. Huber, Toronto, pp 225–238

Aertsen A, Gerstein G, Johannesma P (1985) From neuron to assembly: neuronal organization and stimulus representation. In: Palm G, Aertsen A (eds) Brain theory: proceedings of the first Trieste meeting on brain theory, October 1–4, 1984. Springer, Berlin Heidelberg New York, pp 7–24
Ajmone-Marsan C, Matthies H (eds) (1982) Neuronal plasticity and memory formation. International brain research organization monograph series, vol 9. Raven, New York
Alberts B, Bray D, Lewis J, Raff M, Roberts K, Watson JD (1989) Molecular biology of the cell, 2nd edn. Garland, New York, p 1102
Alkon DL (1987) Memory traces in the brain. Cambridge University Press, Cambridge, pp 135–151
Alkon DL, Farley J (eds) (1984) Primary neural substrates of learning and behavioral change. Cambridge University Press, Cambridge
Anderson OR (1969) An interdisciplinary theory of behavior. J Res Sci Teach 6:265–273
Anderson OR (1971) Quantitative analysis of structure in teaching. Columbia University, Teachers College Press, New York
Anderson OR (1976) The experience of science: a new perspective for laboratory teaching. Columbia University Teachers College Press, New York
Anderson OR (1983) A neuromathematical model of human information processing and its application to science knowledge acquisition. J Res Sci Teach 20:603–620
Anderson OR (1986) Studies on information processing rates in science learning and related cognitive variables, I: some theoretical issues related to motivation. J Res Sci Teach 23: 61–66
Anderson OR, Botticelli S (1987) A quantitative analysis of content structure in some biology texts varying in textual composition. Conference Paper, National Association for Research in Science Teaching. Washington, DC, April 1987. Also (1990) Sci Educ 74:167–182
Anderson OR, Lee MT (1975) Structure in science communications and student recall of knowledge. Sci Educ 59:127–138
Andre T, Phye GD (1986) Cognition, learning and education. In: Phye GD, Andre T (eds) Cognitive classroom learning. Academic, San Diego
Angermeier WF (1984) The evolution of operant learning and memory: a comparative etho-psychology. Karger, Basel
Arbib MA (1972) The metaphorical brain: an introduction to cybernetics as artificial intelligence and brain theory. Wiley-Interscience, New York
Arbib MA (1989) The metaphorical brain 2: neural networks and beyond. Wiley-Interscience, New York
Ashby WR (1952) Design for a brain. Wiley, New York
Asratyan EA, Simonov PV (eds) (1982) The learning brain. Moscow, MIR
Ausubel DP (1963) The psychology of meaningful verbal learning. Grune and Stratton, New York
Avendaño C, Llamas A (1984) Thalamic and nonthalamic direct subcortical projections to association areas of the cat's cerebral cortex. In: Reinoso-Suárez C, Ajmone-Marsan C (eds) Cortical integration. International Brain Research Organization Monograph Series, vol 11. Raven, New York, pp 195–196
Äystö S (1987) Neuropsychological aspects of simultaneous and successive cognitive processes. Jyväskylä studies in education, psychology and social research, 59. (Dissertation)
Äystö S (1988) Neuropsychological aspects of simultaneous and successive cognitive processes. In: Williams JM, Long CJ (eds) Cognitive approaches to neupsychology. Plenum, New York

Bereiter-Hahn J, Anderson OR, Reif W-E (eds) (1987) Cytomechanics: The mechanical basis of cell form and structure. Springer, Berlin Heidelberg New York
Bitterman ME (1965) The evolution of intelligence. Sci Am 212:92–100
Boddy J (1978) Brain systems and psychological concepts. Wiley, New York
Bower GH (1970) Analysis of a mnemonic device. Am Sci 58:496–510
Braitenberg V (1977) On the texture of brains: an introduction to neuroanatomy for the cybernetically minded. Springer, Berlin Heidelberg New York
Braitenberg V (1985) Two views of the cerebral cortex. In: Palm G, Aertsen A (eds) Brain theory: proceedings of the first Trieste meeting on brain theory, October 1–4, 1984. Springer, Berlin Heidelberg New York
Brazier MA (ed) (1979) Brain mechanisms in memory and learning: From the single neuron to man. International brain research organization monograph series, vol 4. Raven, New York
Bullock TH (1984) A framework for considering basic levels of neural integration. In: Reinoso-Suárez C, Ajmone-Marsan C (eds) Cortical integration. International brain research organization monograph series, vol 11. Raven, New York, pp 27–36
Bush RR, Mosteller F (1951) A mathematical model for simple learning. Psychol Rev 58: 313–323
Calvin WH (1987) The brain as a Darwin machine. Nature 330:33–34
Caspari EW (1979) Evolutionary theory and the evolution of the human brain. In: Hahn ME, Jensen C, Dudek BC (eds) Development and evolution of brain size: behavioral implications. Academic, New York, pp 9–28
Changeux J-P, Dennis SG (1982) Signal transduction across cellular membranes. Neurosci Res Program Bull 20:369–370
Chow KL, Leiman AL (1970) The structural and functional organization of the neocortex. Neurosci Res Program Bull 8:153–220
Churchland PS (1986) Neurophilosophy: toward a unified science of the mind-brain. MIT Press, Cambridge, pp 277–313
Churchland PS, Sejnowski TJ (1988) Perspectives on cognitive neuroscience. Science 242:741–745
Clark A (1980) Psychological models and neural mechanisms: an examination of reductionism in psychology. Clarendon, Oxford
Cohen DH (1985) Some organizational principles of a vertebrate conditioning pathways: Is memory a distributed property? In: Weinberger NM, McGaugh JL, Lynch G (eds) Memory systems of the brain: animal and human cognitive processes. Guilford, New York, p 42
Coulter DA, Disterhoft JF, Bank B, Alkon DL (1987) Persistent conditioning-specific changes of post-synaptic membrane currents in *Hermissenda* and hippocampus. In: Milgram NW, Macleod CM, Petit TL (eds) Neuroplasticity, learning, and memory: proceedings of a symposium held at the University of Toronto, Scarborough, Ontario, March 25, 1986. Liss, New York, pp 17–43
Crowder RG (1985) On access and the forms of memory. In: Weinberger NM, McGaugh JL, Lynch G (eds) Memory systems of the brain: animal and human cognitive processes. Guilford, New York, pp 433–441
Das JP, Kirby JR, Jarman RF (1975) Simultaneous and successive syntheses: an alternative model for cognitive abilities. Psychol Bull 82:87–103
Das JP, Kirby JR, Jarman RF (1979) Simultaneous and successive cognitive processes. Academic, New York
Davis WJ (1986) Invertebrate model systems. In: JL Martinez JL, Kesner RP (eds) Learning and memory: a biological perspective. Academic, Orlando, pp 267–297

Dimond SJ, Blizard DA (eds) (1977) Evolution and lateralization of the brain. Ann NY Acad Sci vol 299. The New York Academy of Sciences, New York
Eccles JC (1964) The physiology of synapses. Springer, Berlin Göttingen Heidelberg
Eccles JC (1965) Possible ways in which synaptic mechanisms participate in learning, remembering, and forgetting. In: Kimble DP (ed) Anatomy of memory. Science of Behavior Books, Palo Alto, pp 12–87
Eccles JC (1973) The understanding of the brain. McGraw-Hill, New York
Eccles JC (1977) Part II. In: Popper KR, Eccles JC. The self and its brain. Springer, Berlin Heidelberg New York, pp 225–406
Eccles JC (1978) An instruction-selection hypothesis of cerebral learning. In: Buser PA, Rougel-Buser A (eds) Cerebral correlates of conscious experience. North-Holland, Amsterdam, pp 155–175
Eccles JC, McIntyre AR (1951) Plasticity of mammalian monosynaptic reflexes. Nature 167:466–468
Eccles JC, Robinson DN (1984) The wonder of being human. Free Press, New York
Edelman GM (1981) Group selection as the basis for higher brain function. In: Schmitt FO, Worden FG, Adelman G, Dennis SG (eds) The organization of the cerebral cortex: proceedings of a neurosciences research program colloquium. MIT Press, Cambridge, pp 477–563
Estes WK (1950) Toward a statistical theory of learning. Psychol Rev 57:94–107
Fair CM (1988) Memory and central nervous organization. Paragon House, New York, pp 63–66
Ferraro EJ, Lee MT, Anderson OR (1977) The effects of structure in science communications on knowledge acquisition by students of varying mental maturity. J Res Sci Teach 14:441–447
Fifkova E (1987) Mechanisms of synaptic plasticity. In: Milgram NW, Macleod CM, Petit TL (eds) Neuroplasticity, learning, and memory: proceedings of a symposium held at the University of Toronto, Scarborough, Ontario, March 25, 1986. Liss, New York, pp 61–86
Gazzaniga MS (1989) Organization of the human brain. Science 245:947–952
George FH (1962) The brain as computer. Pergamon, Oxford
Glassman RB (1988) Combinational processes and memory: The power of exponents. In: Markowitsch HJ (ed) Information processing by the brain: views and hypotheses from a physiological-cognitive perspective. Huber, Toronto, pp 207–224
Glees P (1988) The human brain. Cambridge Univ Press, Cambridge
Globus A, Scheibel AB (1967) Pattern and field in cortical structure: the rabbit. J Comp Neurol 131:155–172
Goldman-Rakic PS (1988) Changing concepts of cortical connectivity: parallel distributed cortical networks. In: Rakic P, Singer W (eds) Neurobiology of neocortex. Wiley, New York, pp 177–202
Griffith JS (1967) A view of the brain. Clarendon, Oxford
Grossberg S (1978) A theory of human memory: self-organization and performance of sensory-motor codes, maps, and plans In: Rosen R, Snell F (eds) Progress in theoretical biology. Academic, New York, 5:233–374
Grossberg S (1981) Adaptive resonance in development, perception and cognition. In: Grossberg S (eds) Mathematical psychology and psychophysiology. SIAM/AMS proceedings vol 13. American Mathematical Society, Providence, and Erlbaum Hillsdale, pp 107–156
Grossberg S (1982) Studies of mind and brain. Reidel, Dordrecht
Hahn ME, Jensen C, Dudek BC (eds) (1979) Development and evolution of brain size: behavioral implications. Academic, New York

Hebb DO (1949) The organization of behavior. Wiley, New York
Helige JB (1987) Interhemispheric interaction: models, paradigms and recent findings. In: Ottoson D (ed) Duality and unity of the brain: unified functioning and specialisation of the hemispheres. Plenum, New York, pp 454–466
Hicklin WJ (1976) A model for mastery learning based on dynamic equilibrium theory. J Math Psychol 13:79–88
Hobson JA, Brazier MAB (eds) (1980) The reticular formation revisited: specifying function for a nonspecific system. International brain research organization monograph, vol 6. Raven, New York
Hubel DH, Wiesel TN (1963) Shape and arrangement of columns in cat's striate cortex. J Physiol 165:559–568
Hubel DH, Wiesel TN (1965) Receptive fields and functional architecture in two nonstriate visual areas (18 and 19) of the cat. N Neurophysiol 28:229–289
Hubel DH, Wiesel TN (1968) Receptive fields and functional architecture of monkey striate cortex. J Physiol 195:215–243
Hunt E (1980) Intelligence as an information processing concept. Br J Psychol 71:449–474
Jerison HJ (1976) Paleoneurology and evolution of mind. Sci Am 234:90–101
Jerison HJ (1977) Evolution of the brain. In: Wittrock MC, Beatty J, Bogen JE, Gazzaniga MS, Jerison HJ, Krashen SD, Nebes RD, Teyler TJ (eds) The human brain. Prentice-Hall, Englewood Cliffs, pp 39–62
Jerison HJ (1979) The evolution of diversity in brain size. In: Hahn ME, Jensen C, Dudek BC (eds) Development and evolution of brain size: behavioral implications. Academic, New York, pp 29–57
Johannesma P, Aertsen AH, Van Den Boogaard H, Eggermont J, Epping W (1985) From synchrony to harmony: ideas on the function of neural assemblies and on the interpretation of neural synchrony. In: Palm G, Aertsen A (eds) Brain theory; proceedings of the first Trieste meeting on brain theory, October 1–4, 1984. Springer, Berlin Heidelberg New York, pp 25–47
John ER (1980) A neurophysiological model of purposive behavior. In: Thompson RF, Hicks LH, Shvyrkov VB (eds) Neural mechanisms of goal-directed behavior and learning. Academic, New York, pp 93–115
Johnson V, Lockard D (1985) The effect of kinetic structure and micrograph content on achievement in reading micrographs by college biology students. R Res Sci Teach 22: 713–722
Johnson-Laird PN (1970) The perception and memory of sentences. In: Lyons J (ed) New horizons in linguistics. Penguin, Harmondsworth, pp 261–270
Johnson-Laird PN (1987) Connections and controversy. Nature 330:12–13
Jones EG (1981) Anatomy of cerebral cortex: Columnar input-output organization. In: Schmitt FO, Worden FG, Adelman G, Dennis SG (eds) The organization of the cerebral cortex: Proceedings of a neurosciences research program colloquium. MIT Press, Cambridge, pp 199–235
Kandel ER (1976) Cellular basis of behavior: an introduction of behavioral neurobiology. Freeman, San Francisco
Kandel ER, Schwartz JH (1982) Molecular biology of learning: modulation of transmitter release. Science 218:433–442
Kandel ER, Spencer WA (1968) Cellular neurophysiological approaches in the study of learning. Physiol Rev 58:65–134
Kievit J, Kuypers HGJM (1977) Organization of the thalamo-cortical connexions to the frontal lobe in the rhesus monkey. Exp Brain Res 29:299–322
Kintsch W (1974) The representation of meaning in memory. Wiley, New York, pp 80–89

Kohonen T, Oja E, Lehtiö P (1981) Storage and processing of information in distributed associative memory systems. In: Hinton GE, Anderson JA (eds) Parallel models of associative memory. Erlbaum, Hillsdale, pp 105–143
Kornhuber HH (1973) Neural control of input into long term memory: limbic system and amnestic syndrome in man. In: Zipple HP (ed) Memory and transfer of information. Plenum, New York, pp 1–22
Kornhuber HH (1988) The human brain: from dream and cognition to fantasy, will, conscience, and freedom. In: Markowitsch HJ (ed) Information processing by the brain: views and hypotheses from a physiological-cognitive perspective. Huber, Toronto, pp 241–258
Lambert D, Bramwell M, Lawther G (eds) (1982) The brain: a user's manual. Putnam's, New York
Landahl HD (1961) A note on the mathematical models of interaction of neural elements. Bull Math Biophys 23:91–97
Lasek RJ, Shelanski ML, Brinkley BR, Condeelis JS, Ellisman MH, Goldman RD, Reese TS, Solomon F, Taylor EW (eds) (1981) Cytoskeletons and the architecture of nervous systems. Neurosci Res Program Bull, vol 19
Laszlo E (1972) Introduction to systems philosophy: toward a new paradigm of contemporary thought. Harper and Row, New York
Lawson AE (1986) A neurological model of sensory-motor problem solving with possible implications for higher-order cognition and instruction. J Res Sci Teach 6:503–522
Lenneberg EH (1974) Language and brain: developmental aspects. Neurosci Res Program Bull 12:511–656
Levi-Montalcini R (1965) The nerve growth factor: Its mode of action on sensory and sympathetic ganglia. Harvey Lect 60:217–259
Lindsley DB (1961) The reticular activating system and perceptual integration. In: Sheer DE (ed) Electrical stimulations of the brain. University of Texas Press, Austin, pp 331–349
Luria AR (1966a) Human brain and psychological processes. Harper and Row, New York
Luria AR (1966b) Higher cortical functions in man. Basic Books, New York
Luria AR (1973) The working brain: an introduction to neuropsychology. Penguin, Harmmondsworth
Lynch G (1986) Synapses, circuits, and the beginning of memory. MIT Press, Cambridge, pp 8–70
Lynn R (1967) Attention, arousal and the orientation reflex, Pergamon, Oxford
MacCormac ER (1985) A cognitive theory of metaphor. MIT Press, Cambridge
MacLean P (1972) Cerebral evolution and emotional processes: new findings on the striatal complex. Ann NY Acad Sci 193 (August 25, 1972):137–149
MacLean P (1986) Culminating development in the evolution of the limbic system: The thalamocingulate division. In: Doane BK, Livingston KE (eds) The limbic system: functional organization and clinical disorders. Raven, New York, pp 1–4
Markowitsch HJ (ed) (1988) Information processing by the brain: Views and hypotheses from a physiological-cognitive perspective. Huber, Toronto
Marr D (1970) A theory for cerebral neocortex. Proc R Acad Soc B, London 176:161–234
Mattheis H (1989) In search of cellular mechanisms of memory. Progr Neurobiol 32:277–349
McCulloch W, Pitts W (1943) A logical calculus of the ideas immanent in nervous activity. Bull Math Biophys 5:115–137
McGuinness D, Pribram K (1980) The neuropsychology of attention: emotional and motivational control. In: Wittrock MC (ed) The brain and psychology. New Academic, New York, pp 95–159

Miller GA, Frick FC (1949) Statistical behavioristics and sequences of responses. Psychol Rev 56:311–324
Miller GA, Galanter E, Pribram KH (1960) Plans and the structure of behavior. Holt, New York
Milner PM (1957) The cell assembly: Mark II. Psychol Revs 64:242–252
Minsky ML (1959) Some methods of artificial intelligence and heuristic programming. In: Mechanisation of thought processes. NPL Symposium.
Minsky ML (1968) Semantic information processing. MIT Press, Cambridge
Minsky ML, Papert S (1988) Perceptrons: an introduction to computational geometry. MIT Press, Cambridge
Moché DL (1976) Experiment in matching the structure of instructional modules on astrophysics to cognitive levels among urban community college students. Doctoral Thesis, Columbia University, Teachers College, New York
Montarolo P, Schacher S, Castellucci VF, Hawkins RD, Abrams TW, Goelet P, Kandel ER (1986) Interrelationships of cellular mechanisms for different forms of learning and memory. In: Levi-Montalcini R, Calissano P, Kandel ER, Maggi A (eds) Molecular aspects of neurobiology. Springer, Heidelberg Berlin New York, pp 1–12
Moss DM (1989) Phenomenology and neuropsychology: two approaches to consciousness. In: Valle RS, vonEckartsberg R (eds) Metaphors of consciousness. Plenum, New York, pp 153–166
Mountcastle VB (1957) Modality and topographic properties of single neurons of cat's somatic sensory cortex. J Neurophysiol 20:408–434
Murakami TH (1980) Microtubules and memory: effects of vinblastine on avoidance training. In: Tsukada Y, Agranoff VW (eds) Neurobiological basis of learning and memory. Wiley, New York, pp 165–178
Näätänen R (1982) Processing negativity: an evoked potential reflection of selective attention. Psychol Bull 92:605–640
Näätänen R (1985) Selective attention and stimulus processing: reflections in event-related potentials, magnetoencephalogram, and regional cerebral blood flow. In: Posner MI, Marin OS (eds) Attention and performance XI. Erlbaum, Hillsdale
Näätänen R (1986) Neurophysiological basis of the echoic memory as suggested by event-related potentials and magnetoencephalogram. In: Klix F, Hagendorf H (eds) Ebbinghaus Symposium, North Holland, Amsterdam
Newell A, Simon HA (1972) Human problem solving. Prentice Hall, Englewood Cliffs
Nkpa N (1984) Clarity in biology teaching: student and observer perceptions. Doctoral Thesis, University of London
Noback CR (1959) The heritage of the human brain. James Arthur Lecture on the evolution of the human brain 1959. The American Museum of Natural History, New York
Noback CR (1975) The visual system of primates in phylogenetic studies. In: Luckert WP, Szalay FS (eds) Phylogeny of the primates. Plenum, New York, pp 199–218
Noback CR, Shriver JE (1969) Encephalization and the lemniscal systems during phylogeny. Ann NY Acad Sci 167:118–128
Novak JD (1977) A theory of education. Cornell University Press, Ithaca
O'Keefe J (1979) A review of the hippocampal place cells. Prog Neurophysiol 13:419–439
O'Keefe J, Nadel L (1978) The hippocampus as a cognitive map. Clarendon, Oxford
O'Keefe J, Nadel L (1979) The hippocampus as a cognitive map. Behavioral Brain Sci 2:478–533
Olton DS (1977) Spatial memory. Sci Am 236:82–98
Ottoson D (1983) Physiology of the nervous system. Macmillan, London

Ottoson D (ed) (1987) Duality and unity of the brain: unified functioning and specialisation of the hemispheres. Plenum, New York
Pattee H (ed) (1973) Hierarchy theory: the challenge of complex systems. Braziller, New York
Pompeiano O, Ajmone-Marsan C (eds) (1981) Brain mechanisms of perceptual awareness and purposeful behavior. International brain research organization monograph series, vol 8. Raven, New York
Popper KR, Eccles JC (1977) The self and its brain . Springer, Berlin Heidelberg New York
Preece PF (1982) The fan-spread hypothesis and the adjustment for initial differences between groups in uncontrolled studies. Educational Psychological Measurement 42:759–762
Preece PF (1984) Mathematical modeling of learning. J Res Sci Teach 21:953–955
Rakic P, Singer W (eds) (1988) Neurobiology of neocortex. Wiley, New York
Racine RJ, Kairiss EW (1987) Long-term potentiation phenomena: the search for the mechanisms underlying memory storage processes. In: Milgram NW, Macleod CM, Petit TL (eds) Neuroplasticity, learning, and memory: proceedings of a symposium held at the University of Toronto, Scarborough, Ontario, March 25, 1986. Liss, New York, pp 173–197
Rapp PE (1987) Why are so many biological systems periodic? Prog Neurobiol 29:261–273
Rashevsky N (1945a) A reinterpretation of the mathematical biophysics of the central nervous system in the light of neurophysiological findings. Bull Math Biophys 7:151–160
Rashevsky N (1945b) A suggestion for another statistical interpretation of the fundamental equations of the mathematical biophysics of the central nervous system. Bull Math Biophys 7:223–226
Rashevsky N (1945c) Further contributions to a probabilistic interpretation of the fundamental equations of mathematical biophysics of the central nervous system. Bull Math Biophys 8:51–57
Rashevsky N (1960) Mathematical biophysics. Dover, New York
Rashevsky N (1961) Mathematical principles in biology and their applications. Thomas, Springfield pp 20–28
Riedl R (1984) Biology of knowledge: the evolutionary basis of reason. Wiley, Chichester
Rosenzweig MR, Bennett EI (1976) Neural mechanisms of learning and memory. Massachusetts Institute of Technology Press, Cambridge
Royer JM (1985) Reading from the perspective of a biological metaphor. Contemp Educ Psychol 10:116–123
Russell WR (1959) Brain, memory, learning: a neurologist's view. Clarendon, Oxford Press
Shapiro E, Klein M, Kandel E (1981) Ionic mechanisms and behavioral functions of presynaptic facilitation and presynaptic inhibition in *Aplysia*: a model system for studying the modulation of signal transmission and sensory neurons. In: Ottoson D (ed) Progress in sensory physiology 1. Springer, Berlin Heidelberg New York, pp. 97–137
Shaw GL, Silverman DJ, Pearson JC (1985) Trion model of cortical organization: toward a theory of information processing and memory. In: Palm G, Aertsen A (eds) Brain theory: proceedings of the first Trieste meeting on brain theory, October 1–4, 1984. Springer, Berlin Heidelberg New York, pp 177–191
Simmons ES (1980) The effects of kinetic structure in films on biology student's achievement and attitude. J Res Sci Teach 17:67–74
Smythies JR (1970) Brain mechanisms and behavior, 2nd edn. Blackwell, Oxford
Sokolov EN (1960) Neuronal models and the orienting reflex. In: Brazier MAB (ed) Electrical activity of the central nervous system and behaviour. Macy, New York
Sokolov EN (1963) Perception and the conditioned reflex. Macmillan, New York
Sommerhof G (1974) Logic of the living brain. Wiley, London
Squire LR, Butters N (eds) (1984) Neuropsychology of memory. Guilford, New York

Staubli U, Baudry M, Lynch G (1985a) Olfactory discrimination learning is blocked by leupeptin, a thiol-proteinase inhibitor. Brain Res 337:333–336

Staubli U, Faraday R, Lynch G (1985b) Pharmacological dissociation of memory: anisomycin and leupeptin selectively disrupt different memory tasks. Behav Neural Biol 43:287–297

Sternberg RJ (1985) Beyond IQ: a triarchic theory of human intelligence. Cambridge University Press, Cambridge

Studdert-Kennedy M (1983) Introduction. In: Studdert-Kennedy M (ed) Psychobiology of language. MIT Press, Cambridge, pp 1–6

Studdert-Kennedy M, Lane H (1980) Clues from the differences between signed and spoken language. In: Bellugi U, Studdert-Kennedy M (eds) Signed and spoken language: biological constraints on linguistic form. Dahlem-Konferenzen. Verlag Chemie, Weinheim, pp 29–40

Szentágothai J (1978) The local neuronal apparatus of the cerebral cortex. In: Buser PA, Rougel-Buser A (eds) Cerebral correlates of conscious experience. North-Holland, Amsterdam, pp 131–139

Szentágothai J, Arbib MA (1975) Conceptual models of neural organization. MIT Press, Cambridge

Teyler TJ (1977) An introduction to the neurosciences. In: Wittrock M, Beatty J, Bogen JE, Gazzaniga MS, Jerison HJ, Krashen SD, Nebes RD, Teyler TJ (eds) The human brain. Prentice-Hall, Englewood Cliffs, pp 3–37

Teyler TJ (1986) Memory: electrophysiological analogs. In: Martinez JL, Kesner RP (eds) Learning and memory: a biological view. Academic, Orlando, pp 258–260

Teyler TJ, DiScenna P (1984) The topological anatomy of the hippocampus: a clue to its function. Brain Res Bull 12:711–718

Teyler TJ, DiScenna P (1986) The hippocampal memory indexing theory. Behav Neurosci 100:147–154

Teyler TJ, Perkins AT (1988) The role of hippocampal-neocortical interactions in infor mation processing. In: Markowitsch HJ (ed) Information processing by the brain: views and hypotheses from a physiological-cognitive perspective. Huber, Toronto, pp 199–206

Thompson RF, Berger TW, Berry SD (1980a) An introduction to the anatomy, physiology, and chemistry of the brain. In: Wittrock MC (ed) The brain and psychology. Academic, New York, pp 3–32

Thompson RF, Hicks LH, Shvyrkov VB (eds) (1980b) Neural mechanisms of goal-directed behavior and learning. Academic, New York

Thompson RF, Donegan NH (1986) The search for the engram. In: Martinez JL, Kesner RP (eds) Learning and memory: a biological perspective. Academic, Orlando, pp. 3–52

Trevarthen C (1980) Functional organization of the human brain. In: Wittock MC (ed) The brain and psychology. Academic, New York, pp 33–91

Tucker DM (1987) Hemisphere specialization: a mechanism for unifying anterior and posterior brain regions. In: Ottoson D (ed) Duality and unity of the brain: unified functioning and specialisation of the hemispheres. Plenum, New York, pp 180–193

Ushakova TN (1980) Neurophysiological mechanisms of processing speech information in man. In: Thompson RF, Hicks LH, Shvyrkov VB (eds) Neural mechanisms of goal-directed behavior and learning. Academic, New York, pp 541–554

Waugh NC, Norman D (1965) Primary memory. Psychol Rev 74:89–104

Weinberger NM, Diamond DM (1987) Physiological plasticity in auditory cortex: rapid induction by learning. J Neurophysiol 29:1–55

Weinberger NM, Mcgaugh JL, Lynch GL (eds) (1985) Memory systems of the brain: animal and human cognitive processes. Guilford, New York

Weisenberg RC (1987) The nature and significance of ATP-induced contractions of microtubule gels. In: Bereiter-Hahn J, Anderson OR, Reif W-E (eds) Cytomechanics: the mechanical basis of cell form and structure. Springer, Berlin Heidelberg New York, pp 114–119
Wickelgren WA (1987) Site fragility theory of chunking and consolidation in a distributed associative memory. In: Milgram NW, Macleod CM, Petit TL (eds) Neuroplasticity, learning, and memory: proceedings of a symposium held at the University of Toronto, Scaborough, Ontario, March 25, 1986. Liss, New York, pp 301–325
Will BE, Schmitt P, Dalrymple-Alford JC (eds) (1985) Brain plasticity, learning, and memory. Plenum, New York
Wittrock MC (ed) (1980) The brain and psychology. Academic, New York
Wittrock MC, Beatty J, Bogen JE, Gazzaniga MS, Jerison HJ, Krashen SD, Nebes RD, Teyler TJ (eds) (1977) The human brain. Prentice-Hall, Englewood Cliffs
Young CZ (1964) A model of the brain. Clarendon Press, Oxford.
Young CZ (1966) The memory system. University of California Press, Berkeley
Young CZ (1978) Programs of the brain. Oxford University Press, Oxford
Young CZ (1987) Philosophy and the brain. Oxford University Press, Oxford
Zaidel E (1987) Hemispheric monitoring. In: Ottoson D (ed) Duality and unity of the brain: unified functioning and specialisation of the hemispheres. Plenum, New York, pp 247–279
Zeki S, Shipp S (1988) The functional logic of cortical connections. Nature 335:311–317

Cortical Organization of Language and Verbal Memory Based on Intraoperative Investigations*

G.A. Ojemann

Department of Neurological Surgery, RI-20, University of Washington School of Medicine, Seattle, WA 98195, USA

1 Introduction

The traditional model of the cortical organization of language was developed in the latter half of the nineteenth century by pooling data on the location of brain damage, particularly as a result of strokes, associated with disturbed language. Essential

* The author's research included in this review is supported by NIH grants NS 17111, 21724, and 20482.

language areas were usually lateralized to the left hemisphere. Within that hemisphere, the model included the posterior third of the inferior frontal gyrus as an area for language expression (Broca's area) and a larger area in posterior superior and middle temporal gyri and adjacent parietal operculum for language understanding (Wernicke's area) (Broca 1861; Wernicke 1874). In more recent versions of this model, the Broca and Wernicke language areas are joined by major white matter pathways, where lesions also disturb language (transcortical aphasias). These pathways provide for serial processing of language from decoding in Wernicke's area to expression in Broca's (Geschwind 1970). This model has received support from recent imaging-based studies of brain lesions associated with aphasias (Damasio 1988) and from statistical assessments of formal measure of language deficits with cortical lesions (Kertesz-Phipps 1977). However, other recent studies of these same types of patients have raised questions about major features of the traditional model, including evidence that permanent motor language deficits usually require lesions that extend well beyond Broca's area, into inferior parietal and superior temporal lobe (Mohr 1976), and that language deficits after frontal lesion include not only expressive problems but also a receptive deficit, particularly for syntax (Goodglass 1976).

This paper reviews the organization of language in left lateral cortex from a different perspective, that obtained during neurosurgical operations under local anesthesia in patients with documented lateralization of language to the left hemisphere. This perspective embraces two separate data sets: one set includes localization of language function obtained with a surface electrical stimulation mapping technique and physiologic correlates of that localization derived from electrocorticographic (ECoG) recordings, the other set contains physiologic correlates of language derived from extracellular recording of neuronal activity during appropriate behavioral measures.

Both data sets provide information on cortical organization. The pattern of organization derived from electrical stimulation mapping, like that derived from lesions, depends on the disruption of a behavior when a particular cortical area is functionally inactivated (in the case of stimulation, probably from depolarization blockade evoked by the current). These techniques, then, identify cortical areas that must be active for the behavior, and thus are *essential* for the measured function. On the other hand, neuronal recording, like metabolic measures such as blood flow and oxygen or glucose PET scans, indicate where neurons are active and participating in a behavior, not whether they are essential for it. The patterns of cortical language organization in essential and participatory areas seem to differ, and both suggest the need for revision of the traditional model.

All techniques for investigating language organization in the human brain have limitations as well as advantages. Each technique is applicable only to special populations: the traditional lesion studies of aphasiology, for example, are predominately from the special population of elderly patients with strokes. The studies reported here are from the special populations of patients with diseases requiring craniotomy under local anesthesia, in most cases for medically intractable epilepsy. The extent to which the findings in these special populations can be extrapolated to "normals" is not known, although comparison of patterns of language organization in patients with epilepsy secondary to diseases likely beginning in early life to those in patients with epilepsy secondary to adult acquired disease has shown few differences (Ojemann et al. 1989b), suggesting that this factor, at least, contributes little to the findings reported here. Other clinical considerations also impact on these studies: cortical areas available for stimulation mapping are limited to those subject to the clinically necessary exposure, and invasive studies such as neuronal recording, are restricted to cortical areas that will be subsequently removed. Despite these limitations, the techniques utilized in these intraoperative studies have some special advantages over traditional lesion studies, especially with regard to the reversible nature of the blockades of cortical function, which allows the assessment of function of many small cortical surface areas in an individual subject during measures of different language-related behaviors.

2 Pattern of Essential Areas for Language Within the Left Hemisphere

Applying an electric current to the cortical surface produces both local activation and blockade of neurons and en passage fibers, evoking excitatory or inhibitory effects both locally and at a distance (Ranck 1975). Empirically, threshold stimulation of human cortical surface in the quiet patient evokes responses only from rolandic cortex (movement and localized dysesthetic sensations), primary visual cortex (localized phosphenes) and, rarely, primary auditory cortex (buzzing) (Penfield and Jasper 1954). The remaining cortex is usually silent. Rare exceptions have been reported, such as the illusions and experiential phenomena of Penfield (Penfield and Jasper 1954; Mullan and Penfield 1959; Penfield and Perot 1963). However, there is now evidence that these occur only with afterdischarge, as part of an epileptic manifestation involving medial temporal structures (Gloor et al. 1982).

However, when the patient engages in an ongoing language task, such as naming of objects, stimulation of some of the areas that were silent in the quiet patient, at currents below the threshold for afterdischarges, disturbs naming. Presumably, the behavioral disruption occurs because stimulation evokes depolarization blockage of local function at a site essential for the behavior. This is the technique of electrical stimulation mapping developed by Penfield and his associates (Penfield and Roberts 1959).

Penfield inferred that stimulation and lesions identified the same areas of brain as essential for language. When data from a large series of patients were pooled it was found that many errors were evoked from "silent" cortex only in the left hemisphere, and there from areas that were generally congruent with classical language areas. Subsequently, the technique became a standard method for localizing language during neurosurgical procedures. More direct evidence that stimulation mapping and lesions relate the same areas of cortex to language was provided by Ojemann and Dodrill (Ojemann 1983b), who reported that anterior temporal resections encroaching on sites of repeated stimulation-evoked naming errors were associated with deficits on aphasia tests administered 1 month after operation. These changes were absent when the resection margin avoided those sites, and could not be accounted for by size of resection, degree of seizure control or preoperative verbal ability. Resections that encroached on sites with repeated naming errors were associated with deficits in all aspects of language, not just naming. Technical aspects of stimulation mapping, particularly the selection of appropriate current levels, have been discussed in detail elsewhere (Ojemann 1983a, 1987).

Penfield reported that anomia, the inability to name with retained ability to speak, was the change in naming evoked from cortical sites that most closely corresponded to the traditional language areas. Other types of naming errors – arrest of speech, hesitation and repetition – were evoked from much wider areas including some sites in the nondominant hemisphere. But even anomia, though confined to the dominant hemisphere for sites outside of rolandic cortex, was occasionally evoked outside of traditional language areas (Penfield and Roberts 1959). Van Buren et al. (1978) also evoked anomia from lateral cortical sites far removed from traditional language areas. In addition, Penfield evoked changes in naming from the medial face of the posterior frontal lobe. As a result of this and the effect of excisions he postulated a role for the supplementary motor area in language. Recently, Lueders et al. (1986) reported evoking anomia from the undersurface of the dominant temporal lobe. In addition, there is a single report of a few sites with sporadically evoked anomia from nondominant association cortex (Andy and Bhatnagar 1984). These findings suggest that models of cortical language organization derived from stimulation mapping may differ from the traditional model.

Over the last two decades, the author and his associates have systematically examined the effects of intraoperative cortical stimulation on language functions in the hemisphere established as dominant by previous intracarotid amobarbital perfusion testing. Their recently published experience with stimulation mapping, which identified essential cortical areas for naming in the left, dominant hemisphere of 117 nonaphasic patients, represents the largest series reported to date (Ojemann et al. 1989b). That experience is the basis for the first part of this section. In that study, only a complete failure in naming during stimulation, omission of the name with retained ability to speak, or misnaming were considered naming errors; hesitations were not counted as errors. Although localization of different types of naming errors has been evaluated (Ojemann 1983b), those data have contributed little to an understanding of cortical language organization. Therefore, type of naming errors will not be considered further in this section.

2.1 Extent of Essential Areas for Language in Individual Subjects

When object naming was used as the measure of language, repeatable, statistically significant stimulation-evoked errors in lateral perisylvian cortex of an individual subject were often confined to one or more areas, about 1 cm^2 in surface extent. Figures 1–3 illustrate this in several patients; similar findings in other patients are illustrated in Ojemann (1983a, 1988) and Ojemann et al. (1989c). In the series of 117 patients, an uninterrupted area of cortical surface essential for language of more than 2 cm^2 could be identified in only 39% (Fig. 4). The transition from the localized area where stimulation evokes repeated language errors to the surrounding cortex

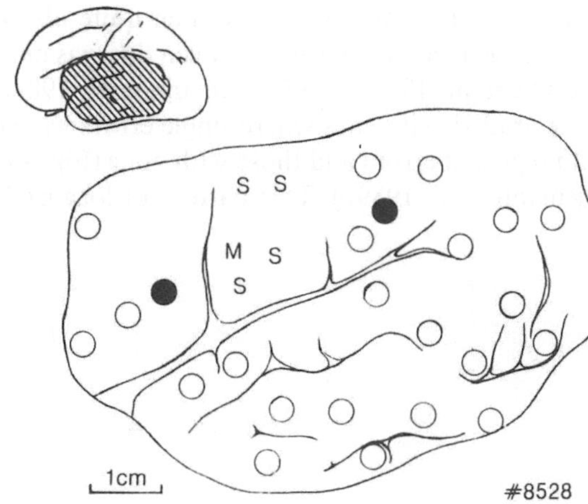

Fig. 1. Localization of essential areas for naming in a 24-year-old female. Circles, sites of stimulation; *filled circles*, statistically significant repeated naming errors. Stimulation at 3 mA. *M,S*, motor and sensory response identifying rolandic cortex. Note localized essential sites for naming in frontal and parietal cortex

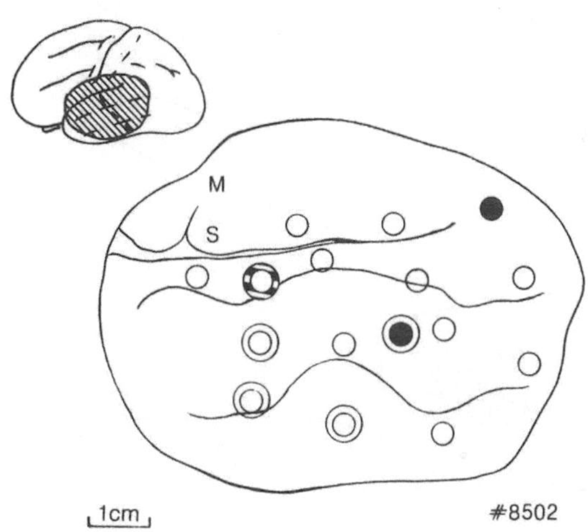

Fig. 2. Localization of sites essential for naming (*small circles*) and sentence reading (*large circles*), presented as in Fig. 5, in a 55-year-old female VIQ 116. Stimulation at 5 mA. Separation of the site essential for reading from that essential for naming is evident. Note also localization of these sites in relation to patient's VIQ (compared to Fig. 5)

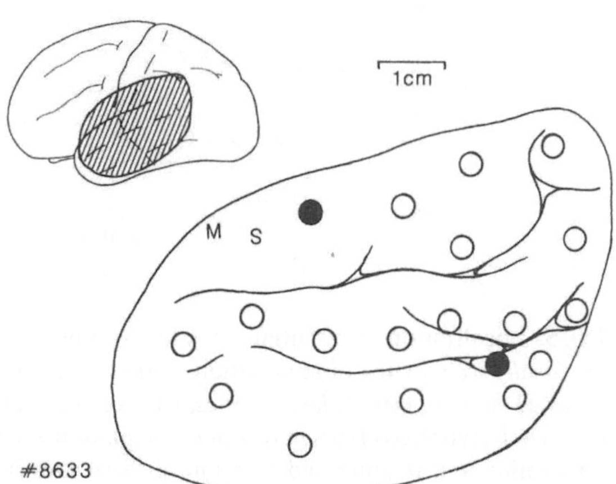

Fig. 3. Localization of sites essential for naming a 70-year-old male; stimulation at 3 mA. Discrete localization similar to that in 4-year-old (Fig. 7). (From Ojemann et al. 1989b)

where no errors are evoked is often quite abrupt, with effects changing when a bipolar electrode with 5 mm between tips was moved a few millimeters (Figs. 2, 3, 5, and Ojemann 1983a, 1988; Ojemann et al. 1989b). Less often, this transition was more graded, with sites where single errors were evoked interposed between areas with repeated errors and those with none (Fig. 4 and Whitaker and Ojemann 1978, Ojemann et al. 1989b). This pattern of localized essential areas often with sharp

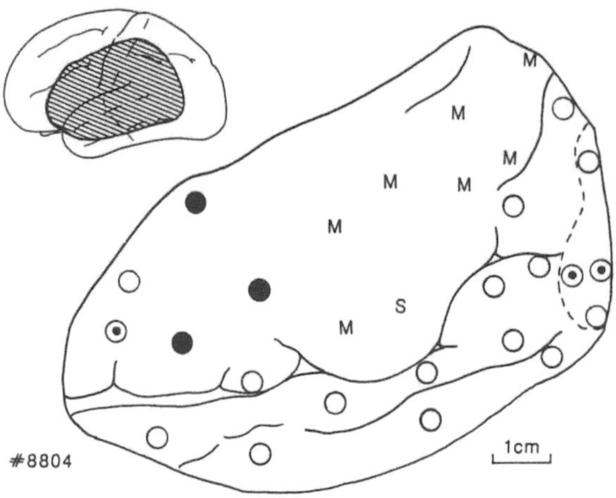

Fig. 4. Localization of essential areas for naming presented as in Fig. 1 in a 37-year-old female, VIQ 99; stimulation at 5 mA. *Dots*, single errors. No posterior language sites were identified. A posterior temporal resection (*dashed line*) did not change language even acutely. (From Ojemann et al. 1989b)

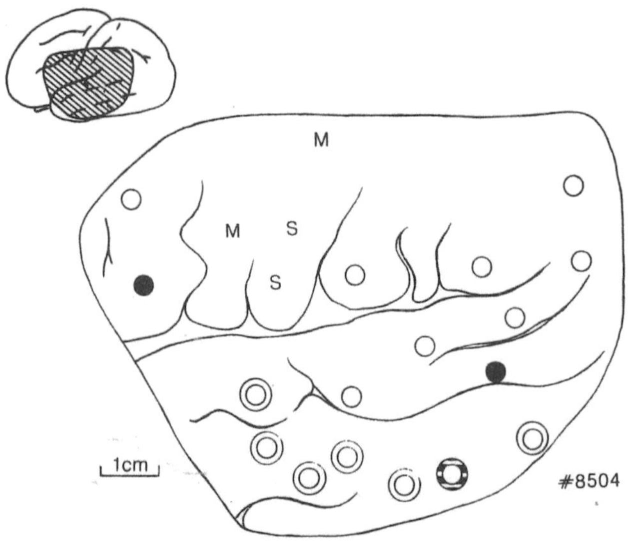

Fig. 5. Localization of essential areas for naming and reading in a 24-year-old male, VIQ 89. *Small circles*, sites of stimulation during naming; *filled circles*, statistically significant repeated naming errors; *large circles*, sites of stimulation during sentence reading; *shaded large circle*, repeated errors in reading. Stimulation at 8 mA. Note separation of sites essential for naming and reading and location of these sites in relation to patient's VIQ (compare to Fig. 2)

boundaries suggests a mosaic or modular organization of essential areas for language in association cortex.

Most subjects have several separate essential modules for language in lateral perisylvian cortex. In two-thirds of the series of 117 patients at least two modules essential for language, separated by cortex not related to language, were identified (Ojemann et al. 1989). In one-quarter of the subjects in that series, more than two separate modules were identified. Usually there was one module in posterior inferior frontal lobe and one or more in temporoparietal cortex (Figs. 1 and 5), but occasionally these sites were present only in parietal cortex (Fig. 1, also Ojemann et al. 1989b, Fig. 3).

The total area of cortical surface identified as essential for language in an individual subject exceeded 6 cm^2 in only 16% of the 117 patient series (Ojemann et al. 1989). The total surface area of cortex related to language in the traditional model is clearly well in excess of 6 cm^2 (cf. Penfield and Roberts 1959). Few individuals then actually have essential areas for language covering the entire area of the traditional model.

Although stimulation mapping usually samples only the cortical surface, the limited information available suggests that essential areas for language are rarely or never located only in the large portion of association cortex that is buried in sulci. The 117 patients with surface sites essential for language identified somewhere in perisylvian cortex came from a series of 119 consecutive left hemisphere dominant nonaphasic individual undergoing stimulation mapping (Ojemann et al. 1989b). Thus surface sites essential for language are present in nearly all patients. These surface sites provide reliable assistance in avoiding language deficits with resections that include both buried and surface cortex (Ojemann 1983b) this would not be so if there were buried essential areas located away from surface sites. On a few occasions, the extension of a surface essential area into a sulcus has been assessed with stimulation mapping. In one case the surface site did not extend into the sulcus, in two others the surface site extended into the immediately adjacent sulcus for about 1 cm (Ojemann and Whitaker 1978b; Ojemann et al. 1989b). Thus there seems to be a preferential

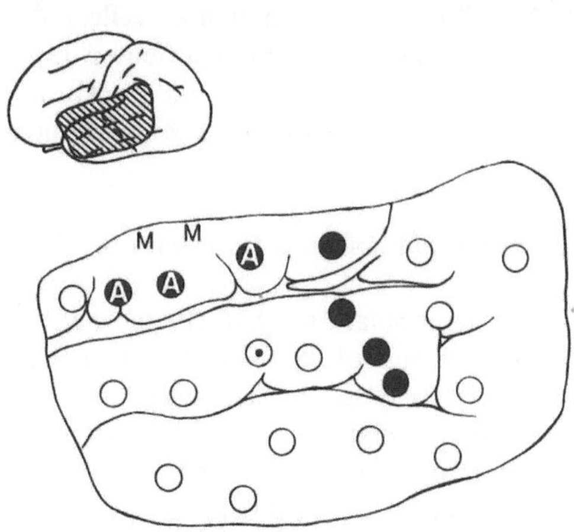

Fig. 6. Localization of sites essential for naming in a 30-year-old male; stimulation at 5 mA. Naming in German; the patient spoke only German. *Filled circles with "A"*, sites of repeated speech arrest. Errors at other sites marked by *filled circles without letters* were anomia. *Dot*, single error. Essential areas in temporal cortex oriented transversely to axis of superior temporal gyrus (Another similar case shown in Fig. 2 of Ojemann et al. 1989b)

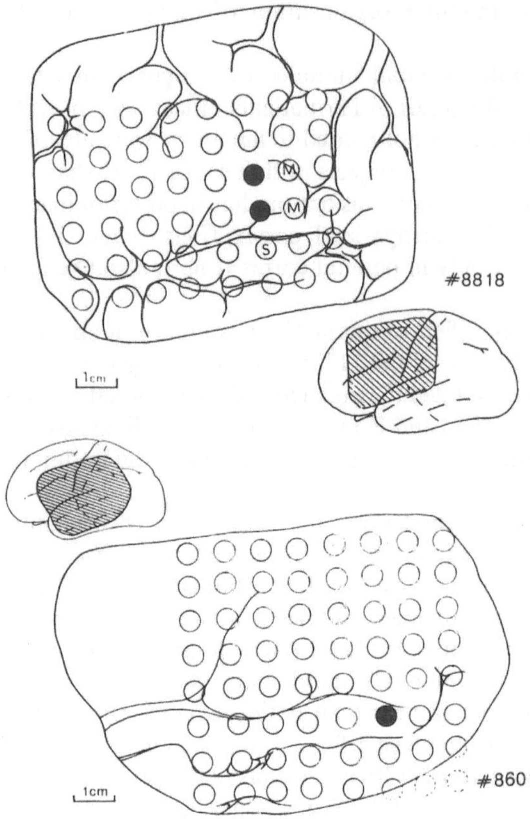

Fig. 7. Localization of essential areas for naming in two 4-year-olds. Stimulation of adjacent contact of subdural electrode grid; contacts stimulated indicated by *circles*. Repeated error evoked only when contacts indicated by *filled circles* included in stimulation. Localized essential areas evident even at this young age. (Case 8601 from Ojemann et al. 1989b)

localization of essential language areas to the surface of gyri. Essential areas were often oriented transversely to the long axis of gyri (Fig. 6; see also Fig. 2 of Ojemann et al. 1989), a pattern that may reflect the connections of these cortical areas, which, in animals, are often oriented in bands perpendicular to the gyral axis (Creutzfeldt O., personal communication). Highly localized essential areas for naming of 1 cm^2 extent have been identified in subjects as young as age 4 (Fig. 7).

2.2 Electrophysiological Correlates of Essential Areas

Neuronal populations that change activity during language measures but not during visuospatial tasks in response to physically identical visual stimuli have been recorded widely in the dominant temporal lobe (Ojemann et al. 1988). Thus the presence of changes in neuronal activity with language, per se, does not distinguish essential from nonessential language cortex. Microelectrode recording is sufficiently invasive that it is not possible to record from essential language areas; thus the presence of any unique patterns of neuronal activity there is unknown. However, there are changes in the surface electrocorticogram (ECoG) recorded during naming that distinguish essential from nonessential language areas (Fried et al. 1981; Ojemann et al. 1989a). ECoG recordings obtained during silent naming showed slow potentials at frontal sites and desynchronization at temporoparietal sites essential for naming, changes that were less intense in surrounding nonessential cortex (Fig. 8). At some, but not all sites essential for naming, these changes were also significantly less intense during a visuospatial task with the same visual stimuli. These frontal and temporoparietal changes occurred in parallel, beginning shortly after the appearance of the item to be named, and lasting about a second, approximately the time required for overt naming (Fig. 8). There was no physiologic evidence for serial processing from posterior to anterior language sites. Thus essential areas for language seem to be organized in separate modules functioning in parallel, analogous to the parallel distributed model for organization of animal association cortex (Mountcastle 1979; Goldman-Rakic 1988).

The slow ECoG potentials occurring during silent naming at frontal sites essential for naming are likely motor prepotentials; similar potentials occur during overt naming. Their presence during a silent language task suggests that the processing of "internal" silent speech proceeds to the level of effector readiness (Fried et al. 1981). Both frontal slow potentials and temporoparietal desynchronization in animals can be signs of selective attention mechanisms, reflecting activity of the thalamocortical activating circuits (Jasper 1960; Skinner and Yingling 1977). Independent evidence of a role for selective attention mechanisms mediated by thalamic circuits has been obtained from observations on the effects of left lateral thalamic stimulation on language and memory measures (Ojemann et al. 1971; Ojemann 1975, 1985). A thalamic selective attention mechanism, then, which simultaneously activates essential frontal and temporoparietal cortical mosaics is likely one physiologic mechanism of language.

2.3 *Variation in Localization of Essential Areas*

Although essential areas for language are often localized to multiple separate mosaics each of 1 cm^2 or so in any individual, the exact location of these mosaics varies greatly between subjects (Fig. 9) (Ojemann 1979, 1983a; Ojemann et al. 1989b; Ojemann and Whitaker 1978b). This variation is so great that there is no perisylvian anatomic location that is essential for language in every left brain dominant subject. As indicated in Fig. 9, in about three-quarters of the 117-patient series the posterior portion of the inferior frontal gyrus was the region where essential areas for language were most consistently found (Ojemann et al. 1989b). Variability in location is particularly great in temporoparietal cortex. Only two-thirds of subjects in that large series had one or more sites essential for language in superior temporal gyrus, and those were distributed from its most anterior to posterior extent. A few subjects had no essential language areas in temporal lobe (Fig. 4); indeed, about 20% of the 117-patient series, all of whom were left dominant, had essential language areas identified in only frontal or temporoparietal lobes, but not both (Ojemann et al. 1989b). Previous stimulation mapping studies implied variability in localization, though it was not systematically identified. Penfield recorded some sites essential for

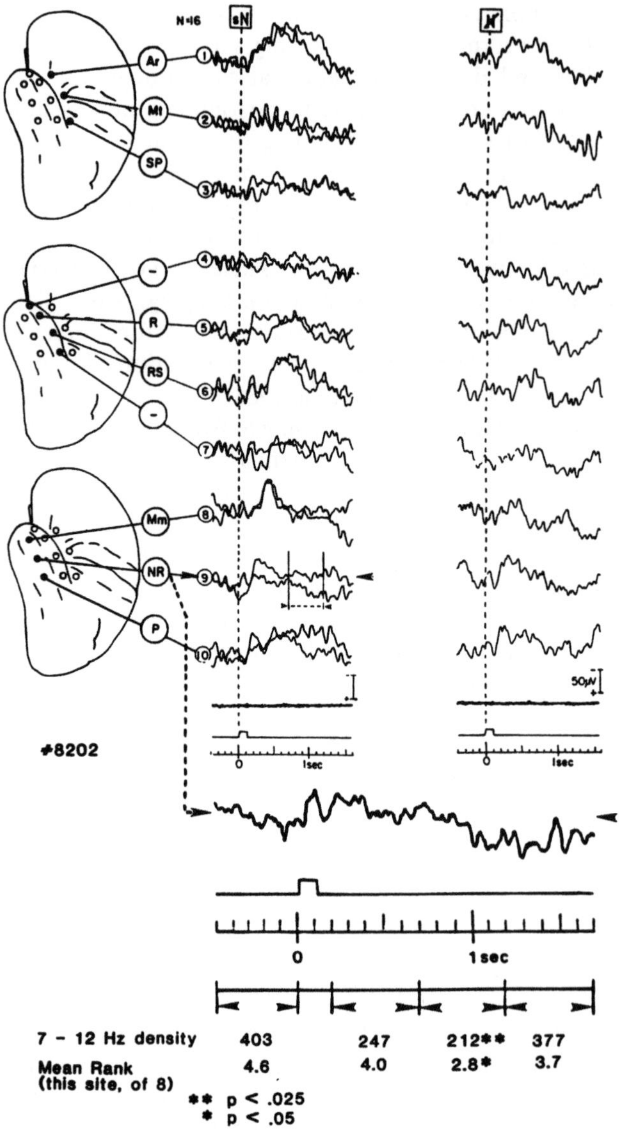

Fig. 8. Electrocorticographic (ECoG) changes during naming, at 10 sites with relation to naming established by stimulation mapping, in a 24-year-old male, VIQ 113. *Above left*: recording sites with relation to naming indicated. *Ar*, arrest of all speech; *N*, anomia; *Mt*, motor. Stimulation effect on sentence reading, recent verbal memory, phoneme identification and orofacial movements also determined: *R*, reading errors; *MN*, memory errors; *P*, phoneme errors; *S*, errors on mimicry of sequences of orofacial movements; *lines*, no errors. *Center, right*: ECoG average for 16 trials of 500 ms before and 2000 s after 100-ms presentation of slides containing object pictures and lines. *Center*: two separate averages for slides silently named. *Right*: average when line angles perceived. In each trial accuracy of silent task was checked 4 s later by overt production of the name, or matching of angle.

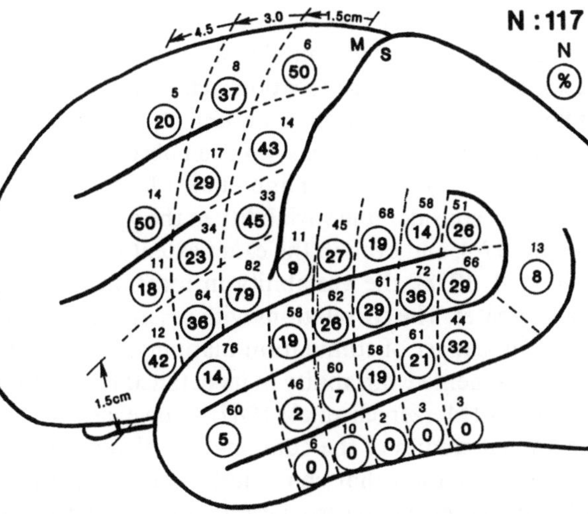

Fig. 9. Variability in language localization in 117 patients, all with left brain language. Individual maps aligned to motor cortex and end of sylvian fissure, and cortex divided into zones identified by *dashed lines*. *Upper number* in each zone, number of patients with a site in that zone; *lower number in circle*, percentage of those patients with sites of significant evoked naming errors in that zone. (From Ojemann et al. 1989b)

naming outside of the traditional language areas, without specific comment (Penfield and Roberts 1959), while Van Buren et al. (1978) discussed the unusual localization of essential areas for naming in several patients. Lesser et al. (1984) noted variable localization in frontal language sites, and Lueders et al. (1986) the variable presence of basal temporal language sites. In the author's series, substantial variability has been evident, even when there were only a few cases (Ojemann and Whitaker 1978b).

This substantial variability in the location of essential areas for language has some relation to subjects' sex and verbal abilities as measured by the verbal IQ (VIQ). Essential language sites in parietal cortex were significantly more frequent in males with lower VIQ than in females with lower VIQ, although there was no difference for the higher VIQ subjects (Ojemann et al. 1989b). Males with essential language sites in superior temporal gyrus had significantly lower VIQ than those without essential

◀**Fig. 8** (*continued*). Note simultaneous slow potential at frontal naming site (*1*) and desynchronization at temporoparietal naming site (*9*), changes that are less evident during the angle matching task. *Below*: a single trial ECoG segment recorded during silent naming from temporal naming site *9*. Beneath that, the 7–12 Hz spectral densities for 500 ms before slide presentation, and the 200–700, 700–1200, 1200–1700 ms epoch of all silent naming trials recorded at that site. Below that, for each epoch the relative rank of the density at that site among the eight temporoparietal recording sites (3–10). Note the significant decrease in 7–12 Hz density at 700–1200 ms, and the decreased rank of the naming site, quantitative evidence of desynchronization at the naming site, during naming that is greater than in surrounding cortex not related to naming. (Adapted from Ojemann and Creutzfeldt 1987 and Ojemann et al. 1989a)

sites there, while males with essential sites in middle temporal gyrus and all subjects with sites widely distributed in frontal lobe, tended to have higher VIQ than subjects without essential sites in those locations (cf. Figs. 2 and 5) (Ojemann et al. 1989b). No age differences in localization have been reported; the highly localized modular pattern for essential language areas has been observed in subjects as young as 4 and as old as 80 (Figs. 3 and 7) (Ojemann et al. 1989b).

Although there is considerable variability in the gyral patterns in perisylvian cortex (Rubens et al. 1976), and even in the exact extent of cortical cytoarchitectonic area (Galaburda et al. 1978), the variability in location of essential language areas is substantially greater. The correlations with sex and VIQ reduce the chance that this variability is an artifact of stimulation mapping. Recent studies of primate motor cortex have documented variability in its detailed localization between members of the same species (Merzenich et al. 1987). However, there seems to be no direct parallel in animal studies of the combination of very localized modules with high variance in their location that characterizes the organization of essential cortical areas for human language in association cortex. The presence of the high variance does explain the origin of the traditional maps of language localization. The wide extent of language cortex in these traditional maps is an artifact of pooling lesion location in a population of aphasic patients, each with essential language areas in only a part of the map.

2.4 Localization of Multiple Language Functions

The author and his associates extended the stimulation mapping technique to the investigation of localization of other language functions besides naming. Those studies provide considerable evidence that different language functions often have different essential modules. In the limited experience with stimulation mapping localization of naming of the same object pictures in two different languages in bilingual subjects, sites essential for only one of the two languages were identified in every case (Figs. 10 and 11; also Ojemann and Whitaker 1978a; Ojemann 1983a; Black and Ronner 1987; Rapport et al. 1983; Cawthon et al. 1987). This differential localization has been identified for both Oriental and Indo-European languages (Rapport et al. 1983). No consistent differences in localization of one or another language have been observed, but only localization of naming has been investigated. Some subjects also had additional sites essential for both languages. When these were present they were often near to rolandic cortex, with sites essential for only one or the other language further away. Essential areas for naming with manual communication systems are also partially separate from those for oral language (Mateer et al. 1982, 1984).

Similarly, assessment of localization of several language functions in an individual subject often identifies different modules as essential for different language functions. When the effects of stimulation on both naming and reading of simple sentences were assessed at 479 lateral cortical sites in 55 patients, 23% of sites showed changes in only one or the other, and an additional 13% changes in both, with no changes evoked at the remaining sites (Ojemann 1989). In an individual subject, the total area

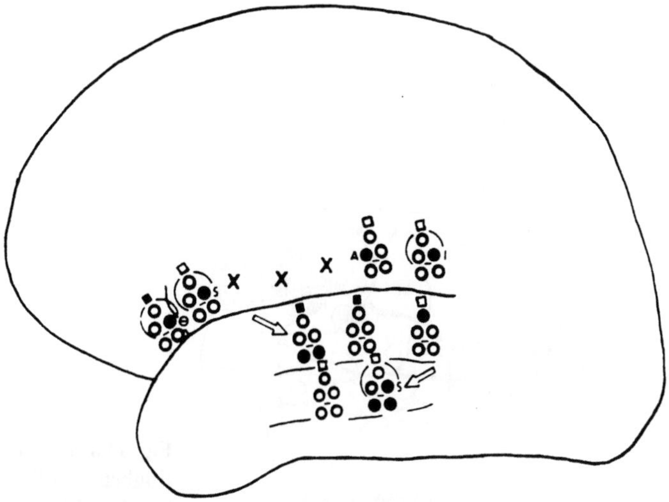

Fig. 10. Essential areas for naming (*top circle*), sentence reading (*left circle, second row*), recent verbal memory (*right circle, second row*), phoneme identification (*left circle, third row*), mimicry for single or sequential orofacial movements (*right circle, third row*), and naming in a second language (*square*) at nine sites in a 30-year-old female VIQ 103. Stimulation at 3 mA. *Filled symbols* indicate significant error. Except for the association between phoneme identification and orofacial movement, individual sites were usually related to single functions, including separate sites related to naming in the two languages. (From Ojemann 1983c)

of cortex essential for sentence reading was usually slightly larger than that for naming (Fig. 12). As with naming, there was considerable individual variability in the location of essential areas for sentence reading (Ojemann 1989). The patterns of localization of temporoparietal sites essential only for naming or only for sentence reading were highly correlated to the VIQ (Ojemann 1989). Subjects with sites for only naming in middle temporal gyrus and sites for only sentence reading in superior temporal gyrus were likely to have high VIQ (Figs. 2, 12), while those with the reverse pattern, sites essential for only naming in superior temporal gyrus and only sentence reading in the middle temporal gyrus were likely to have low VIQ (Fig. 5). Localization of sites essential for both naming and reading was not related to the VIQ. A possible explanation for this rather unexpected finding, involving the important relationship of reading to phoneme identification mechanisms and the extent of cortex that was previously occupied during acquisition of naming, has been discussed in detail elsewhere (Ojemann 1989).

Further evidence of the separation of modules for different language related functions was derived from assessment of the effect of stimulation on measures of naming, reading, and recent verbal memory at all of 118 sites in 14 subjects, and additionally from stimulation effects on phoneme identification and motor control for single and sequential orofacial movements at 80% of those sites. Changes in only one of the measured functions were evoked at 44% of the sites, in multiple functions

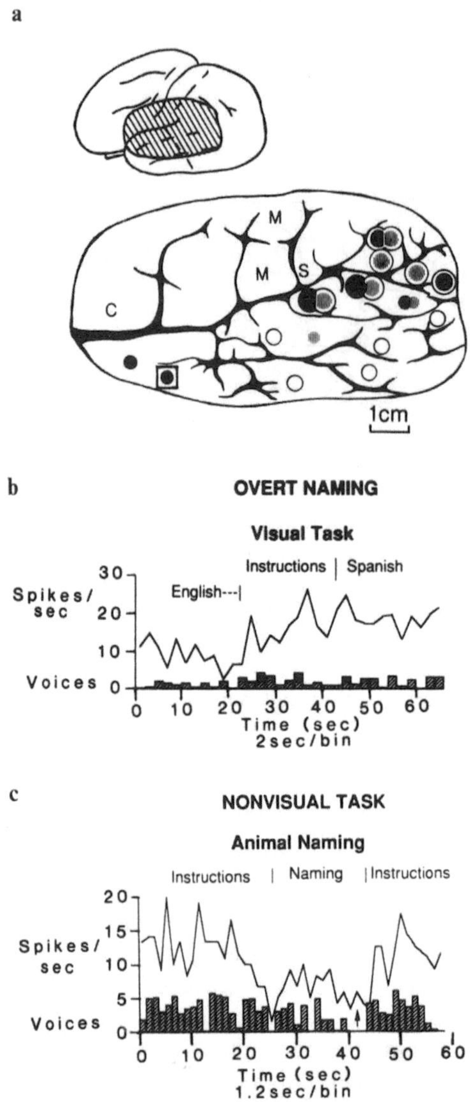

Fig. 11 a–c. Naming in two languages, English and Spanish, in a 27-year-old female. **a** Identification of essential areas with stimulation mapping at 6 mA during naming of same object pictures. *Large filled circles*, English errors; *large shaded circles*, Spanish errors; *large conjoined circles*, errors in both at same site. *M*, face motor response; *S*, face sensory response. **b** Neuronal activity recorded at middle temporal site identified by *square* in **a**, at transition from naming in English to naming (of same objects) in Spanish. Activity increases abruptly with instructions to name in Spanish. **c** Activity remained at higher level. About 10 min later, the subject was requested to produce names of animals in English. Activity abruptly decreases during this task. *Bars* below on **b** and **c**, voice activity on audio channel. (From Cawthon et al. 1987 and unpublished observations)

at 33%, and at the remaining 23% no changes in any measured function were evoked (Ojemann 1983b). Figures 8 and 10 illustrate this separation for two subjects of that study.

A model of the organization of language related functions in perisylvian cortex was derived from the patterns of the behaviors altered by stimulation of a particular site in those cases (Ojemann 1983a). One set of sites where multiple functions were altered by stimulation was essential for naming, reading, output from recent memory, and motor control of even single orofacial movements. That pattern suggests a final motor pathway for speech, and those sites were most often in

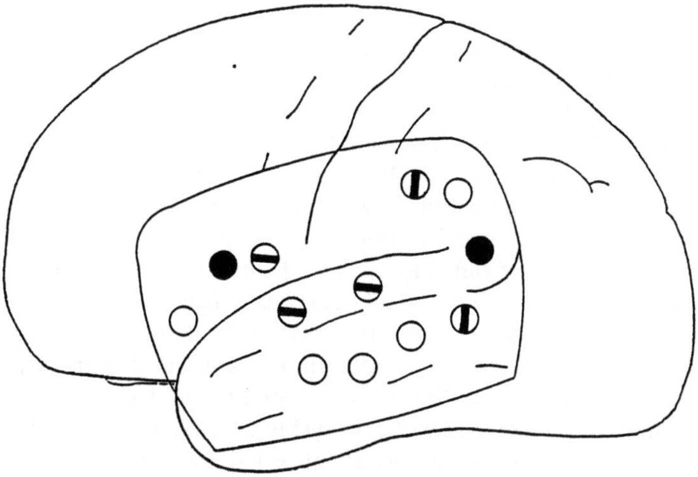

Fig. 12. Localization of sites essential for naming and/or sentence reading in 33-year-old female with VIQ of 108; stimulation at 6 mA. *Filled circle*, essential for both naming and reading; *vertical line*, only naming; *horizontal line*, only reading. (From Ojemann 1989)

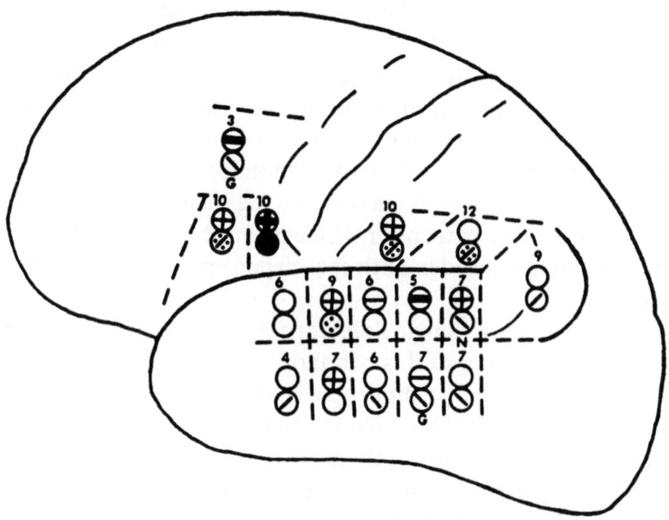

Fig. 13. Organization of perisylvian language cortex, based on localization of essential areas of naming, reading, recent verbal memory, phoneme identification and control of orofacial movement at 118 sites in 14 patients. For each zone, the number indicates how many patients had stimulation sites there. *Upper circle*, extent of essential areas for naming and reading: *thin line*, present at 50% or more of sites; *heavy line*, 70%–80% or more; *vertical line*, naming; *horizontal line*, reading. *Lower circle*, zones with frequent changes representing one of the components of the model of language organization: *filled circle*, final motor pathway; *dots*, sequential motor system. Phoneme identification error frequent at zones with either of these two symbols and infrequent elsewhere. *Slash top to right*, recent verbal memory; *slash top to left*, only reading or naming: *N*, only naming; *G*, only reading errors of syntactic type. (Adapted from Ojemann 1983b)

posterior inferior frontal cortex immediately anterior to face motor cortex. A second set of sites where multiple functions were altered was also essential for naming and/or reading, and motor control for sequences, but not for single orofacial movements. These sites were most often located more anteriorly in inferior frontal gyrus, in parietal operculum, and in superior temporal gyrus (Fig. 13). Stimulation at these sites usually did not alter recent verbal memory, including stimulation at the time of retrieval, indicating, among other things, that perception of the visual test stimuli was not altered by stimulation. Thus the portions of cortex essential for motor functions of language included not only the traditional Broca posterior inferior frontal area, but also a larger area of frontal lobe, inferior parietal lobe and superior temporal lobe. This is the area that Mohr's studies indicate must usually be destroyed to produce a permanent motor aphasia (Mohr 1976).

The same cortical sites that were essential for orofacial motor control were often essential for phoneme identification (Figs. 10, 13; also Ojemann and Mateer 1979; Ojemann 1983a, b). In the 14-patient series, about three-quarters of sites essential for control of orofacial movements, either singly or in sequence, were also essential for phoneme identification, and this accounted for over three-quarters of the sites essential for that latter measure. The phoneme identification test in that study was a test of phoneme perception, for stimulation occurred only during the time the phoneme was heard, with a response period without stimulation. Thus the motor speech disturbance per se would not account for the finding that a common site was often essential for both functions. Particularly in a context where two-thirds of sites were identified as essential either for one or none of the measured functions, this finding suggests some common mechanism for both orofacial motor control and speech sound identification located in nonrolandic cortex immediately above and below the sylvian fissure. One such mechanism is described by the motor theory of speech perception (Liberman et al. 1967). Another is a mechanism of precise timing utilized in both speech perception and expression (Tallal 1983; Calvin 1983).

Sites essential for single language or verbal memory functions surround the perisylvian location of this common cortex for speech production and perception, both frontally and temporoparietally (Fig. 13) (Ojemann and Mateer 1979; Ojemann 1983a, b). Those closest to the perisylvian area were usually essential for one language function – only naming or only reading. Errors at sites essential for only reading tended to involved the grammatical aspects of sentences, rather than the semantic aspects, suggesting that these sites might be especially essential for syntax (Ojemann 1983a).

Peripheral to sites essential for single language functions were cortical sites essential for only recent verbal memory (Ojemann 1978, 1983a, b; Ojemann and Dodrill 1985, 1987). These sites have been shown to be important in memory deficits with temporal lobe resections (Ojemann and Dodrill 1985, 1987). The location of these sites essential for recent memory, like the sites essential for naming and those essential for reading, shows substantial variability between subjects (Ojemann and Dodrill 1985). Thus these stimulation studies also alter the usual model of brain mechanisms for recent memory to include not only medial temporal structures, especially hippocampus, but also lateral neocortex. Dominant temporal neocortical sites essential for recent verbal memory were also identified by Fedio and Van Buren (1974).

Outside of the final motor pathway for all speech output, sites essential for recent memory rarely overlap with sites essential for language (Ojemann 1978, 1983a; Ojemann and Dodrill 1985). The behavioral memory distinction between "episodic" and "semantic" (Tulving 1972), or "declarative" and "procedural" (Squire 1987) types likely reflect this anatomic separation of essential areas for recent memory from those for language.

There is evidence of lateralized motor control systems and memory systems in animals (Trevarthan 1974; Dewson 1977). The evolutionary development of language, then, may involve the lateralized development of a perisylvian area for orofacial motor control, and surrounding memory system, with the more specialized language functions developing at the interface between them.

Apart from the frontally located final motor pathway for speech, the systems identified in the model of lateral cortical language organization derived from these studies have similar processes in frontal and temporoparietal cortex. This pattern, of course, differs from the traditional model of language cortex that assigns expressive functions to frontal lobe and receptive functions to temporoparietal areas. In the stimulation mapping study, sites common to both control of orofacial sequential motor movements and phoneme identification were found both frontally and temporoparietally. Sites for only naming were found in both locations. Sites with only sentence reading changes suggestive of a role in syntax were found in both locations. Sites related to recent verbal memory were also present in both locations. However, there is some evidence that the frontal memory sites have a different role than the temporoparietal sites. The memory measure used in these studies (Ojemann 1978, 1983a, b) indicated separation of stimulation effects according to the time of stimulation, whether during entry of information into memory, during the time the memory was stored, or when it was retrieved. Stimulation during memory entry or storage altered performance at 27% of frontal memory sites, but at 62% of temporal and at 64% of parietal memory sites (Ojemann 1983b). Thus temporoparietal sites were related to memory entry or storage, while frontal sites were related to memory retrieval. This has suggested that the frontal-temporoparietal separation for many language related functions may represent a retrieval-storage dichotomy.

2.5 Changes in Localization Related to Experience and Injury

On the few occasions when language mapping has been repeated after an interval without an intervening injury or resection, the maps have been quite similar (Ojemann 1983a). There is only limited information on the changes in localization with different experiences with language, or with injury. Patterns of localization seems to be similar in the very young, during language acquisition, to those of the adult or the very old, all showing the discrete modular organization (Ojemann et al. 1989b). There is some inferential evidence on the effects of experience, derived from the size of essential areas for each language in bilingual subjects (Ojemann and Whitaker 1978a; Cawthon et al. 1987). Larger essential areas seemed to be associated with languages of lesser fluency acquired later in life, the most recent study of this suggesting that later acquisition is a more important factor than fluency. That study identified somewhat different patterns of stimulation evoked errors related to fluency in a language, with complete blockade of all speech more likely in the more fluent language. These few data suggest that the size of essential areas for language may change with experience, being larger for less fluent, later acquired language. Moreover, subjects with poorer language skills as reflected in the VIQ tended to have larger cortical areas essential for language than those with high VIQ, although this difference did not reach statistical significance (Ojemann et al. 1989b).

Similarly, there are only limited data on the effects of injury in dominant hemisphere lateral cortex on language organization. Essential sites for language immediately surrounded the area of cortical injury in both frontal and temporal lobes in a subject who had partially recovered from a severe global aphasia following trauma decades previously (and in whom there was no evidence for language in the other hemisphere on intracarotid amytal testing) (Ojemann G., unpublished observations). In several subjects who developed expressive aphasias on progression of malignant gliomas, remapping during the progression demonstrated the disappearance of a few essential sites as the aphasia worsened; the remaining essential areas were somewhat more widespread, but in the same general location. There was little evidence for the appearance of new distant sites (Ojemann G., unpublished observations). Thus there is presently evidence for little or no reorganization or essential language areas following injury in adult life.

2.6 *Summary of Organization of Essential Language Areas*

Electrical stimulation mapping identification of essential language areas in an individual suggests that these are organized in several modules or mosaics usually at least one frontally and one or more temporoparietally, with a surface area often as small as 1 cm^2. The total surface area of these modules in an individual subject is rarely as great as that of the traditional Broca and Wernicke language areas. Modules are preferentially located at least partially on the surface of gyri. Somewhat different modules are generally essentially for different language-related fuctions. There is substantial variability between subjects in the location of modules for any one language-related function. Some patterns of localization differ by sex and language skills as measured by the VIQ. Modules can be distinguished from surrounding cortex by the appearance of ECoG desynchronization and slow potentials, suggestive of mechanisms of selective attention, that appear in parallel at frontal and temporoparietal sites during the appropriate language function. These studies provide no evidence for serial processing from posterior to anterior language areas. However, they do provide evidence that language cortex is organized in the parallel distribution pattern described for primate association cortex (Mountcastle 1979; Goodman-Rakic 1988). The prominent individual variability in location of the language system is nevertheless, a feature little described in animals. There are some hints that modules for language may decrease in size with greater language skill or earlier acquisition, but there is little evidence for changes after injury to language cortex in the adult.

From the location of essential areas for multiple language related functions has been derived a model of lateral cortical organization for language that differs from the traditional Broca–Wernicke model. Perisylvian cortex, both frontally and temporoparietally is an orofacial motor control system. This same cortical area is essential for identification of speech sounds. It is surrounded both frontally and temporoparietally by sites essential for only single language functions, including only naming, and only syntactic aspects of reading. These in turn are surrounded by neocortical sites essential for recent verbal memory, with the temporoparietal portion of those sites related to memory input and storage, the frontal portion more to memory retrieval. The identification of this neocortical memory system also necessitates a revision in the traditional model of recent memory, namely addition of the neocortical system to the more traditional medial temporal lobe system.

These findings have been based on a technique that identifies essential cortical areas for a function. The next section discusses the organization of language and memory in cortex that is not essential for those functions.

3 Neuronal Activity Related to Language and Memory in Nonessential Cortex

There are only a few investigations of neuronal activity during language and memory – those of Bechtereva (1987) from frontal cortex, striatum and thalamus, Halgren et al. (1978) from medial temporal structures, and the investigations of the author and his associates reviewed in this section. These latter investigations represent initial findings from microelectrode recordings at nonessential lateral temporal neocortical sites during two set of measures that separate perception, attention, language processing, speech output and, in the second set, recent verbal memory. The first set were auditory language tasks, listening to series of words with varying phonemic and syllabic structure, and then listening to the same set a second time with the instruction to repeat each word aloud. The inclusion of pure tones within the range of speech frequencies provided controls for auditory perception apart from speech perception. In a few recordings additional controls were provided by the use of degraded speech and speech backward (Creutzfeldt et al. 1989a). The second set of behavioral measures included visually presented items, physically identical stimuli used to elicit several tasks: a language task requiring silent and overt naming or reading, a visuospatial task requiring matching of line angles, and as part of a recent verbal memory measure (Ojemann et al. 1988). In a few bilingual cases recordings were obtained when these identical visual stimuli were used to elicit the same linguistic function in two languages (Cawthon et al. 1987). As with findings from stimulation mapping, the clinical situation where these recordings can be accomplished limits them to the special population of patients with medically intractable epilepsy undergoing resective neurosurgery under local anesthesia. An additional limitation is that microelectrode recordings are restricted to those

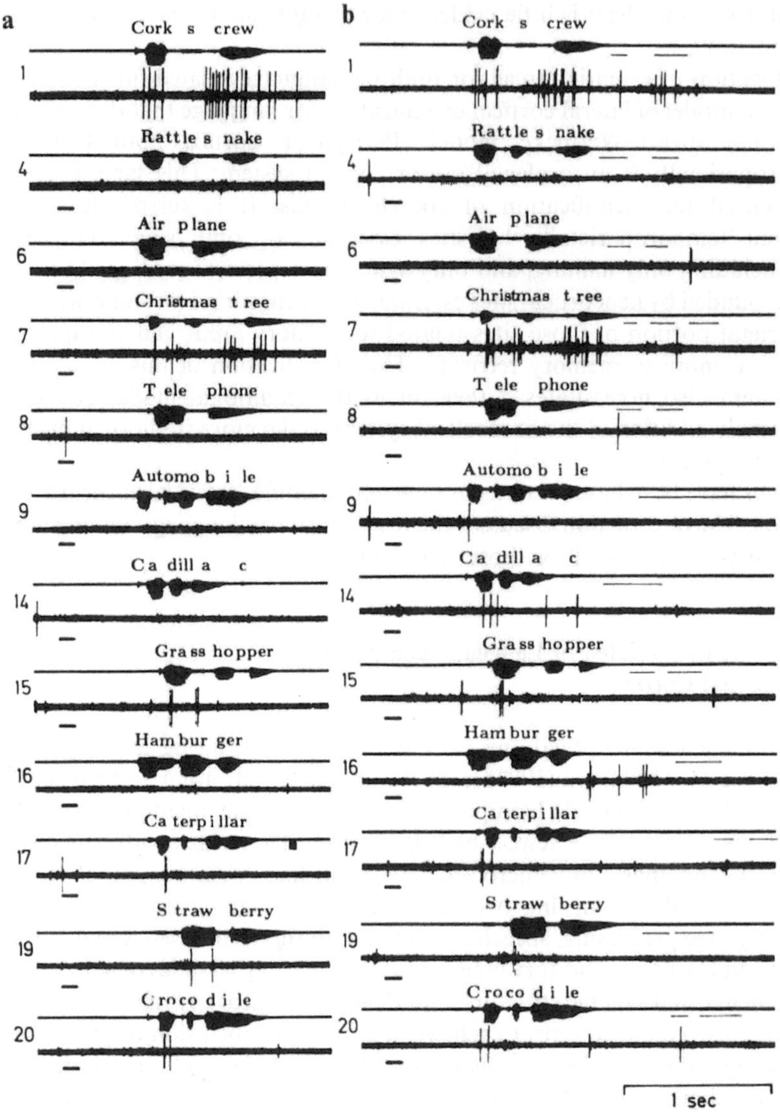

Fig. 14 a, b. Activity of right, nondominant *superior* temporal gyrus neuronal population during listening to specific words. **a** Listening to tape-recorded words, preceded by 1000-Hz tone (*left, heavy bar*). **b** Several minutes later, the same words and tones were presented again with the instruction to repeat the word aloud (*right, light bar*). Patterns of activity are similar with each presentation of the same word, but differ between words, suggesting response to specific class of phonemes. *Above*, audio tape channel; *below*, microelectrode indicating word presentation. Case 8512. (From Creutzfeldt et al. 1989a)

portions of exposed cortex that are subsequently resected. However, recordings were obtained from the portion of resected cortex with the least evidence of epileptic activity on ECoG, and the data considered here were from neuronal populations with little or no evidence of the burst firing that has been associated with animal and human epileptic neurons (Calvin et al. 1973). Thus subjects' epilepsy probably contributed little to the findings discussed here.

The extracellular recordings of neuronal activity commonly obtained from a microelectrode in temporal neocortex included activity from several cells often distinguishable by amplitude of action potentials (Figs. 14–17; Creutzfeldt et al. 1989a, b; Ojemann et al. 1988). The average frequency of activity of an individual neuron was often quite low, less than one discharge per second. Upon activation, activity rarely increased beyond 50 discharges per second. Both statistically

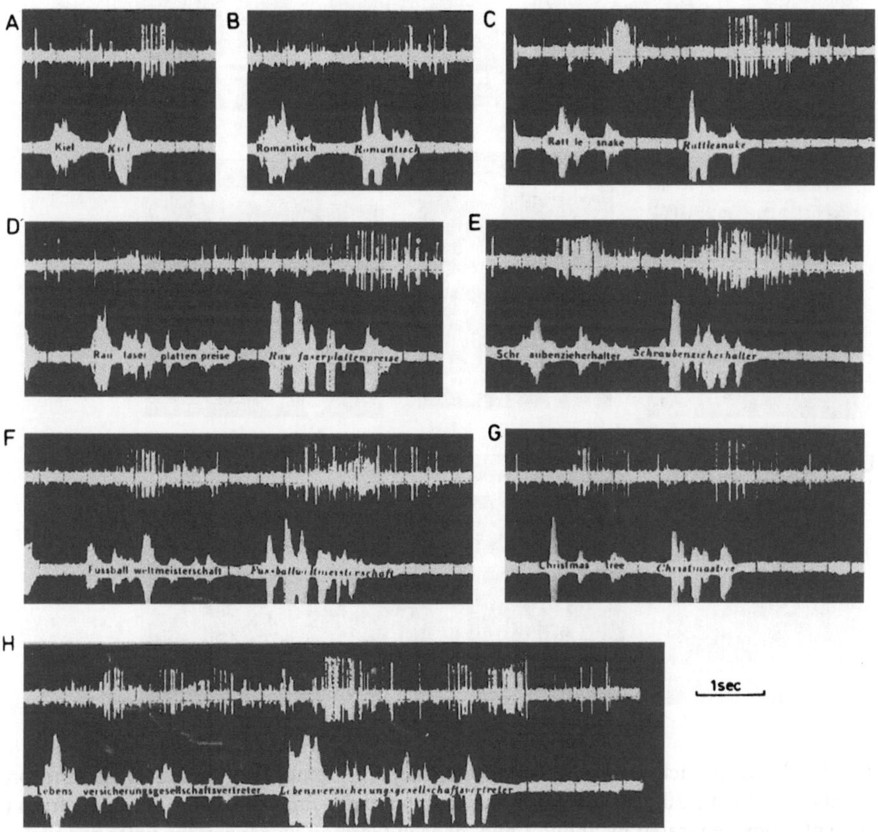

Fig. 15. Activity of left, dominant superior temporal gyrus neuron during listening to and repeating tape-recorded German words. A *large unit* represents activity with later syllable of word repetition, a *mid sized unit*, activity with both listening and repeating, in similar pattern for same word (especially *C, D, E, F*). *Above*, microelectrode; *below*, audio tape channel. Data from case 8742, a German-speaking male. (From Creutzfeldt et al. 1989b)

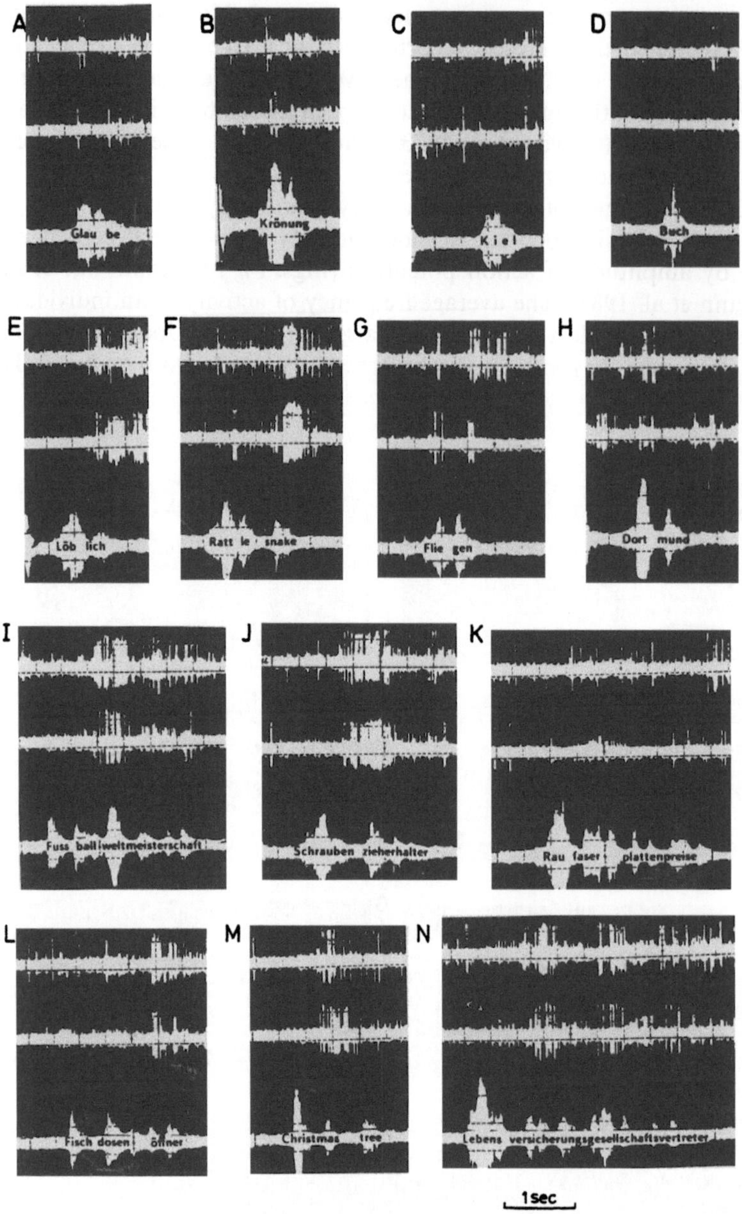

Fig. 16. Neuronal activity from left, dominant superior temporal gyrus during tape-recorded word listening in a 30-year-old German-speaking male. *Top and middle*, recording from two presentations, separated by about 3 min, of each German word; *bottom*, voice channel with approximate corresponding syllables. Note similar patterns of activity with each presentation. Case 8742. (From Creutzfeldt et al. 1989a)

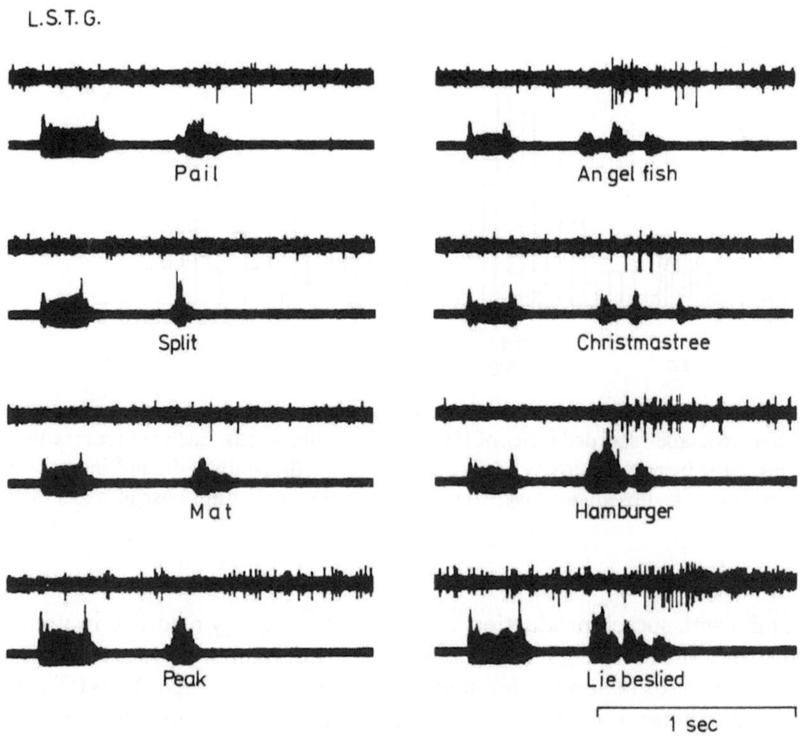

Fig. 17. Neuronal activity from left, *non*dominant superior temporal gyrus in a 14-year-old female during tape-recorded word listening. *Left*, recording during monosyllabic words; *right*, recording during multisyllabic words. Note activity appearing with later syllables on multisyllabic words and little or no activity on single syllable words. Case 8531. (From Creutzfeldt et al. 1989a)

significant excitation and inhibition have been recorded during the language and memory measures.

3.1 Changes Are Recorded Widely Beyond Essential Areas

Although the invasive nature of microelectrode recording prevents its use in essential areas for language, changes in neuronal activity during language but not during control tasks have been frequently recorded from many cortical areas, including the nondominant hemisphere. Two-thirds of neuronal populations recorded in nondominant temporal lobe changed activity during some portion of the auditory language tasks (Creutzfeldt et al. 1989b). Changes in activity during speech perception were identified in six of the eight initial recordings in nondominant superior temporal gyrus (Creutzfeldt et al. 1989a, b). Figure 18 provides another example. Some of these changes recorded from nondominant temporal lobe had quite complex relationships to speech perception, as illustrated in Fig. 14.

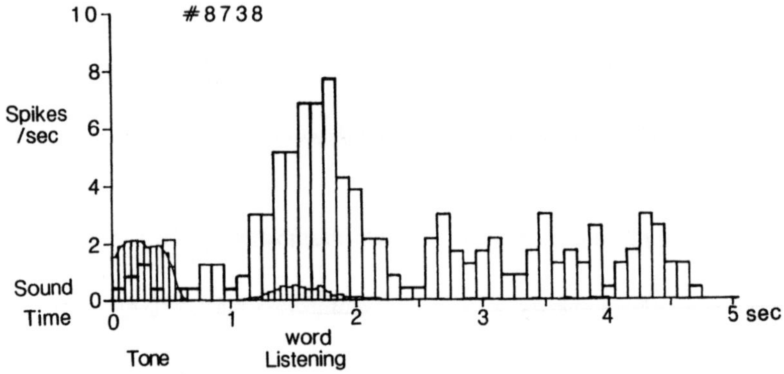

Fig. 18. Activity of right, nondominant *superior* temporal gyrus neuronal population during listening to a tape-recorded series of 1000-Hz tones and words. Each *bar* represents activity in 100 ms. Note increased activity when listening to words. Activity did not increase with word production in a subsequent task. Case 8738. (Ojemann J., unpublished observations)

Changes with speech production were even more widely recorded in nondominant cortex, in seven of the eight nondominant superior temporal gysus recordings and in about two-thirds of nondominant middle temporal gyrus recordings (Fig. 19), few of those recordings showed changes with speech perception. Some nondominant temporal lobe neuronal populations that increased activity with speech production demonstrated maximal increases in activity when the speech was part of spatial tasks, rather than object naming. This suggests that nondominant visuospatial systems include neurons related to language effector mechanisms, perhaps representing the interface between visuospatial and language functions (Ojemann et al. 1989c).

Recordings from nonessential sites in the dominant hemisphere demonstrated a slightly greater proportion of changes with speech perception or production, with changes during speech production similarly recorded more widely (Creutzfeldt et al. 1989a, b). Changes in activity related to naming or reading, compared to activity during a visuospatial task in response to physically identical stimuli, were recorded at 6 of 17 nonessential sites (in 13 patients) in the dominant hemisphere (Ojemann et al. 1988). Changes related to recent verbal memory were identified at 12 of these sites. Changes in activity with naming in only one of two languages were identified at a nonessential dominant temporal lobe recording site in each of three bilingual subjects (Cawthon et al. 1987). Thus changes in neuronal activity that seem to be quite specific to certain language tasks can be recorded well beyond essential areas, even in the nondominant hemisphere. Changes in neuronal activity at frontal sites likely not essential for language, as well as widely in striatum and thalamus, have been reported by Bechtereva (1987). Investigation of blood flow changes that reflect where neurons are working during language tasks has often shown a somewhat similar pattern, with symmetrical flow changes widely in both temporal lobes (Ingvar 1983). Models of the organization of language must include this very

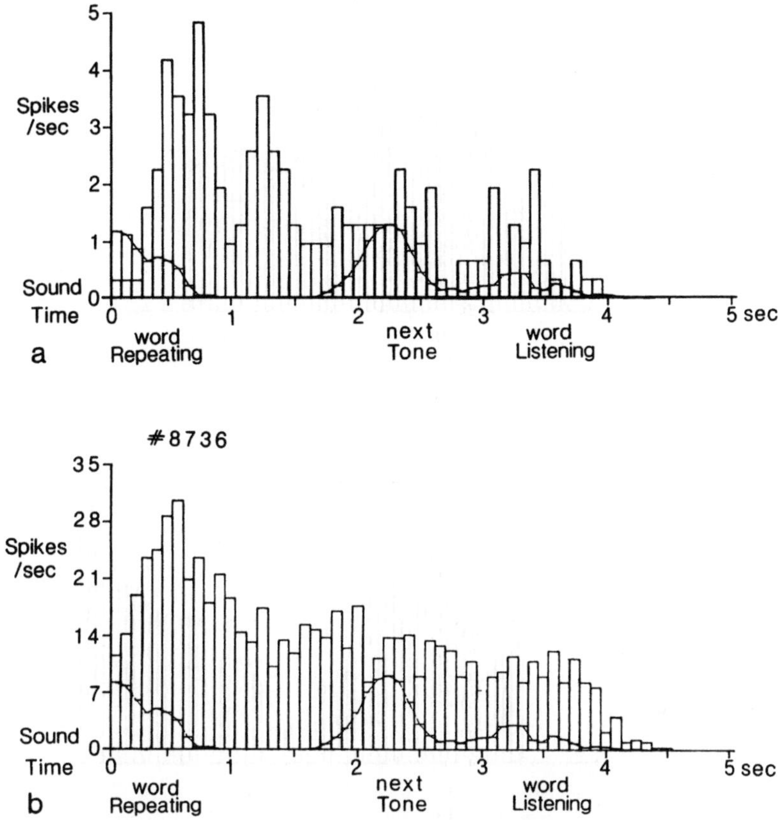

Fig. 19 a, b. Single neuron (**a**) and multiunit (**b**) activity from right, nondominant middle temporal gyrus during listening to and repeating of words. Activity is aligned to onset of patient's voice during repetition. *Below*, audio channel. A 1000-Hz tone preceding tape-recorded words acts as control for auditory perception. Each bin is 100 ms. Note increased activity with repetition with also onset, but sustained after end of word. Case 8736. (Ojemann J., unpublished observations)

widespread nonlateralized neuronal activity, as well as the lateralized, highly localized modular essential areas.

3.2 Different Neuronal Populations Are Often Related to Different Functions

When simultaneous activity of several neurons is separated by amplitude of action potentials, nearby neurons frequently show different responses during a language task (Fig. 15; Creutzfeldt et al. 1989b, Fig. 5). One of the most consistently observed separations was in responses to speech perception and production. As a general rule, neuronal populations in nonessential areas of temporal lobe responded either to

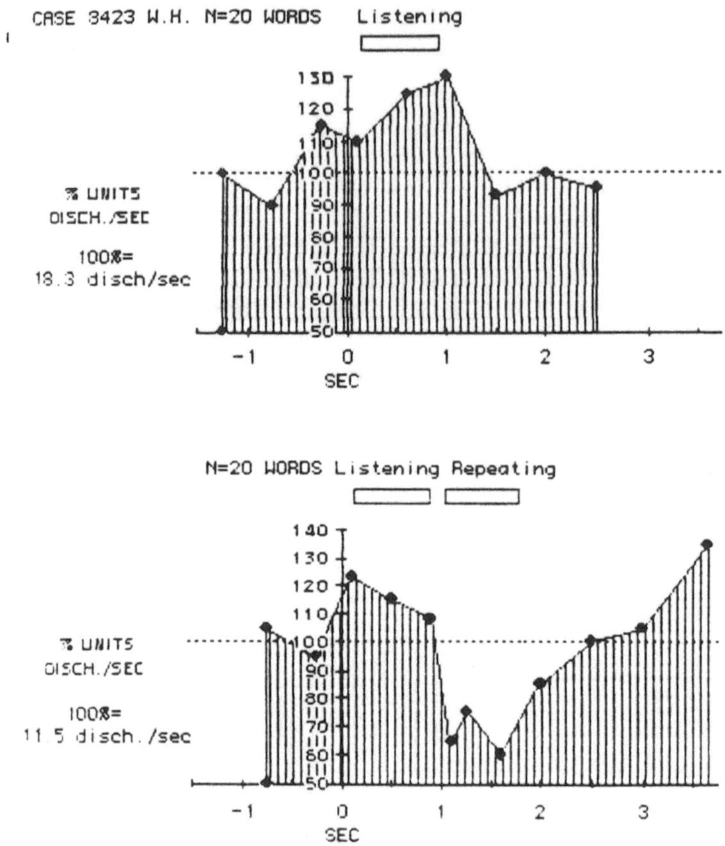

Fig. 20. Activity of left dominant middle temporal gyrus neurons during listening to and repeating of 20 tape-recorded words. Activity increases with listening and is inhibited during repetition. Case 8423. (Creutzfeldt, Ojemann, Lettich, unpublished observations)

speech production or to perception but not to both, and if they responded to both, the response was most often in opposite directions, as illustrated in Fig. 20. The pattern illustrated in the figure, increased activity with speech perception and inhibition during speech production, was particularly common in neuronal populations responding to both speech perception and production. When inhibition was observed, a rebound excitation commonly followed. Changes in temporal neuronal activity during speech production have also been identified in visually presented language and memory tasks (Ojemann et al. 1988). This general separation of neural activity during speech production and perception is another feature that must be incorporated into a model of cortical language organization. The mechanisms for perceiving externally generated speech seem to be generally inoperative when one's own speech is produced; the differentiation between perceiving our own speech and that of others is, then, based on activity of different cortical physiological mechanisms.

A role for temporal neocortex in the motor aspects of language was inferred from stimulation mapping effects. The finding of widespread changes in neuronal activity there during speech production provides additional evidence of such a motor role for temporal lobe. An unusual aspect of this motor role for temporal neocortex was suggested by recordings from the dominant hemisphere anterior temporal lobe in several patients. Changes in activity occurred at the latency of overt responses to naming or reading but were present only when these tasks were done silently and not when they were done overtly (Ojemann et al. 1988). These neurons may be involved in suppressing overt speech production.

These neuronal populations, and recorded human temporal lobe neurons in general, did not show response to auditory perception (tones, random noises) (Creutzfeldt et al. 1989a) or visual perception (Ojemann et al. 1988). Additional evidence of the specificity of neuronal activity to language was detected in recordings during degraded speech (Creutzfeldt et al. 1989a). A few neuronal populations changed activity (either a significant increase or decrease) only when perceiving natural speech, and not speech backward, speech in various degraded forms or with speech production. Such neurons seem to be very specific to speech perception. A single neuronal population was identified that responded only to natural speech and speech backwards, suggesting that it was specifically tuned to respond to speech frequencies, but not content.

Separate neuronal populations in nonessential cortex seem to be related to different languages even when the same linguistic function was performed, complementing the stimulation mapping derived separation of essential areas for the same function in different languages. Each of five neuronal populations recorded from three bilingual patients demonstrated a different level of activity with naming of the same pictures in one language compared to the other language, as illustrated in Fig. 11 (Cawthon et al. 1987 and unpublished observations). One population from one patient showed changes to both languages but in opposite directions. Changes were specific not only to the language, but also to the task, being more marked for naming than reading in two populations recorded from two cases, and for overt than silent naming in two other populations from two cases.

Similar task specificity was seen in 17 recordings obtained from 13 patients during measures of naming, reading and recent verbal memory (Ojemann et al. 1988). One population was related only to reading. Another six populations showed significant changes only with the recent verbal memory measure. Some overlap in responses to language and memory was evident in five of these six populations, however. But even in those populations there was often a differential response to language or memory tasks, as illustrated in Fig. 21, which shows that a relative inhibition of activity is evident when a memory is stored during a reading distracting task, compared to activity during that reading task alone.

Thus, the neuronal system for a particular language behavior includes anatomically separate essential modules and separate neurons distributed widely across non-essential association cortex, where they are in close proximity to neurons related to a different system. Few neurons seem to be part of more than one system. There is little evidence for "general purpose" neurons.

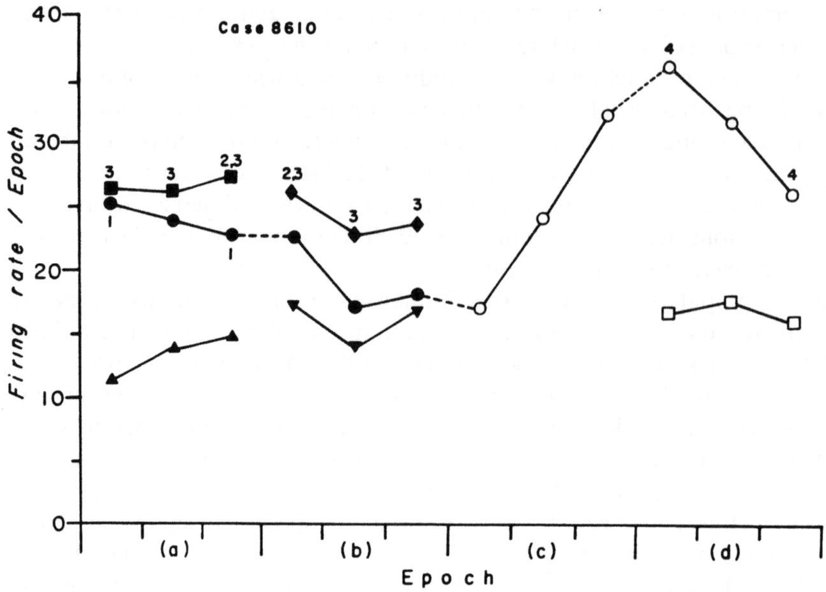

Fig. 21. Neuronal activity in left, dominant middle temporal gyrus during naming, word reading and recent verbal memory. Memory measure (*circles*) has four phases: (*a*) entry (input) of silently named information; (*b, c*) storage of that information during two reading tasks, word reading (*b*) and sentence reading (*c*); and (*d*) overt retrieval of the name entered in (*a*). Activity during silent naming (*filled squares*) and angle matching (*filled upright triangles*) provides controls for memory entry (*a*): physically identical visual stimuli are used for memory entry, the same language task, silent naming, but without the recent memory component, and for the spatial task, angle matching. Activity during silent word reading (*diamonds*) and angle matching to the same reading stimuli (*inverted triangles*) provide the same controls for the first half of memory storage (*b*). Activity during overt naming (*open squares*) is the control for memory recall (*d*), with the same words overtly produced in each task. Each task is divided into three 1.3-s epochs of equal length. Each point is the average of activity of six trials of that measure. *Numbers* indicate statistically significant differences in activity between memory and control tasks: *1* between memory entry or storage and analogous angle matching, *2* between memory entry or storage and analogous language task, *3* between language task and analogous angle matching, *4* between memory retrieval and overt naming. This case shows sustained increased activity during naming and reading, with relative inhibition with recent verbal memory storage (*a*, last epoch; *b*, first epoch). (From Ojemann et al. 1988)

In this context of neuronal populations usually related to only one of several measured language functions in only one language, the presence of a few neuronal populations responding to several language functions takes on added significance. Several populations, usually in the superior temporal gyrus of dominant hemisphere, responded in the same way to both speech production and perception (Fig. 15, 22). The neoronal populations represented by midsized action potentials in the recording illustrated in Fig. 15 are of particular interest. The activity in that neuron seemed to have specific patterns during both perception and production of

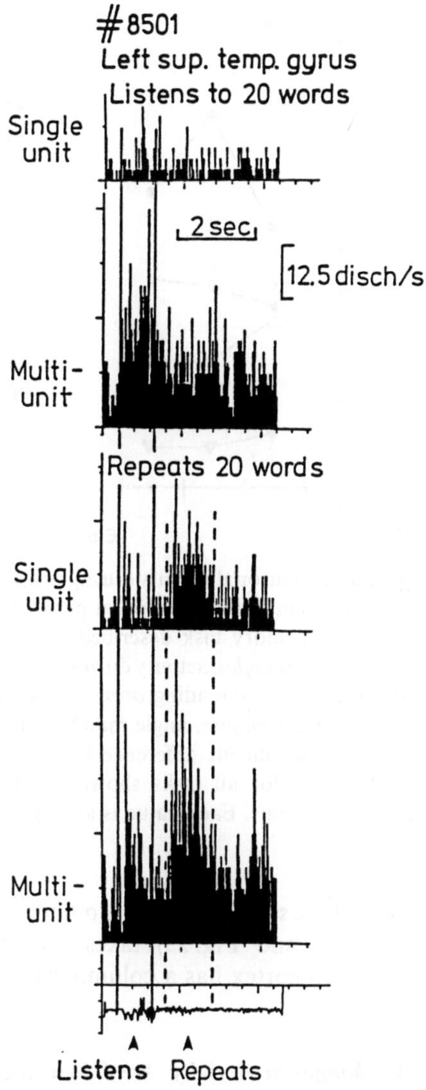

Fig. 22. Neuronal activity in left, dominant superior temporal gyrus during listening to and repeating of a tape-recorded word list. In multiunit recordings, increased activity during both listening to and repeating of word is evident, an unusual pattern compatible with activity appropriate to the motor theory of speech perception. The more usual pattern is response to only listening, or repeating (as seen for the single unit activity here) or differential responses to listening or repeating as in Fig. 20. (From Creutzfeldt et al. 1987)

specific words (compare Fig. 15*C* to Fig. 16*F*), though changes during production were partly masked by the activity of another neuron (large action potential) that was active only with speech production. The neuron illustrated in Fig. 14 from a nondominant temporal lobe also seemed to have some similar response to the same phoneme during speech perception and perception (Creutzfeldt et al. 1989b). These neuronal populations would be the appropriate substrate for the common cortex related to orofacial motor control and speech sound identification by stimulation mapping, and an appropriate neural substrate for the motor theory of speech perception (Liberman et al. 1987).

Similarly, most recordings have shown different responses in simultaneously recorded neuronal populations. Thus the finding that the few recordings at different

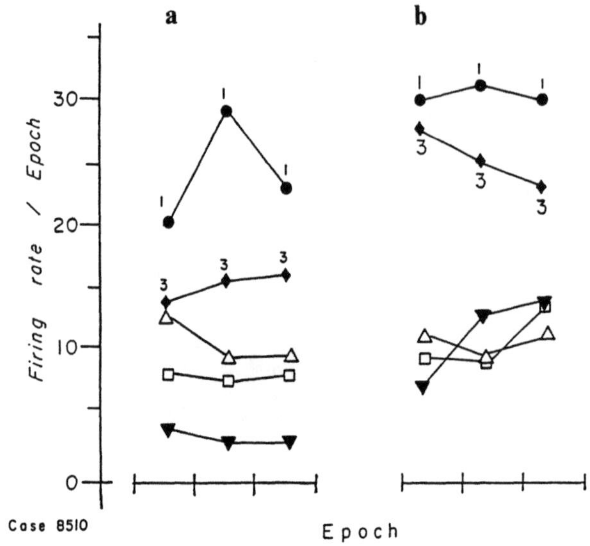

Fig. 23 a, b. Neuronal activity during word reading at two depths (**a** 2 mm, **b** 4 mm) of one site in left, dominant middle temporal gyrus. *Filled circles*, activity with silent word reading as part of the memory task described in Fig. 21; *filled diamonds*, silent word reading alone; *inverted filled triangles*, activity during angle matching using physically identical visual stimuli to that used for word reading; *open symbols*, activity during naming task and controls; *squares*, silent naming; *triangles*, angle matching to stimuli identical to those for naming, *numbers*, statistically significant difference between reading tasks and analogous angle matching. Visual stimuli for all tasks shown for 4s; activity during that time divided into three equal 1.3-s epochs. Each value is average of activity on six trials. (From Ojemann 1989)

depths at the same cortical site often showed similar patterns of responses (Ojemann et al. 1988 and Fig. 23) takes on added significance, suggesting that human association cortex has a columnar organization.

3.3 Changes in Activity Are Often Suggestive of Mechanisms of Selective Attention

Many of the changes in activity related to specific language tasks were sustained shifts in the frequency of discharges, lasting seconds or minutes. Figures 11, 21 and 23 show such changes. In the recording illustrated in Fig. 11, the sustained shift in activity was evident with the transition from a low level of activity with naming in English to a higher level with naming in Spanish; the shift to a lower level of activity reappeared with a second measure of naming in English (Cawthon et al. 1987). In the recording shown in Fig. 21, the level of activity through the entire period of a naming task is elevated compared to that during visuospatial task in response to the same visual stimuli. The actual time required for naming is, however, less than half of the 4-s epoch over which this change is sustained. Similar prolonged shifts in activity recorded during word reading are illustrated in Figs. 21 and 23. The transition from one level of activity to another was often quite abrupt, occurring with instructions for the new task (Fig. 11; Cawthon et al. 1987). Although the number of samples is still quite small, there is a suggestion that inhibitory tonic shifts in activity are more likely at sites more distant from essential areas than at sites with sustained increases in activity.

The mechanisms for these sustained shifts in activity is unknown. They are, however, suggestive of activity of a mechanism of local selective activation, such as that mediated by thalamocortical activating activity in animals. If such a mechanism accounts for these neuronal changes in non essential cortex, and for ECoG desynchronization that differentiates essential from noneessential sites during a

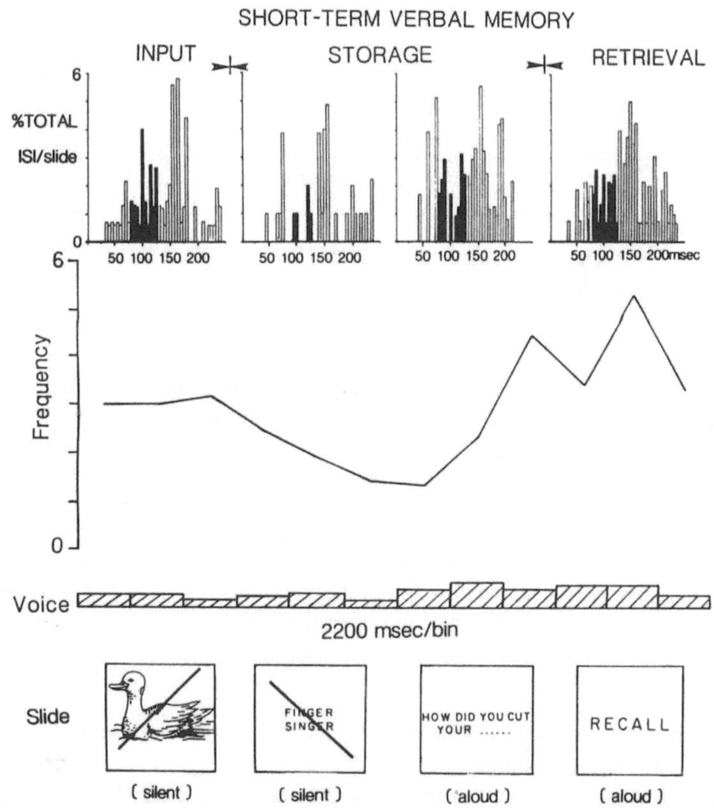

Fig. 24. Neuronal activity during recent verbal memory measure, recorded from left, dominant inferior temporal gyrus. Memory measure same as used in Fig. 21. *Middle*, frequency of activity averaged over five trials of memory measure; note increase with memory input and retrieval. This was a well-isolated single unit. *Above*, interspike intervals of activity averaged over the five memory trials during each phase of that task. The interspike intervals from 75 to 125 ms are significantly ($p < .003$) less likely during the first half of memory storage than in other phases. Case 8503. (From Cawthon et al. 1988 and unpublished observations)

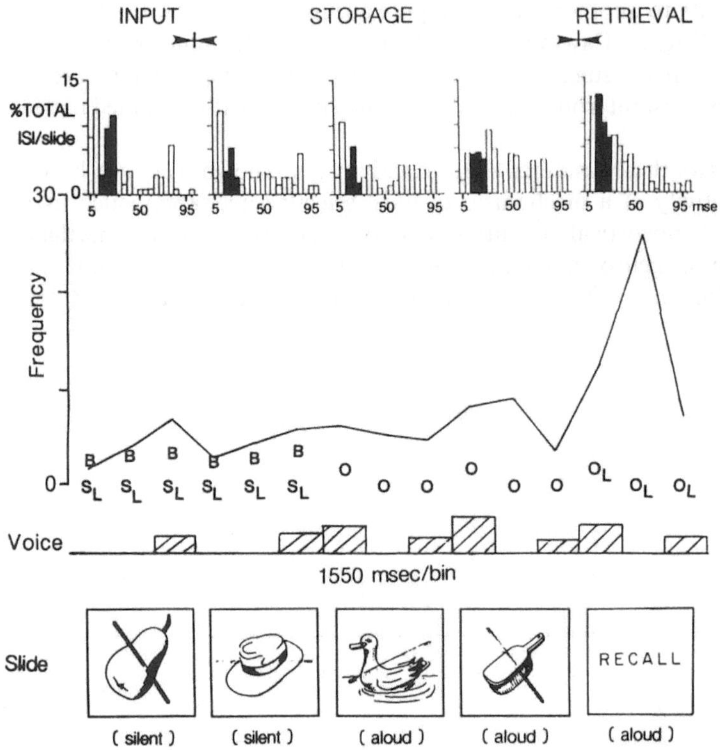

Fig. 25. Neuronal activity during recent memory measure recorded from left, dominant superior temporal gyrus. Memory measure somewhat modified from previous figure. *Middle*, average frequency on four trials of memory measure, along with activity during control measures for each phase of test. *s*, silent naming alone; *O*, overt naming alone; *L*, angle matching to same visual stimuli. Activity is significantly increased at memory retrieval. This too was well-isolated activity of a single neuron. *Above*, interspike intervals during various phases of memory task. Intervals between 10 and 25 ms are significantly ($p<.005$) over-represented during memory input and retrieval, compared to storage. Data from case 8742. (From Cawthon et al. 1988 and unpublished observations)

language task, then thalamocortical activity may be quite widespread, but more intense at essential sites.

Prolonged alterations in activity, lasting seconds, also characterized the changes related to recent verbal memory in recordings from temporal neocortex (Ojemann et al. 1988). Those prolonged changes, however, did not last throughout the memory task, but only for some seconds at the time information entered the memory, and again at its retrieval, and in each case they outlasted the time required for identifying the entering verbal information, or its actual retrieval. Figs. 21, 24 and 25 provide examples. Perhaps this period of prolonged activity at memory entry and retrieval represents the recent memory trace. Its reappearance with retrieval may be the mechanism sustaining recent memories. Changes related to memory were evident in 12 of 17 neuronal populations recorded from dominant temporal neocortex. This represents evidence beyond that derived from stimulation mapping and discussed previously (Ojemann 1978, 1983a; Ojemann and Dodrill 1985) of the importance of this cortex in memory mechanisms. The prolonged nature of these neocortical neuronal changes related to memory also suggests mechanisms of selective attention, possibly mediated through thalamocortical activating circuits. A memory role for those circuits has been defined from the effects of thalamic stimulation (Ojemann et al. 1971; Ojemann 1975, 1985).

Slow sustained changes in activity have also been recorded during auditory tasks. Sustained reductions in activity have been a frequent response of temporal neuronal populations to music (Creutzfeldt and Ojemann 1989). Responses to speech production are also often relatively slow by neuronal standards. These often began 100 ms or more after production of the word, and outlasted the motor output by some hundreds of milliseconds (Fig. 19).

3.4 Patterns of Neuronal Activity Specific to Aspects of Language and Memory

Although the sustained tonic changes were usually highly specific to particular behaviors, the pattern of the detected changes was a nonspecific shift in overall frequency. However, a few other recordings have also shown patterns of neuronal activity specific to particular aspects of language and memory.

Neuronal populations with activation only on the perception of second and subsequent syllables of multisyllable words, with no activity, or at most a single action potential at the onset when perceiving single syllable words, have been recorded from both dominant and nondominant temporal lobe (Fig. 17; Creutzfeldt et al. 1989a). One of these populations from the nondominant hemisphere seemed to respond to the second syllable only when it was part of a compound word. The activity of these neurons is likely related to *prosody* of speech. Several populations in superior temporal gyrus of either hemisphere seemed to increase activity with certain *phonemes* (Fig. 14, 16; Creutzfeldt et al. 1989a). One population showed increased activity to both certain phoneme and syllable combinations (Fig. 16). As a result, this population seemed to have a particular pattern of activity each time a specific word was perceived, a pattern that also seemed to be present with production of that word (Fig. 15). Thus some neurons may produce a specific pattern of activity each time a *particular word* is encountered. Whether these specific patterns of response to prosody, phonemes or words and sustained tonic changes in activity with language are characteristics of the same neuronal population or of separate populations is not presently known. Patterns of neuronal activity specific to particular words and semantic clauses, recorded from frontal cortex, striatum and thalamus have been reported by Bechtereva and her associates (Bechtereva 1987).

Changes in the pattern of neuronal activity accompany some of these tonic shifts in frequency of activity related to memory entry and retrieval. When microelectrode recordings have shown well isolated activity from a single neuron, this has been analyzed for changes in the intervals between action potentials related to different behavioral conditions (Cawthon et al. 1988). As indicated in Figs. 24 and 25, the proportion of certain intervals between action potentials is significantly over represented in activity recorded during memory entry and retrieval, compared to the period when the memory is stored. Although the over represented intervals are different for the few neurons suitable for these analyses, the presence of some intervals related to both memory entry and retrieval suggest that the pattern of neural activity then differs from that during storage: That different pattern may represent that neuron's coding of the memory trace.

3.5 Summary

Neuronal activity related to language or memory can be recorded widely in the dominant and the nondominant hemisphere, well beyond the essential modules identified by stimulation mapping. The change in neural activity in nonessential cortex was often highly specific to one language task, even to one language. This high degree of specificity is similar to that identified for essential modules by stimulation mapping, suggesting that throughout both essential and nonessential cortex there are separate neuronal systems for specific language functions. This specificity of neuronal activity in nonessential cortex included considerable separation in activity during speech perception or production, indicating that different neural circuits subserve perception of external speech than perception of one's own speech. Neuronal responses to speech production were recorded widely in the temporal lobe, well beyond the extent of responses to speech perception, providing confirmation of the temporal lobe role in motor speech activity.

The change in neuronal activity related to language was most often a prolonged shift in the level of activity, suggestive of mechanisms of selective attention. Thus selective attention mechanisms seem to be functioning in both nonessential and essential cortex, although the ECoG changes at essential language sites, during the language task, suggest that those mechanisms are more active at essential sites. A few neuronal populations showed specific patterns of activity related to prosody, phonemes, or even specific words.

Neuronal activity related to recent verbal memory was also recorded widely from temporal neocortex, evidence additional to that derived from stimulation mapping and the effects of lesions for a role of this cortex in memory processes. The change in neuronal activity related to memory was most often a prolonged increase in activity at the time of entry of information into memory and with retrieval. This increase lasted much longer than the time required for retrieval; it may be the activity that sustains the memory trace. Within this period of increased activity, increases in specific interspike intervals have been identified in a few recordings of unusually well isolated single neuron activity, suggesting that there may be specific patterns of activity during the memory trace.

4 References

Andy OJ, Bhatnagar (1984) Right-hemispheric language evidence from cortical stimulation. Brain Lang 23:159–166

Bechtereva N (1987) Some general physiological principles of human brain functioning. Int J Psychophysiol 5:235–251

Black P, Ronner S (1987) Cortical mapping for defining the limits of tumor resection. Neurosurgery 20:914–919

Broca P (1861) Remarques sur le siege de la faculte du language articule, suivies d'une observation d'aphemie (perte de la parole). Bull Soc Anatol 36:330–357

Calvin WH (1983) Timing sequencers as a foundation for language. Behav Brain Sci 6:210–211

Calvin W, Ojemann G, Ward A Jr (1973) Human cortical neurons in epileptogenic foci: comparison of interictal firing patterns to those of "epileptic" neurons in animals. Electroencephalog Clin Neurophysiol 34:337–351

Cawthon D, Lettich E, Ojemann G (1987) Human temporal lobe neuronal activity: inhibition during naming in only one of two languages. Soc Neurosci Abstr 13:839

Cawthon D, Lettich E, Creutzfeldt O, Ojemann G (1988) Memory-specific changes in interspike intervals in dominant human temporal lobe. Soc Neurosci Abstr 14:861

Creutzfeldt O, Ojemann G (1989) Neuronal activity in human lateral temporal lobe. III. Activity changes during music. Exp Brain Res 77:490–498

Creutzfeldt O, Ojemann G, Lettich E (1987) Single neuron activity in right and left human temporal lobe during listening and speaking. In: Engel J Jr et al. (eds) Fundamental mechanisms of human brain function. Raven, New York

Creutzfeldt O, Ojemann G, Lettich E (1989a) Neuronal activity in human lateral temporal lobe. I. Responses to speech. Exp Brain Res 77:451–475

Creutzfeldt O, Ojemann G, Lettich E (1989b) Neuronal activity in human lateral temporal lobe. II. Responses to the subject's own voice. Exp Brain Res 77:476–489

Damasio A (1998) Concluding remarks: neuroscience and cognitive science in the study of language and the brain. In: Plum F (ed) Language, communication and the brain. Raven, New York

Dewson JH III (1977) Preliminary evidence of hemispheric asymmetry of auditory function in monkeys. In: Harnad S (ed) Lateralization in the nervous system. Academic, New York

Fedio P, Van Buren JM (1974) Memory deficits during electrical stimulation of speech cortex in conscious man. Brain Lang 1:29–42

Fried I, Ojemann GA, Fetz EE (1981) Language-related potentials specific to human language cortex. Science 212:353–356

Galaburda A, Sanides F, Geschwind N (1978) Human brain. Cytoarchitectonic left-right assymetries in the temporal speech region. Arch Neurol 35:812–817

Geschwind N (1970) The organization of language and the brain. Science 170:940–944

Gloor PA, Olivier A, Quesney LF, Andermann F, Horowitz S (1982) The role of the limbic system in experimental phenomena of temporal lobe epilepsy. Ann Neurol 12:129–144

Goldman-Rakic PS (1988) Topography of cognition: parallel distributed networks in primate association cortex. Annu Rev Neurosci 11:137–156

Goodglass H (1976) Agrammatism. In: Whitaker H, Whitaker HA (eds) Studies in neurolinguistics, vol 1. Academic, New York, pp 237–260

Halgren E, Babb TL, Crandall PH (1978) Activity of human hippocampal formation and amygdala neurons during memory testing. Electroencephalogr Clin Neurophysiol 45:585–601

Ingvar D (1983) Serial aspects of language and speech related to prefrontal cortical activity. A selective review. Hum Neurobiol 2:177–189
Jasper HH (1960) Unspecific thalamocortical relations. In: Handbook of physiology: neurophysiology. American Physiological Society, Washington DC
Kertesz A, Phipps J (1977) Numerical taxonomy of aphasia. Brain Lang 4:1–10
Lesser RP, Lueders H, Dinners DS, Hahn J, Cohen L (1984) The location of speech and writing functions in the frontal language area. Results of extraoperative cortical stimulation. Brain 107:275–291
Liberman AM, Cooper FS, Shankweiler DP, Studdert-Kennedy M (1967) Perception of the speech code. Psychol Rev 74:431–461
Lueders H, Lesser RP, Hahn J et al. (1986) Basal temporal language area demonstrated by electrical stimulation. Neurology 36:505–510
Mateer CA, Polen SB, Ojemann GA et al. (1982) Cortical localization of finger spelling and oral language: a case study. Brain Lang 17:46–57
Mateer CA, Rapport RL, Kettrick C (1984) Cerebral organization of oral and signed language responses: case study evidence from amytal and cortical stimulation studies. Brain Lang 21:123–135
Merzenich MM, Nelson RJ, Kaas JH et al. (1987) Variability in hand surface representations in areas 3b and 1 in adult owl and squirrel monkeys. J Comp Neurol 258:281–296
Mohr JP (1976) Broca's area and Broca's aphasia. In: Whitaker H, Whitaker HA (eds) Studies in neurolinguistics. Academic, New York
Mountcastle V (1979) An organizing principle for cerebral function: the unit module and the distributed system. In: Schmitt FD, Warden FG (eds) The neurosciences: fourth study program. MIT Press, Cambridge MA
Mullan S, Penfield W (1959) Illusions of comparative interpretation and emotion produced by epileptic discharge and by electrical stimulation in temporal cortex. Arch Neurol Psychiatr 81:269–284
Ojemann GA (1975) Language and the thalamus: object naming and recall during and after thalamic stimulation. Brain Lang 2:101–120
Ojemann GA (1978) Organization of short-term verbal memory in language areas of human cortex: evidence from electrical stimulation. Brain Lang 5:331–348
Ojemann GA (1979) Individual variability in cortical localization of language. J Neurosurg 50:164–169
Ojemann GA (1983a) Brain organization for language from the perspective of electrical stimulation mapping. Behav Brain Sci 6:189–206
Ojemann GA (1983b) Electrical stimulation and the neurobiology of language. Behav Brain Sci 6:221–230
Ojemann GA (1983c) Interrelationship in the brain organization of language-related behaviors: evidence from electrical stimulation. In: Kirk U (ed) Neuropsychology of language, reading and spelling. Academic, New York
Ojemann GA (1985) Enhancement of memory with human ventrolateral thalamic stimulation: effect evident on a dichotic listening task. Appl Neurophysiol 48:212–215
Ojemann G (1987) Intraoperative functional mapping at the University of Washington, Seattle. In: Engel J Jr (ed) Surgical treatment of the epilepsies. Raven, New York
Ojemann GA (1988) Effect of cortical and subcortical stimulation on human language and verbal memory. In: Plum F (ed) Language, communication and the brain. Raven, New York
Ojemann GA (1989) Some brain mechanisms for reading. In: Von Euler C (ed) Brain and reading. Macmillan, New York

Ojemann GA, Creutzfeldt OD (1987) Language in humans and animals: contribution of brain stimulation and record. In: Plum F (ed) Handbook of physiology, Sect 1, Vol V, Part 1, Higher functions of the brain. Oxford University Press, New York

Ojemann GA, Dodrill CB (1985) Verbal memory deficits after left temporal lobectomy for epilepsy. J Neurosurg 62:101–107

Ojemann G, Dodrill G (1987) Intraoperative techniques for reducing language and memory deficits with left temporal lobectomy. In: Wolf P et al. (eds) Advances in epileptology: the 16th epilepsy international symposium. Raven, New York

Ojemann GA, Mateer C (1979) Human language cortex: localization of memory, syntax and sequential motor-phoneme identification systems. Science 205:1401–1403

Ojemann GA, Whitaker HA (1978a) The bilingual brain. Arch Neurol 35:409–412

Ojemann GA, Whitaker HA (1978b) Language localization and variability. Brain Lang 6:239–260

Ojemann GA, Blick KI, Ward AA (1971) Improvement and disturbance of short-term verbal memory with human ventrolateral thalamic stimulation. Brain 94:225–240

Ojemann GA, Creutzfeldt O, Lettich E et al. (1988) Neuronal activity in human lateral temporal cortex related to short-term verbal memory, naming and reading. Brain 111:1383–1403

Ojemann GA, Fried I, Lettich E (1989a) Electrocorticographic (ECoG) correlates of language: I. Desynchronization in temporal language cortex during object naming. Electroencephalog Clin Neurophysiol 73:453–463

Ojemann G, Ojemann J, Lettich E, Berger M (1989b) Cortical language localization in left, dominant hemisphere. J Neurosurg 71:316–326

Ojemann J, Ojemann G, Creutzfeldt O, Lettich E (1989c) Neuronal activity related to faces and matching in human right, nondominant temporal cortex. Soc Neurosci Abst 15:247

Penfield W, Jasper H (1954) Epilepsy and the functional anatomy of the human brain. Little, Brown, Boston MA

Penfield W, Perrot P (1963) The brain's record of auditory and visual experience – a final summary and discussion. Brain 86:595–696

Penfield W, Roberts L (1959) Speech and brain mechanisms. Princeton University Press, Princeton NJ

Ranck JB Jr (1975) Which elements are excited in electrical stimulation of mammalian central nervous system: a review. Brain Res 98:417–440

Rapport RL, Tan CT, Whitaker HA (1983) Language function and dysfunction among Chinese- and English-speaking polyglots: cortical stimulation, Wada testing and clinical studies. Brain Lang 18:342–366

Rubens AB, Mahowald MW, Hutton JT (1976) Asymmetry of the lateral (Sylvian) fissures in man. Neurology 26:620–624

Skinner J, Yingling C (1977) Central gating mechanisms that regulate event-related potentials and behavior: a neural model for attention. Prog Clin Neurophysiol 1:30–69

Squire L (1987) Memory: neural organization and behavior. In: Plum F (ed) Handbook of physiology, Sect. 1, Vol V, Part 1. Higher functions of the brain. Oxford University Press, New York

Tallal P (1983) A precise timing mechanism may underlie a common speech perception and production area in the perisylvian cortex of the dominant hemisphere. Behav Brain Sci 6:219–220

Trevarthen C (1974) Functional relations of disconnected hemispheres with the brain stem, and with each other: monkey and man. In: Kinsbourne M, Smith WL (eds) Hemispheric disconnection and cerebral function. Thomas, Springfield IL

Tulving E (1972) Episodic and semantic memory. In: Tulving E, Donaldson W (ed) Organization of memory. Academic, New York
Van Buren JM, Fedio P, Frederick GC (1978) Mechanism and localization of speech in the parietotemporal cortex. Neurosurgery 2:233–239
Wernicke C (1874) Der aphasische Symtomen-Komplex. Cohen and Weigart, Breslau
Whitaker HA, Ojemann GA (1978) Graded localisation of naming from electrical stimulation mapping of left cerebral cortex. Nature 270:50–51

Subject Index

MIX
Papier aus verantwortungsvollen Quellen
Paper from responsible sources
FSC® C105338

If you have any concerns about our products,
you can contact us on
ProductSafety@springernature.com

In case Publisher is established outside the EU,
the EU authorized representative is:
Springer Nature Customer Service Center GmbH
Europaplatz 3, 69115 Heidelberg, Germany

Printed by Libri Plureos GmbH
in Hamburg, Germany